MATHEMATICS IN TRANSPORT PLANNING AND CONTROL

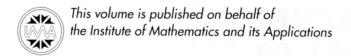 *This volume is published on behalf of
the Institute of Mathematics and its Applications*

Related Pergamon books

BELL
Transportation Networks: Recent Methodological Advances
DAGANZO
Fundamentals of Transportation and Traffic Operations
ETTEMA & TIMMERMANS
Activity Based Approches to Travel Analysis
GÄRLING, LAITILA & WESTIN
Theoretical Foundations of Travel Choice Modeling
STOPHER & LEE-GOSSELIN
Understanding Travel Behaviour in an Era of Change

Related Pergamon journals

Transportation Research Part A: Policy and Practice
Editor: Frank A. Haight

Transportation Research Part B: Methodological
Editor: Frank A. Haight

Free specimen copies of journals available on request

MATHEMATICS IN TRANSPORT PLANNING AND CONTROL

Proceedings of the
3rd IMA International Conference on
Mathematics in Transport Planning and Control

Edited by

J.D. Griffiths
University of Wales, Cardiff, Wales

1998
PERGAMON
An imprint of Elsevier Science
Amsterdam – Lausanne – New York – Oxford – Shannon – Singapore – Tokyo

ELSEVIER SCIENCE Ltd
The Boulevard, Langford Lane
Kiflington, Oxford OX5 1GB, UK

First edition 1998

Library of Congress Cataloging in Publication Data
A catalog record from the Library of Congress has been applied for.

British Library of Cataloguing in Publication Data
A catalogue record from the British Library has been applied for.

ISBN: 0 08 043430 4

⊗ The paper used in this publication meets the requirements of ANSI/NISO Z39.48-1992
(Permanence of Paper).
Printed in The Netherlands.

PREFACE

This volume contains forty of the papers presented at the International Conference on Mathematics in Transport Planning and Control which was held at Cardiff University from 1-3 April, 1998. This was the third such conference run under the auspices of the Institute of Mathematics and Its Applications, the first being held at the University of Surrey in 1979, and the second at Cardiff University in 1989. Over fifty papers were submitted for presentation at the conference, and the papers included in these Proceedings are those selected following a strict refereeing procedure.

It will be clear from the contents that mathematical ideas and methodologies continue to play a prominent part in the description and solution of the many and varied problems that are being currently investigated in quite diverse areas of transport research. There are papers relating to modes of transport such as road, rail, air and shipping. Applications will be found on transport planning, congestion, assignment, networks, signalling, road safety, and environmental issues. It is hoped that the reader, whether an 'old hand" or a newcomer to the subject area, will find much to interest him/her. For those with little previous knowledge of the subject area, the Plenary papers presented by Professor R.E. Allsop (University College London) and Dr. R. Kimber (Transport Research Laboratory) are recommended as starting points.

I wish to express my personal thanks to the members of the organizing committee (Professor M. Maher, Dr. B.G. Heydecker, Dr. N. Hounsell, Dr. J.G. Hunt, Professor C. Wright), who also acted as Associate Editors, and to the referees who gave freely of their time and expertise. Finally, my thanks are due to my secretary, Ms June Thomas, without whose sterling efforts these Proceedings would not have seen the light of day.

J.D. Griffiths
Editor

CONTENTS

ANALYSIS OF TRAFFIC CONDITIONS AT CONGESTED ROUNDABOUTS

R.E. Allsop, University of London, Centre for Transport Studies, University College London

ABSTRACT

At roundabouts where entering traffic is required to give way to traffic circulating in the roundabout, the traffic capacity of each entry is a function of the flow of traffic circulating past it. This relationship has previously been analysed in two main ways: using a linear relationship based on regression and using a non-linear relationship based on a model of entering drivers' acceptance of gaps in the circulating traffic. The linear analysis has previously been extended to estimation of the reserve capacity or degree of overload of the roundabout as a whole in relation to a given pattern of approaching traffic. The non-linear analysis is extended similarly in this paper.

The relationships between entry capacity and circulating flow imply in turn that the capacity of each entry is a function of the entering flows and the proportions of traffic making various movements through the junction from some or all of the entries. Equations are established for determining derivatives of capacity or delay on each entry with respect to the demand flow for each movement. In particular, it is shown that when the roundabout is overloaded the capacity of an entry can depend upon the demand flow on that same entry, giving rise to a corresponding term in the derivative of the delay-flow relationship for the approach concerned.

1. GENERAL PRINCIPLES

Fundamental to the effectiveness of the roundabout as a road junction in resolving conflicts between the various traffic movements is the *offside priority rule*, whereby drivers of vehicles wishing to enter the roundabout are required to wait for gaps in, or *give way* to traffic already circulating in the roundabout. Having given way at this point, drivers then have priority for the rest of their passage through the junction. All merging and crossing conflicts between vehicular movements are thus resolved by requiring each driver to be ready to give way at one point, and only diverging conflicts remain for drivers to handle as they leave the circulating traffic at their chosen exits. Conflicts with pedestrian movements usually occur just before vehicles enter the junction or just after they leave it.

Vehicular traffic approaching a roundabout can be divided into identifiable sets called *streams* such that queuing theory can be applied to the traffic in each stream. With the offside priority rule, the *capacity* Q of a stream can be expected to be a decreasing function $F(q_c)$ of the *circulating flow*

q_c of traffic to which it should give way. The parameters of this function can depend upon the characteristics of the circulating traffic, and certainly depend upon the geometry of the junction, including the detailed layout of the entry concerned, but the latter is assumed to be such that the circulating flow for a stream does not include any traffic from other streams using the same entry. The value of q_c will thus depend on the approaching flows q_a and capacities in the streams at the preceding entries, and the proportions of traffic in these streams making each relevant movement through the junction. For each stream it is assumed that the *entering flow* is the lesser of the approaching flow and the capacity, and that the proportions of the entering vehicles making each movement are the same as those of the approaching vehicles. Flows and capacities are measured in passenger car units (pcu)/unit time, so that their values are approximately independent of the composition of the traffic.

The capacity of each stream is thus determined by the interaction between the whole pattern of traffic and the geometry of the junction, and for a given pattern of traffic the capacities of the various streams and the corresponding amounts of delay to traffic in them can be influenced by adjusting the geometrical layout.

Compliance with the offside priority rule should maintain free movement of circulating traffic so long as the exits from the roundabout do not become blocked. It is not, however, a guarantee of acceptable distribution of capacity and delay among the traffic streams, and can in extreme cases lead to disproportionate queues in certain streams as a result of lack of gaps in the corresponding circulating traffic. In such cases, it may be helpful to introduce signal control.

There are two main ways in which capacity can be analysed for a given geometrical layout.

(a)   For any given pattern of traffic, to estimate the resulting value of Q for each stream, thus enabling the queue-length and delay for the given approaching flow to be estimated for each stream by means of applied queueing theory. The results enable the given layout to be evaluated in terms of provision for the given pattern of traffic, and provide indications where attempts should be made to adjust the layout to provide more appropriate levels of capacity in different streams. They also enable the derivatives of capacity and delay in any one stream with respect to the amount of traffic wishing to make any particular movement through the junction from any stream to be estimated.

For this purpose, it is necessary to calculate the circulating flows past the various entries, each of which depends on the entering flows at preceding entries. Calculation of circulating flows is straightforward when the approaching flow in every stream is within capacity, because each entering flow is then equal to the corresponding approaching flow and is therefore known. When one or more streams is overloaded, however, the corresponding entering flows are equal to the as yet undetermined capacities of the corresponding streams, and a stepwise calculation is required.

(b)   For a given pattern of traffic, to calculate at what common multiple T^* , say, of the given approaching flows the most heavily loaded stream will have an approaching flow that is as close as is desirable in practice to its capacity. This is relevant to the question for how long, in a scenario of traffic growth, the junction will continue to function satisfactorily, and for which stream or streams difficulty will first arise.

For this purpose, suppose there are m vehicular traffic streams, and let the given approaching flow in stream i be q_{ai} ($i = 1,2,...,m$), Then using the indicator of junction capacity first introduced by the author (Allsop, 1972) in the context of traffic signal control, and applied to roundabouts by Wong (1996), the junction is said to be working *at practical capacity for approaching flows proportional to the* q_{ai} when the arrival rates are T^*q_{ai} , and the *practical capacity of the junction for approaching flows proportional to the* q_{ai} is T^* times these approaching flows.

2. SPECIFICATION OF THE PATTERN OF TRAFFIC

For each stream i , let q_{ci} be the circulating flow to which traffic in stream i should give way. The flows q_{ai} and q_{ci} are flows that arise in the operation of the roundabout from the rates at which drivers wishing to make the various possible movements through the junction approach it in the various streams. To specify the demand for movement of vehicles through the junction in these terms, let a *movement* comprise traffic wishing to enter the junction in a particular stream and leave by a particular exit. Suppose that there are n distinct movements, and let the approaching flow in movement j be q_j ($j = 1,2,...,n$) .

Then following Wong (1996), but with a different choice of symbols, let **a** be the m×n matrix (a_{ij}) such that

$$a_{ij} = \begin{cases} 1 & \text{if movement } j \text{ contributes to } q_{ai} \\ 0 & \text{if not} \end{cases}$$

and let **c** be the m×n matrix (c_{ij}) such that

$$c_{ij} = \begin{cases} 1 & \text{if movement } j \text{ contributes to } q_{ci} \\ 0 & \text{if not} \end{cases}$$

Thus **a** and **c** specify which movements contribute to the approaching flow and circulating flow for each stream, and by the definition of a movement, **a** has just one non-zero element in each column. It also follows that

$$q_{ai} = \sum_j a_{ij} q_j \tag{2.1}$$

but the relationship between the q_{ci} and the q_j is less straightforward because each q_j can contribute to the relevant q_{ci} only to the extent that it can enter the roundabout.

Let q_{ej} be the rate at which q_j can enter. Then to reflect the assumption that vehicles making different movements from each stream can enter at rates proportional to those at which they approach, for each j

$$q_{ej} \;=\; \min\!\left(q_j \,,\; q_j \sum_i a_{ij} Q_i / q_{ai}\right)$$

(2.2)

It then follows that

$$q_{ci} \;=\; \sum_j c_{ij} q_{ej}$$

(2.3)

When analysing the effect of the flow in any one movement upon capacity and delay for the various streams, a convention concerning the numbering of streams and movements may be helpful.

Suppose there are p exits from the roundabout, so that each stream may comprise up to p movements (including U-turns, but assuming that every driver succeeds in taking the intended exit at the first opportunity). The following notation allows for all p movements to be present in each stream. Absence of a movement is then represented by setting the corresponding q_j to zero.

Let the streams be numbered in order round the roundabout in the opposite direction to the circulation of the traffic, starting with the nearside stream at one of the entries. Then let the p movements within each stream be numbered in the order of the exits in the direction of circulation of the traffic, starting with the first exit available to the traffic in the stream after it enters. Thus the movements forming stream i are movements

$$(i-1)p + k \qquad\qquad (k = 1,2,...,p)$$

and to avoid expressions as subscripts, let

$$q_{ik} = q_{(i-1)p+k} \qquad\qquad ((i,k) = (1,1),...,(m,p))$$

and $\qquad q_{eik} = q_{e,(i-1)p+k}$

so that $\qquad q_{ai} = {}_{;k}\, q_{ik}$

and $\qquad q_{ei} = {}_{;k}\, q_{eik}$

To obtain corresponding expressions for the circulating flows, it is helpful to suppose that entries (each of which may be used by one or more streams) and exits are placed alternately around the roundabout. Situations where this is not the case can then be represented by inserting dummy entries or exits and setting the flows in the corresponding movements to zero.

The numbers of the streams using each entry are consecutive and the numbers of all m streams form p sets

$$S_k = \{ i_{k-1} + 1 , ..., i_k \} \qquad (k = 1,2,...,p) ,$$

where $i_o = 0$ and $i_p = m$, such that S_k contains just the numbers of the streams using entry k.

For a stream in S_1 the circulating traffic comprises all the traffic from streams in S_2 except that which takes the first exit, all that in S_3 except that which takes the first two exits, and so on. Thus

$$q_{ci} = \sum_{k=2}^{p} \sum_{l \in S_k} \sum_{r=k}^{p} q_{elr} (i \; \varepsilon \; S_1)$$

(2.4)

and the circulating flows for streams in the other sets S_K are given by the same expression with S_k replaced by $S_{k+K-1(mod\,p)}$.

This convention implies corresponding dimensions for and arrangements of non-zero elements in the matrices **a** and **c** .

3. PRACTICAL CAPACITY OF THE ROUNDABOUT AS A WHOLE

The estimation of the practical capacity of the roundabout for flows proportional to a given set of flows was defined in Section 1 in terms of finding a certain common multiplier T^* applied to the approaching flows q_{ai} in the streams. Since each q_{ai} is a linear combination of movement flows q_j , this is equivalent to finding the same common multiplier regarded as applied to the pattern of traffic demand represented by the q_j . Since the operational constraint on capacity applies at the level of the stream, the calculation will be expressed first in terms of stream flows.

Let the capacity of each stream i be regarded as $F_i(q_{ci})$, where *for this particular purpose only* each q_{ci} is calculated as if all the flows q_{ai} were within the capacities of their respective streams, thus making $q_{ci} = ;_jc_{ij}q_j$. For all sufficiently small values of T , the flows Tq_j will satisfy this condition and the corresponding capacities will be correctly given by $F_i(Tq_{ci})$.

For the flows Tq_{ai} to be within the respective *practical capacities*, it is deemed to be necessary that

$$Tq_{ai} \leq p_i F_i(Tq_{ci}) \qquad \text{for } i = 1,2,...,m \qquad (3.1)$$

where p_i is the maximum acceptable ratio of approaching flow to capacity, or *degree of saturation* for stream i .

Since the left hand side of each of these inequalities increases with T whilst the right hand side decreases as T increases, each determines a largest value T_i, say, of T for which that inequality is satisfied. Each T_i can be found by solving the corresponding inequality as an equation.

Now $T*$ is the largest value of T for which all m inequalities are satisfied, so

$$T^* = min_i\{T_i\}$$

and the critical stream is the stream i for which T_i is least.

It should be noted that with approaching flows T^*q_{ai}, there is in general only one stream that is at practical capacity with degree of saturation $= p_i$, namely the one for which $T_i = T^*$. The other streams have degrees of saturation $< p_i$ and therefore have spare capacity, but this can be utilised only by changing the pattern of traffic.

4. LINEAR DEPENDENCE OF CAPACITY UPON CIRCULATING FLOW

The work of Kimber (1980) provides an empirical basis for analysing the capacity of roundabouts on the basis that traffic using each entry forms a single stream, and that for each such stream i

$$F_i(q_{ci}) = A_i - B_i q_{ci} \tag{4.1}$$

where the A_i and B_i can be estimated from the geometry of the junction in a manner demonstrated by Kimber, on the basis of regression analysis of extensive observations.

In this case the equation obtained from (3.1) for each T_i is linear and as Wong has documented (1996) but in the present notation

$$\mu^* = \min_i \left(\frac{p_i A_i}{\sum_j a_{ij} q_j + p_i B_i \sum_j c_{ij} q_j} \right) \tag{4.2}$$

The marginal effect of changes in junction geometry upon $T*$ over the range within which the same stream remains critical in determining $T*$ can be estimated from this expression through the effect of the changes on the A_i and B_i for the critical stream.

The treatment of each entry as a single stream has long been debated internationally, notably by Akçelik and Troutbeck (e.g. 1991), and has recently been questioned under British conditions by Chard (1997), but the analysis presented here is not affected by entries being used by more than one stream, provided that $F_i(q_{ci})$ takes the form (4.1) for each stream.

5. DEPENDENCE OF CAPACITY UPON GAP ACCEPTANCE

The gap acceptance model of the behaviour of drivers at the head of a queue of vehicles waiting to enter or cross a priority stream of traffic to which they have to give way has been analysed extensively. Its basis is that the driver at the head of the queue compares each *gap*, or time-headway between arrivals of successive vehicles, in the priority stream in turn with a *critical gap*. The driver will *accept*, that is enter or cross through, the first gap that exceeds the critical gap. There follows a *move-up time* during which any immediately following driver moves forward to the give-way position and at the end of which this driver begins to assess gaps, starting with the remainder of the gap during which the move-up time ends. The model is thus specified in terms of the critical gap, the move-up time, and the distribution of durations of gaps, that is time-headways.

Of the several versions of this model that have been used extensively in the analysis of operation of roundabouts (Akçelik, 1998), the only one that will be referred to here is one (AUSTROADS, 1993) in which the headway distribution is due to Cowan (1975). In this distribution, a given circulating flow is regarded as comprising a known proportion of vehicles travelling in bunches with a known common headway within bunches, and a remainder whose headways have a shifted exponential distribution with parameters depending upon the quantities already specified.

Let q_c = the circulating flow being considered

c = the critical gap

u = the move-up time

P = the proportion of circulating vehicles travelling in bunches

$\setminus$ = the mean headway between bunched vehicles

$>(h)$ = the distribution function of the headways h in the circulating traffic

where $>(h)$ = $\begin{cases} 0 & \text{for } h < \therefore \\ \{ 1 - (1-P)\exp\{-S(h-\setminus)\} \} & \text{for } h \geq \setminus \end{cases}$

and S = $(1-P)q_c/(1-\setminus q_c)$ $\qquad$ (5.1)

with $P<1$ and hence $\setminus q_c<1$ in all realistic cases.

Then it can be shown that the capacity of a stream that gives way to this circulating flow is

Q = $F(q_c)$ = $(1-\Pi)q_c \exp\{-\Sigma(c-.:\}/(1-\exp(-\Sigma u))$ $\qquad$ (5.2)

It follows from (5.2) that the equation obtained from (3.1) for stream i in the estimation of the practical capacity of the roundabout as a whole becomes (with subscript i omitted from all quantities for the time being to simplify the notation and q_c calculated for this purpose as described in Section 3)

$$Tq_a \quad = \quad (1-\Pi)\,Tq_c\,\exp\{-\Sigma(c\text{-.}:)\}/(1-\exp(-\Sigma u)) \tag{5.3}$$

where by (5.1)

$$\Sigma \quad = \quad (1-\Pi)\,Tq_c/(1-.:Tq_c) \tag{5.4}$$

Equation (5.3) reduces to an equation for S which can be written

$$f(\Sigma) \quad = \quad q_a(1 - \exp(-\Sigma u)) - p(1-\Pi)q_c\,\exp\{-\Sigma(c\text{-.}:)\} = 0 \tag{5.5}$$

When this has been solved for Σ, then T is given by (5.4) as

$$T \quad = \quad \Sigma/q_c(1 - \Pi + \Sigma.:) \tag{5.6}$$

Equation (5.5) can be seen to have a unique positive solution S because as S increases indefinitely from zero, the first term in $f(S)$ increases from zero towards q_a whereas the absolute value of the second term decreases from $p(1-P)q_c$ towards zero (since $c>1$). The equation is therefore favourable to solution by Newton's method and a good starting point is the value of S given by (5.1), which corresponds to $T=1$. At each iteration, the current S is increased by

$$\Delta\lambda = -\frac{f(\lambda)}{f'(\lambda)} = \frac{(1-\theta)\,pq_c\exp\{-\lambda(c-\tau)\} - q_a(1-\exp(-\lambda u))}{(c-\tau)(1-\theta)\,pq_c\exp\{-\lambda(c-\tau)\} + uq_a\exp(-\lambda u)} \tag{5.7}$$

All this has assumed that c, u, P and $\setminus$ are independent of T, but in the application of this gap acceptance model to the analysis of roundabout operation (Akçelik and Troutbeck, 1991; AUSTROADS, 1993), the values of these parameters for any stream are obtained from empirically based look-up tables in which an important determinant is the circulating flow, i.e. Tq_c.

Fortunately, this can be accommodated in the solution of (5.5) by Newton's method by using (5.6) to calculate the value of T corresponding to the current value of S at each iteration, and updating the values of c, u, P and $\setminus$ to the current value of Tq_c by reference to the look-up tables.

In that analysis of roundabout operation, traffic on each lane of a multilane entry is regarded as forming a separate stream. On such entries, the lane having the highest approaching flow is called the *dominant* stream, and the other streams are called *subdominant*. On a single-lane entry the one stream is regarded as dominant. Some of the look-up tables for the gap acceptance parameters give different values according as the stream concerned is dominant or subdominant.

The analysis of capacity requires an assumption about the distribution of approaching traffic on any multilane entry among the lanes. As the calculation of T proceeds, the plausibility of the assumed distribution of traffic for the current approaching flows (i.e. the original ones multiplied by the current value of T) among the various streams $i\,\hat{I}\,S_k$ for any one entry k should be kept under review. For example, if the current approaching flows and capacities would lead to very different queue-lengths in adjacent lanes where drivers would in practice choose lanes to make the queues about equal, the distribution of approaching traffic between the lanes should be adjusted to reflect this. The same will be true of any extension of the analysis described in Section 4 to allow for entries being used by more than one stream.

Equations (5.5) and (5.6) are solved in the foregoing way to give T_i for each stream i, and hence T^*.

6. DEPENDENCE OF STREAM CAPACITIES UPON DEMAND FLOWS

When the approaching flows in all streams are within their respective capacities, that is when, with $p_i=1$ for all i, $T^*\geq1$, the dependence of stream capacities on demand flows follows directly from (2.2), in which $q_{ej} = q_j$ for all j, so that for all i

$$Q_i \quad = F_i(q_{ci}) \quad = \quad F_i(\sum_j c_{ij}q_j)$$

and provided that $T^*>1$ with $p_i=1$ for all i,

$$\frac{\partial Q_i}{\partial q_j} = c_{ij} F_{i'}(q_{ci})$$

(6.1)

It then follows from (2.4) that $\partial Q_i/\partial q_j = 0$ for all movements j in stream i or in any other stream using the same entry as stream i.

When the approaching flows in one or more streams exceed their respective capacities, however, the demand flows q_j forming part of those overloaded streams are replaced in terms of their effects upon the capacity of any particular stream by the corresponding q_{ej}. These are determined jointly by the q_j concerned and by the capacities of the overloaded streams, as indicated in (2.2).

Let A be the set of movements that form part of overloaded streams, so that

$$A \quad = \{ j : q_{al} > Q_l \text{ for the stream } l \text{ such that } a_{lj} = 1 \}$$

Then for any stream i, by (2.1), (2.2) and (2.3)

$$q_{ci} = \sum_{s\notin\Omega} c_{is}q_s + \sum_{s\in\Omega} c_{is}q_s \sum_l (a_{ls}Q_l / \sum_t a_{lt}q_t)$$

(6.2)

It follows first that the relations $Q_i = F_i(q_{ci})$ lead to m simultaneous equation for the stream capacities Q_i. These equations are linear in the case of linear dependence of capacity upon circulating flows, but both in that case and in the case of dependence of capacity upon gap acceptance, their formulation requires knowledge of which streams are overloaded.

In practice the author has found that the Q_i can be calculated by the following stepwise procedure. This terminates after a small number of steps, usually about twice as many as there are streams, but the author is not aware of a proof that this must always be the case, or that the solution is unique.

Step 1   Suppose there is no circulating traffic.

    2   Choose the entry with the highest approaching flow as the *current entry*, and estimate the capacities of streams on this entry in the absence of circulating flow.

    3   Calculate the entering flows for streams using the current entry.

    4   Make *current* the next entry in the direction of the circulating traffic, update its circulating flow and calculate the capacities of streams using this entry.

    5   Repeat steps 3 and 4 until no entering flows have changed during one complete circuit of the roundabout.

    6   Stop: the current values of capacity, entering flows and circulating flows are mutually consistent and $q_{ej} = \min\{q_j, q_j; _i a_{ij} Q_i/q_{ai}\}$ for all j.

This stopping criterion is conservative - with practice it is usually possible to stop one or two steps short of this.

When equation (5.2) is used for $F(q_c)$, the value of $F(q_c)$ corresponding to $q_c = 0$, as required in Step 2, is $1/u$, as can be shown by taking the limit as $q_c \circledR 0$ of expression (5.2) after substituting for S in terms of q_c from (5.1).

Having identified A by calculation of the Q_i, equation (6.2) can be used further to derive simultaneous linear equations for the $\P Q_i/\P q_j$ as follows.

For any i, differentiation of (6.2) using (2.1) and substitution in the equation

$$\frac{\partial Q_i}{\partial q_j} = F_{i'}(q_{ci}) \frac{\partial q_{ci}}{\partial q_j}$$

shows that

(a)   for $j \circledcirc A$

$$\frac{\partial Q_i}{\partial q_j} = F_i'(q_{ci}) \left\{ c_{ij} + \sum_{s \in Q} c_{is} q_s \sum_1 \left(\frac{a_{ls}}{q_{al}} \frac{\partial Q_1}{\partial q_j} - \frac{a_{lj} a_{ls} Q_1}{q^2_{al}} \right) \right\} \tag{6.3}$$

(b)   for $j \in A$

$$\frac{\partial Q_i}{\partial q_j} = F_i'(q_{ci}) \left\{ c_{ij} \sum_1 \frac{a_{lj} Q_1}{q_{al}} + \sum_{s \in Q} c_{is} q_s \sum_1 \left(\frac{a_{ls}}{q_{al}} \frac{\partial Q_1}{\partial q_j} - \frac{a_{lj} a_{ls} Q_1}{q^2_{al}} \right) \right\} \tag{6.4}$$

These equations may well yield non-zero values of $\partial Q_i/\partial q_j$ even for movements j in stream i or in other streams using the same entry as stream i.

7. DEPENDENCE OF QUEUEING AND DELAY UPON DEMAND FLOWS

There is an extensive literature concerning approximate expressions, applicable to time-dependent queueing of vehicles, for queue-length L, average delay per unit time D over a given period, and average delay d to a vehicle arriving in a given period. The expressions most relevant to roundabout analysis are those of Kimber and Hollis (1979) and Akçelik and Troutbeck (1991). For a timeslice beginning at time $t = 0$ and during which the arrival rate or approaching flow is q and the capacity is Q, each regarded as independent of t within the timeslice, these expressions are of the general form

$$E \text{ (queue-length at time } t) = L(q,Q,L_0,t)$$

$$E \text{ (average delay/unit time over } (0,t)) = D(q,Q,L_0,t)$$

$$E \text{ (average delay to a vehicle joining queue in } (0,t)) = d(q,Q,L_0,t)$$

where L_0 is the queue-length at time $t=0$.

The implications of Section 6 for each of these expressions are similar, and are presented here for the last of them, because of its relevance to traffic assignment.

The average delay to a vehicle joining the queue in stream i at the roundabout in the period $(0,t)$ is estimated by

$$d_i = d(q_{ai}, Q_i, L_{0i}, t)$$

where L_{0i} is the value of L_0 for stream i.

Then for any movement j

$$\frac{\partial d_i}{\partial q_j} = a_{ij} \frac{\partial d}{\partial q_{ai}} + \frac{\partial d}{\partial Q_i} \frac{\partial Q_i}{\partial q_j}$$

(7.1)

In the modelling and estimation of the assignment of traffic to any network that includes the roundabout, each stream of traffic entering the roundabout is represented by a link in a node-link representation of the network. In this connection, the derivatives $\partial d_i/\partial q_j$ represent derivatives of link travel times with respect to sums of path flows. It has long been recognised that the second term of expression (7.1) is non-zero for many flows that enter the roundabout in streams other than stream i. The results of Section 6 show that if the roundabout is overloaded the second term can be non-zero even for flows j that enter the roundabout in stream i. The mechanism for this is that, whereas in the absence of overload, movements entering by the same entry as stream i can affect capacities of streams only up to the previous entry in the direction of circulation, if any of these entries is overloaded, the consequent effect on the entering flows of movements entering there can in turn extend to capacities of entries up to the entry previous to that one, and can therefore extend to the capacity of stream i itself.

8. EXAMPLE

The following description of a 4-arm roundabout and a peak-hour pattern of traffic demand is sufficient to enable each of the techniques described by Kimber (1980) and AUSTROADS (1993) to be applied to it. Traffic drives on the left.

The largest circle that can be inscribed within the outer kerblines of the roundabouts has a diameter of 50 m and the central island a diameter of 30 m so that circulating traffic travels in 2 lanes in a circulating carriageway width of 10 m. Each approach road has 2 lanes 4 m wide leading towards the roundabout, widening to 5 m per lane over the last 25 m of the approach. For each entry, the smallest radius of the nearside kerbline where the traffic enters the roundabout is 25 m and the paths of entering and circulating vehicles merge at an angle of $X/8$ radians.

The maximum acceptable degree of saturation is 0.85 for each stream. The pattern of approaching traffic in pcu/h is:

from:	to (movement number): South	West	North	East	in lane: Left	Right	Both together	Entry number
East	200 (1)	400 (2)	100 (3)	0	400	300	700	1
North	1200 (5)	800 (6)	0	100 (4)	1100	1000	2100	2
West	800 (9)	0	100 (7)	400 (8)	500	800	1300	3
South	0	200 (10)	1200 (11)	100 (12)	800	700	1500	4

All left-turners approach in the left lane, all right-turners in the right lane, and the balance of traffic in each lane goes straight ahead.

For the application of Kimber's technique, suppose that all traffic on each entry forms a single stream and the stream numbers are those of the entries.

Since there are no U-turners, these movements can be omitted from the numbering of movements, and with this amendment, the convention of Section 2 yields, with one stream per entry, the movement numbers shown in brackets in the foregoing table.

Then the matrix **a** is

$$(a_{ij}) = \begin{pmatrix} 1 & 1 & 1 & 0 & 0 & 0 & 0 & 0 & 0 & 0 & 0 & 0 \\ 0 & 0 & 0 & 1 & 1 & 1 & 0 & 0 & 0 & 0 & 0 & 0 \\ 0 & 0 & 0 & 0 & 0 & 0 & 1 & 1 & 1 & 0 & 0 & 0 \\ 0 & 0 & 0 & 0 & 0 & 0 & 0 & 0 & 0 & 1 & 1 & 1 \end{pmatrix}$$

and the matrix c is

$$(c_{ij}) = \begin{pmatrix} 0 & 0 & 0 & 0 & 1 & 1 & 0 & 0 & 1 & 0 & 0 & 0 \\ 0 & 0 & 0 & 0 & 0 & 0 & 0 & 1 & 1 & 0 & 0 & 1 \\ 0 & 0 & 1 & 0 & 0 & 0 & 0 & 0 & 0 & 0 & 1 & 1 \\ 0 & 1 & 1 & 0 & 0 & 1 & 0 & 0 & 0 & 0 & 0 & 0 \end{pmatrix}$$

8.1 Practical Capacity of the Roundabout as a Whole for Demand Flows Proportioned to the Given Ones

In equation (4.1) the A_i and B_i given by Kimber's formulae are $A_i = 3010$ and $B_i = 0.867$ for all i, and (4.2) gives

$$T^* = \min \{0.926, 0.837, 1.097, 1.041\} = 0.837$$

with the stream entering from the north as clearly the critical stream.

This outcome implies that (unless the two techniques were grossly inconsistent) in using equations (5.5) and (5.6) to calculate T^* using the AUSTROADS technique, it is necessary only to calculate T for the two lanes (each forming a separate stream in that technique) entering from the north.

First suppose that the allocation of approaching flow among the two lanes is fixed at 1100 and 1000 pcu/h in the left and right lanes respectively.

In calculating T for the dominant left lane, the first iteration begins with $T = 1.0$ and $Tq_c = 1300$. For the given layout and approaching flows and for this circulating flow, the look-up tables give $T = 2.04$, $c = 2.24$, $P = 0.521$ and $\setminus = 1.0$.

It follows from (5.4) that $S = 0.271$ and from (5.7) that $-S = -0.050$, giving a new S of 0.221, so that by (5.6) the second iteration starts with $T = 0.874$ and $Tq_c = 1137$.

In the second iteration, S is first increased to 0.237 as a result of revision of the parameter values given by the look-up tables, and then reduced to 0.222 because $-S = -0.015$.

The third iteration therefore starts with $T = 0.836$ and $Tq_c = 1087$, and S is increased first to 0.226 and then by $-S = 0.003$ to 0.229 to give $T = 0.843$, where calculation for the left lane is terminated.

When the values $T = 0.836$, $Tq_c = 1087$ are taken as the starting point for calculations for the subdominant right lane, the values of $-S$ in two iterations are -0.027 and -0.002, at which point the calculation is terminated with $T = 0.758$.

The right lane is thus critical and $T^* = 0.758$. This result is not comparable with the value $T^* = 0.837$ given by Kimber's technique because the flow of 1000 pcu/h in the right lane has been constrained to remain there despite the resulting degree of saturation being 112 per cent compared with only 99 per cent in the left lane.

If instead the traffic is reallocated at each iteration to balance the currently estimated degrees of saturation of the left and right lanes, the allocation of flows stabilises at about 1150 and 950 pcu/h in the left and right lanes respectively after two iterations with a common value of T. A third iteration in each lane separately then results in $-S = -0.001$ in each case, resulting in values of 0.821 and 0.832 for T, which are close to the value of 0.837 given by Kimber's method. The two techniques are thus consistent in their estimation of junction capacity in this example, when consistent assumptions are made about the allocation of traffic among the lanes on the same entry.

8.2   Stream Capacities with the Given Demand Flows

With the Kimber technique it can be seen from the values of T_i with $p_i = 0.85$ obtained in Section 8.1 that the stream entering from the north is the only one for which $q_{ai} > Q_i$, so that in equation (6.2) $A = \{4,5,6\}$ and the equations $Q_i = F_i(q_{ci})$ become

$$Q_1 = 3010 - 0.867\,(q_9 + (q_5 + q_6)Q_2/2100)$$
$$Q_2 = 3010 - 0.867\,(q_8 + q_9 + q_{12})$$
$$Q_3 = 3010 - 0.867\,(q_3 + q_{11} + q_{12})$$
$$Q_4 = 3010 - 0.867\,(q_2 + q_3 + q_6 Q_2/2100)$$

which on substituting the values of the q_j give

$$Q_1 = 762 \quad Q_2 = 1883 \quad Q_3 = 1796 \text{ and } Q_4 = 1955$$

The same result is obtained without knowledge of A by the stepwise procedure, starting with the entry from the north, after 7 executions of steps 3 and 4.

According to the AUSTROADS technique, equations (5.2) and (6.2) give the following capacities lane by lane.

Entry number	Capacity in pcu/h		Total
(from)	**Left lane**	**Right lane**	
1 (east)	656	408	1064
2 (north)	1048	867	1915
3 (west)	792	1025	1817
4 (south)	1076	899	1975

The total for each entry is similar to the capacity estimated by Kimber's technique for the entry regarded as a single stream except for the entry from the east, for which the AUSTROADS method estimates a substantially higher capacity.

For the values of the Q_i given by the Kimber technique, calculation of $\partial Q_i/\partial q_j$ using equations (6.3) and (6.4) results in the matrix

$$
\left(\frac{\partial Q_i}{\partial q_j}\right) = \begin{pmatrix}
0 & 0 & 0 & 0.740 & -0.037 & -0.037 & 0 & 0.716 & -0.151 & 0 & 0 & 0.716 \\
0 & 0 & 0 & 0 & 0 & 0 & 0 & -0.867 & -0.867 & 0 & 0 & -0.867 \\
0 & 0 & -0.867 & 0 & 0 & 0 & 0 & 0 & 0 & 0 & -0.867 & -0.867 \\
0 & -0.867 & -0.867 & 0.296 & 0.296 & -0.481 & 0 & 0.286 & 0.286 & 0 & 0 & 0.286
\end{pmatrix}
$$

in which the elements can be seen to be of the expected signs and relative magnitudes. The most interesting is element (4,12), which shows that the capacity of stream 4 is increased by an increase in its own right-turning movement. This is because the right-turning entering flow from stream 2, which contributes to the circulating flow for stream 4, is thereby reduced.

ACKNOWLEDGEMENTS

A good deal of the content of this paper arises from time spent by the author in 1997 as Visiting Erskine Fellow in the Department of Civil Engineering at the University of Canterbury, Christchurch, including helpful discussions with Dr Alan Nicholson. The author is also grateful to his colleagues Dr. Benjamin Heydecker and Dr. S.C.Wong for useful comments relating to Section 6.

REFERENCES

Akçelik R (1998) Roundabouts: capacity and peformance analysis. *Research Report ARR 321.* Vermont South: ARRB Transport Research.

Akçelik R and R Troutbeck (1991) Implementation of the Australian roundabout analysis method in SIDRA. In U.Brannolte, ed: *Highway Capacity and Level of Service*. Rotterdam: Balkema, 17-34.

Allsop R E (1972) Estimating the traffic capacity of a signalized road junction. *Transportation Research*, **6**(3), 245-255.

AUSTROADS (1993) Roundabouts. *Guide to traffic engineering practice, Part 6*. Sydney: Australian Association of Road and Traffic Authorities.

Chard B (1997) ARCADY health warning: Account for unequal lane usage or risk damaging the Public Purse. *Traffic Engineering and Control*, **38**(3), 122-132.

Cowan R J (1975) Useful headway models. *Transportation Research*, **9**(6), 371-375.

Kimber R M (1980) The traffic capacity of roundabouts. *TRRL Laboratory Report LR942*. Crowthorne: Transport and Road Research Laboratory.

Kimber R M and E M Hollis (1979) Traffic queues and delays at road junctions. *TRRL Laboratory Report LR909*. Crowthorne: Transport and Road Research Laboratory

Wong S C (1996) On the reserve capacities of priority junctions and roundabouts. *Transportation Research*, **30B**(6), 441-453.

CURRENT AND FUTURE RESEARCH STRATEGIES AT TRL

Rod Kimber, Research Director, Transport Research Laboratory

1. INTRODUCTION

Transport is at a turning point, and the role of transport research is more than ever vital. While car dominance races ahead, it is easy to see that widespread aspirations for clean, safe, and sustainable living environments present a real challenge. Governments worldwide, and our own UK Government in particular, are turning away from "predict and provide" policies for roads, and strongly embracing integration of transport modes and provision. The scope for benefits from top quality research is enormous. New technology will enable all kinds of new processes to aid integration. But how will people react when confronted with automatic systems, charging involving mode and route choice, better but still imperfect information, and so on?

TRL has been at the forefront of transport research for many years. When it was privatised in 1996, the successful bid came from the Transport Research Foundation, a non-profit distributing Foundation currently of 85 Sector Members drawn from all parts of the transport and transport research community. The Foundation guarantees impartiality and independence. My task here is to outline the sort of research we currently do, and how I see it developing. TRL's range of expertise covers all of the major disciplines: from civil engineering, environmental issues, and traffic and transport operations through to the behaviour and safety of road users and their vehicles. Mathematical and statistical techniques are used widely. I can only touch on a few areas, but I will develop a couple in a little more detail later on. What we are greatly interested in is the practical interface: mathematical modelling becomes extremely powerful when coupled with careful testing and experimentation. Some obvious examples are the growth in finite element modelling of a wide range of interactions: from bridges and other civil engineering structures, through vehicle-infrastructure and vehicle-vehicle collision processes, ultimately to mixed biomechanical systems involving hard and soft tissue.
Mathematical models of such systems are extremely powerful provided that they can be pinned down by the results of practical tests.
Similarly, statistical models for accident occurrence are utterly dependent on sound and usually extensive data, often underpinned by designed experiments. The same applies to transportation and route assignment models.

2.   THE FUTURE

Many areas of transport research are likely to undergo change from the confluence of new technologies and policies to limit or reduce the adverse impact of travel - congestion, pollution, and accidents. Coupled with the potential policies and practices needed to contain growth in car usage - including pricing and similarly challenging mechanisms - there is likely to be much argument about how far particular technologies could be used. The automatic enforcement of traffic offences is an obvious example. It is perfectly feasible to use IT and communications technologies to detect and record offences, and to levy charges or penalties. The trend in that direction is obvious from the successful introduction of speed and red light cameras; but going substantially further technologically and institutionally, involves key decisions about the traditional role of the police. How such systems could work, and what people's behavioural responses would be are important seams of research.

Telematics - much hyped, but genuinely mould-breaking at least in the longer run - should enable radical changes to take place in operations. Whether driverless cars really are viable is an open question (the obvious answer is that they could be at some stage, but is it in the next quarter century?). But there is still enormous potential even if that question is left aside. Communications and control could enable much more efficient uses of network capacity - road, rail, bus, tram, walk, cycle - and could couple multi-mode journeys far better than at present. Developments in automatic banking and payment could begin to bring in *transactional integration:* through- ticketing in a very broad sense, encompassing road and parking charges as well as payments for the public transport stages of a journey. Telematics could also be used for example to apply vehicle speed control in built up areas; and this would extend the benefits of traffic calming enormously, without the physical intrusion.

These things all involve large unknowns of one sort and another, but unknowns that research is perfectly capable of answering. The benefits of such research go way beyond the potential benefits of the technology itself - large though those might be. Enabling the best policy and practical decisions to be made carries enormous benefits in itself. Evidence-based decision making carries impact and moves policies forward extremely positively.

Technology and improved management systems using IT and telecommunications could also be brought to bear on the physical transport infrastructure itself. Roads, bridges, tunnels, rail, and so on could all be operated more efficiently and cheaply if the infrastructure were self monitoring. Intelligent structures could diagnose, on-line, their own fatigue history, and their safety margin before the need for serious repair or rebuilding. Given the scale of the transport infrastructure - roads alone in the UK have a replacement cost upwards of £100bn of which some £60bn is in the trunk road network - and given the uncertainties in the load factors and reserve capacity in much of it, the scope for improved management is extremely large. Moreover maintenance and repair always involves disruption to moving traffic, with the large economic penalties that carries. If we could build-in widespread intelligent sensing and communications links the scope for savings is very big.

3. TWO SPECIFIC EXAMPLES

To illustrate a small sample of TRL's current activities, I will present two examples. Both come from safety and accident analysis: the first describes some work we have been doing on area wide accident prediction and reduction, and the second on the application of computer modelling to vehicle safety research.

3.1 Network Wide Casualty Evaluation and Reduction

The movement of traffic in road networks has been much studied, and poses intriguing problems. Traffic assignment models predict the routes drivers will take across a network of roads to get from origin to destination, according to the conditions they encounter *en route*, the length of the route, and other factors. Two well known examples are CONTRAM (Taylor, 1990) and SATURN (van Vliet, 1982 *et seq*). Such models can be used to assess the effect of a traffic management scheme on congestion, taking account of re-routing effects. But although they can estimate the changes in flows that will occur, and the area-wide effects of a scheme on queues, delays, and vehicle operating costs, they do not take account of the potential effects on safety. In contrast, programmes for the design of individual road junctions, such as ARCADY (Webb & Peirce, 1992), PICADY (Webb & Peirce, 1993), and OSCADY (Crabtree & Harrison, 1993) can take account both of delay and safety at individual roundabouts, priority junctions, and traffic signal junctions respectively.

Some two-thirds of all casualties occur on networks of roads in built up areas, where traffic flow patterns are very dependent on the state of congestion as a function of time. It would therefore be extremely valuable to be able to predict across a whole network the effects of traffic management measures on casualties. To do this two components are needed: (a) an assignment model to predict the traffic flows on all road links and junction turning movements, and (b) a set of relationships between the magnitude and type of intersecting flows (including pedestrian flows) and the accident risk, ie the number of casualties of a given severity per unit time. Over the past several years, TRL in conjunction with Southampton University and others, have pioneered the development of accident risk models of this type, and much is now known about the best form and parameters for the predictive equations required (Maher and Summersgill 1995). The general form of these predictive relationships is:

$$A = kQ_a^\alpha Q_b^\beta \exp(\Sigma \gamma_{ij} D_{ij} + \Sigma \delta_i G_i)$$

where A is the accident frequency per unit time, Q_a and Q_b are the traffic flows, or functions of them, D_{ij} and G_i are dummy and continuous variables defining the junction layout and operation, and k, α, β, γ, δ are parameters to be estimated. This expression relates to accidents arising out of collisions between vehicles in the traffic streams a and b. Similar relationships apply for pedestrian-vehicle interactions, but with pedestrian flows substituted for one of the traffic flows. Recent work has combined these relationships with CONTRAM. Thus for every element within a road network, the traffic flows are predicted by means of CONTRAM, from a (time dependent) origin destination matrix of trips, and the casualties are predicted by applying equations of the above form to the sets of interacting flows. This produces a combined model which can (i) predict how traffic will redistribute after the implementation of a new traffic management scheme or safety scheme, (ii) estimate the resulting accident pattern, (iii) reduce the risk of wasting resources on unsuccessful measures.

3.1.1 Case Study: a Town in Southern England

Figure 1a shows the sketch plan for a largely residential and shopping area connecting to a town centre, for which we have looked at a hypothetical traffic management scheme. The objective was to remove through-traffic from unsuitable routes. There were two main parallel roads running east-west, leading from out-of-town sources (XY) into the town centre. The more northerly, A2, is a wide single carriageway A road. The other main east-west road, B1, is a B road with mixed residential use and shopping. Speeds are high, but there is a lot of pedestrian activity. Minor residential streets run between the A2 and B1 roads. Other roads are A1, a fast dual carriageway connecting to a near-by motorway, B2 bringing traffic from the southwest, and B3 connecting the A2 and B1 on the edge of the built up area.

One of the main problems in the network is that besides local traffic, B1 is carrying a substantial flow of traffic to and from the town. Much of it is travelling at speed and would be better suited to A2. Thus an objective of the proposed scheme is to discourage through-traffic from using B1 and the roads C between A2 and B1. To achieve that the main treatments suggested were: (i) to close the junctions of the residential streets C1 and C3 with A2 and B1, and to place speed humps along the whole length of C2 and C4, making these roads unattractive as links between B1 and A2; (ii) to convert the junction of C2 with A2 from major/minor priority control to a roundabout, so as to reduce accidents to vehicles turning right from C2; (iii) to convert other major/minor junctions along A2 and B1 to mini-roundabouts which should have lower accident rates; (iv) to place speed cushions along the section of B1 where the shops lead to a particularly high level of pedestrian activity; (v) to place speed humps along C6 to reduce speeds and deter through-traffic from C6 and B4 as an alternative to the more appropriate A1; (vi) to make C5 a "residents only" street, introducing a barrier at the mid-point.

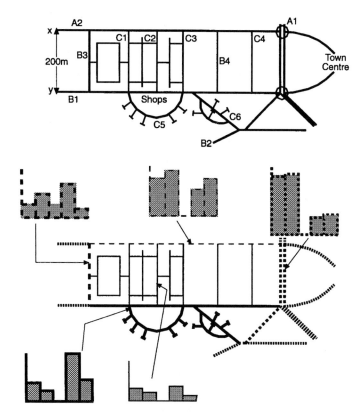

Figure 1: Town plan (a: upper, b: lower). Each histogram represent delay (left-hand pair) and accidents (right-hand pair). Each pair is before (left) and after (right)

Modelling. The prior OD matrix of trips was not available for this exercise, and was therefore synthesized to produce link flows consistent with those observed on the physical network . This matrix (time dependent) was held unchanged throughout the modelling: ie it was assumed that the introduction of the traffic management measures would not cause any significant changes in the origin-destination pattern, the mode of travel, or the time of day at which journeys were made. *Before* and *after* runs were then made using CONTRAM. The main changes between them were to represent the measures (i) to (vii) by : (a) removing the connections between some links to represent the junction closures, (b) changing the specification of controlled and controlling links to represent the change in priority from major/minor to roundabout, and (c) reducing the speeds along the calmed links to represent the effects of the humps and cushions. These changes in speed were based on those observed at similar sites in practical experiments.

Results. The main results from modelling this scheme are shown in Figure 1(b). The aggregate journey time changed little (an increase of about 1% from 844,000 to 850,000 hrs pa, whilst the queuing time reduced by 6% from 92,000 to 84,000 hrs pa, the reduction being offset by the increased journey distances on the new network. In aggregate, personal injury accidents across the network however reduced by some 32%, from 50 before to 34 pa after. Much of the improvement was predicted to come from a reduction in accidents along the B1 because of reduced flows, reductions in accidents on the calmed and closed links, and reductions in accidents at the junctions converted from major/minor priority control. Offsetting this was a small increase in accidents along the A2, reflecting the higher flows. The traffic flow along C2 was predicted to increase, but because this traffic was restrained by the road humps, there was little change in the predicted number of accidents: the one effect balancing out the other. The essential source of the benefit was from removing traffic from the other roads.

This of course is only one rather simple case study. The effects of changes depend very much on the particular configuration and properties of the network itself. But what this example illustrates is how models of this type can be used to predict shifts in accident frequency aggregated across whole networks, and how the pattern of accidents might be expected to change.

3.2   Second Example. Computer Modelling Applied to Vehicle Safety

The second example belongs to *secondary* safety - devising ways of predicting the injury consequences of collisions, and reducing them. Structural modelling, and particularly the *finite element method*, has transformed engineering design over the last decade or so. Shapes of more or less any complexity can now be analysed, and materials with highly non-linear and unusual characteristics can also be included, even liquids. The work falls into three main areas: *Vehicle Engineering* - vehicle structural behaviour, including the interaction of anthropomorphic dummies with the vehicle, airbags and restraint systems; *Biomechanics* - understanding how injuries occur in vehicle accidents by modelling the human body; and *Roadside Structures* - the impact performance of safety barriers, bridge parapets, and other roadside furniture.

3.2.1. Principles of the Method

The objective is to predict the physical displacements, and their derivatives, of the parts of complex mechanical structures in response to the applied forces which result from impacts, and the associated stress and strain fields. This is achieved by (i) dividing the structure up into elements sufficiently small that the shape of the displacement field across them can be approximated by simple functions; (ii) establishing the governing equation, which becomes a summation of integrals over elements, taking the form for each element:

$$\int_v \sigma^t \varepsilon dv = \int_v p_v^t u dv + \int_s p_s^t u ds$$

where t denotes transpose, σ, ε, p_v, p_s,u are vectors respectively of: the stress and strain components at any point, the body and surface forces, and the displacements, and v and s denote volume and surface integrals; (iii) approximating the behaviour of each element using *shape* functions which describe the form of the displacement field, ensure continuity between elements, and account for the degrees of freedom from element to element; the magnitudes of these shape functions then become the unknowns in the governing equation; (iv) integrating each element numerically by fitting a polynomial to the shape function and employing gaussian integration; (v) incorporating the summation of integrals into the governing equation and rearranging it in terms of the virtual displacements; (vi) equating the coefficients of the virtual displacements to zero, rearrange the terms into the form **K.u** =**R** where **K** is the stiffness matrix and **R** the vector of applied loads; (vii) applying the loads and boundary conditions before solving for **u** and hence the internal forces and stresses.

To account for dynamic movement the governing equation **K.u** =**R** is modified to include accelerations, velocities, and time-dependence in the forces: **K.u** + **C.v** + **M.a** = **R**(t), where **C** and **M** are the damping and mass matrices. This means a repetitive process has to be introduced to calculate **C** and **M** as well as **K**. Gross displacements can be very high relative to local deformations and this can have numerical implications. If calculations are carried out relative to a fixed "global" set of axes, then local deformations can be lost through rounding errors due to the very high global displacements. Local element frames of reference therefore have to be updated to account for global displacements.

These methods when used for simulating real life impacts require extensive numerical calculation. Properly applied and cross checked against physical measurements on test experiments involving full-scale impact, the method is an extremely powerful way of exploring a wide range of parameters and situations.

3.2.2 Vehicle Engineering

By developing a verified model of a specific test, additional 'data' can then be generated to those physically measured during the test itself, for instance about the energy absorbed by a specific component, or the acceleration of a point on the structure that was not instrumented during the test. Figure (2) shows just one example: the simulated movement of anthropomorphic dummies in a minibus.

Figure 2: Minibuses: unrestrained rear passenger

The seatbelt anchorages on coaches and minibuses are generally attached to the seat itself rather than the vehicle structure. This means that the seat and its fixings to the floor take the full loading induced by the restrained occupant. When the occupant of the seat behind is unrestrained as in the Figure, additional loading is caused by contact with the rear of the seat. As part of a research programme for DETR, TRL is carrying out full-scale tests to determine the practical limits of seat belt anchorage strength in minibuses. To assist the development of optimised seat and floor systems, typical floors and seats have been modelled using the finite element method. Models were first validated against the results of full scale tests and then used to explore the effects of different floor stiffness, occupant size, and seats facing on the occupants. Output data included the effects on the impact locations, forces and velocities of the occupant's knees and head on the seat in front, and forces in the seat and floor structures.

3.2.3 Biomechanics

Aided by the recent advances in finite element techniques, particularly the development of improved material models, biomechanics is a rapidly growing area. TRL's interest is in the study of injuries, the precise conditions for their occurrence, and the development of practical means for determining the likelihood of injury in a given car impact. Full scale physical testing is complicated by the range of sizes and physiques of human subjects. Modelling can therefore be used to extend the results, or to "normalise" results for non standard cases. Models are also used to help in determining the injury mechanism. By building a model of a part of the human body and verifying it against test data as far as possible, the stresses, forces, movements, pressures, etc can be deduced under a range of circumstances. By comparison with observed injuries under similar conditions, assessments can be made about the causes. As an example, Figure 3 shows the *simulated* pattern of stresses (shaded areas) in the lower leg when subjected to a pendulum impact to the ball of the foot - equivalent to some forms of footwell injury in frontal collisions.

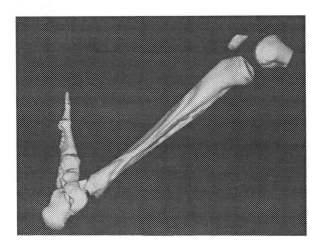

Figure 3: Lower leg: impact to ball of foot. Shading represents stress

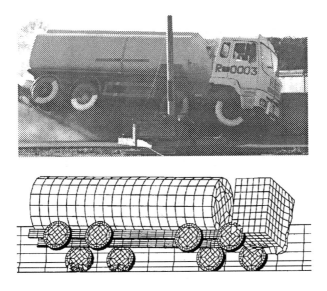

Figure 4: Tanker/barrier impact

3.2.4 Roadside Restraint Systems

Vehicle impacts with roadside structures are different in that the overall timescale of an event - car to safety barrier for example - is of the order of seconds, whereas the key events in a frontal vehicle impact or biomechanical impact happen on a timescale of milliseconds. Relatively high displacements are typical of impacts with safety barriers, and trajectories and gross motions are of particular interest, rather than detailed studies of the deforming vehicle structure. Mass, inertia, and frictional effects therefore dominate over stiffness effects. Figure (4) is one example of many investigations: it is the representation of a bulk-liquid-carrying tanker in impact with a safety barrier. This case was an interesting one, since the first full-scale test resulted in instability in the tanker which bucked and finally rolled over onto the side opposite the barrier. This unusual behaviour might have arisen from liquid surge within the tanker, which was unbaffled. Full-scale physical tests of this sort are not trivial, and before repeating the test, we modelled the tanker-barrier interaction carefully, over a range of parameters. One particular case is shown in the lower part of Figure (4). In no cases did the tanker roll over. Careful mechanical inspection of the original tanker revealed that the forward tank mounting had failed - which would not normally happen. The inference was that this was probably the cause of a large torsional deformation of the chassis during the anomalous test. Finally a further full-scale physical test using another tanker - in which there was no mounting failure - conformed extremely closely to the modelling predictions, and the tanker did not roll over.

4.   ACKNOWLEDGEMENTS

I am grateful to Ian Burrow and Charles Oakley for their valuable contributions and help in assembling the examples of Section 3, and to George Washington University for the supply of the original lower leg model.

5.   REFERENCES

Crabtree M R and Harrison S (1993). OSCADY User Guide. AG22, Transport Research Laboratory.

Maher M J and Summersgill I (1996). A comprehensive methodology for the fitting of predictive accident models. *Accid. Anal. and Prev.*, 28, 3 pp281-296.

Taylor N B (1990). CONTRAM 5 - an enhanced traffic assignment model. RR 249

Van Vliet D (1982). SATURN - a modern assignment model. *Traffic Engineering and Control*, **24,2**

Webb P J and Peirce J R (1992). ARCADY User Guide. AG17. Transport Research Laboratory.

Webb P J and Peirce J R (1993). PICADY User Guide. AG18. Transport Research Laboratory.

BILEVEL OPTIMISATION OF TRANSPORTATION NETWORKS

J. Clegg and M.J. Smith, York Network Control Group, Department of Mathematics, University of York, England.

ABSTRACT

The need to reduce traffic congestion is becoming increasingly important. The means of achieving this aim involves optimising parameters such as traffic signal green-times, road prices and public transport fares. This paper will describe a new bilevel method of optimising traffic signals and prices. The method uses the steepest descent direction together with projections in order to define a descent direction which will reduce the objective function subject to the over-riding necessity to be in equilibrium.

The paper will provide a description of the bilevel method together with results on two simple problems. Optimisation is performed on two functions simultaneously; the equilibrium function E (which must have value zero for equilibrium) and the objective function Z which is minimised subject to the constraint that E is zero. For most traffic problems equilibrium is not mathematically well behaved and therefore the method approaches equilibrium in stages, at each stage it minimises Z whilst avoiding the difficult equilibrium region.

1. INTRODUCTION

This paper is concerned with a bilevel optimisation procedure which is particularly geared towards problems in traffic networks. The need to reduce urban traffic congestion is becoming increasingly important; and present thinking is that this reduction in congestion may be assisted by using appropriate traffic signal settings, parking charges, road prices and public transport fares, so as to manage the network. The "optimal control problem" central to this paper has attracted much theoretical attention and is also an important and perhaps urgent day-to-day concern for traffic engineers and traffic planners in real life.

This paper proposes a steepest descent, bilevel method of calculating control variables which seeks to take correct account of drivers' route-choices (and travellers' mode-choices), in the sense that travellers are supposed to choose routes and modes which cause the overall traffic distribution to be in equilibrium as specified essentially by Wardrop (1952). The method is designed to produce locally optimal controls in a variety of steady state models. The steepest descent method described here relies in a fundamental manner on Lyapunov techniques similar to those demonstrated in Smith (1984a,b,c).

The following section of this paper provides an outline of the method while the subsequent sections supply two computational results to illustrate the method and a variant. The first problem is an "artificial" problem; the second is to optimise travel time plus construction costs by choosing appropriate link capacities

2. A NATURAL DIRECTION FOR BILEVEL PROBLEMS

In the following, link flows, costs and delays will be represented by a vector z and controls represented by a vector λ. The function $E(z, \lambda)$ will be a measure of disequilibrium; E will be non-zero and positive away from equilibrium and zero if and only if equilibrium has been reached. The function $Z(z, \lambda)$ is the objective function and contains all quantities (for example total travel time) which we may wish to minimise (under the constraint that the equilibrium function E is zero). Possible equilibrium functions E are given in Smith (1984,a,b,c). Let the vector $x = [z, \lambda]$ represent values of all variables and we require to travel in the direction of a vector δ which reduces E towards zero whilst reducing Z, or reducing any increase in Z, as much as possible.

Naturally -grad E is the direction which reduces E most rapidly. Consider the half-space $\{\delta; \delta \cdot (\text{grad } E) \le 0\}$ of locally non-increasing E. Travelling in any direction within this half-space will not increase E. The method is essentially to follow at each point the direction

$$- \text{grad } E + \text{Proj} (- \text{grad} Z), \qquad (2.1)$$

where Proj (- grad Z) is -(the gradient of Z) projected onto the half-space $\{\delta; \delta \cdot (\text{grad } E) \le 0\}$ to which grad E is normal. The first half of equation (2.1) uses the steepest descent direction, - grad E, to reduce E; whereas the second half of (2.1) attempts to reduce Z, but instead of taking the direction - grad Z this gradient is projected onto the half space of non-increasing E. The direction takes account of the bilevel nature of the problem since reducing E always gets priority. Once E is reduced below some initial target value, say $E < \varepsilon$ (where $\varepsilon > 0$), an attempt is made to reduce the function Z more quickly by emphasising the term second term in (2.1). This second direction is followed reducing Z subject to $E < 2\varepsilon$ (say). This two stage method can be described using the following single direction which ensures that within both parts of the trajectory the direction followed changes in a continuous manner:

$$\max\left(\left(E/\varepsilon-1\right),0\right)\left(-\text{grad } E\right)+\max\left(\left(1-E/\varepsilon\right),0\right)\left(-\text{grad } Z\right)+E/\varepsilon\,\text{Proj}_{\perp gradE}\left(-\text{grad } Z\right). \qquad (2.2)$$

To reduce E some initial target value of ε is chosen; to then reduce Z a larger value of ε (say 2ε) is chosen. This is repeated with $\varepsilon/2$, $\varepsilon/4$, $\varepsilon/8$,....

It proves beneficial to define desc E to be the direction of steepest descent of E, $-gradE/\|gradE\|$, to define desc Z similarly and to change direction (2.2) to (putting x in explicitly)

$$\delta_\varepsilon(x)=\max((E(x)/\varepsilon-1),0)\text{desc } E(x)+\max((1-E(x)/\varepsilon),0)\text{desc } Z(x)+E(x)/\varepsilon\,\text{desc}\,(Z,E)(x),\,(2.3)$$

where $\text{desc}(Z,E)(x)$ is the steepest descent direction desc $Z(x)$ projected onto the half-space defined by desc E.

3. AN ALGORITHM FOR SOLVING THE BILEVEL PROBLEM

Let D be defined by:

$$D(x)=\|\text{desc }(Z,E)(x)\|\text{ for all } x.$$

The proposed algorithm depends on the two functions $E(.)$ and $D(.)$ where E is as above.

The direction $\delta_\varepsilon(x)$ in (2.3) is related to these two functions as follows. If $E(x)>\varepsilon$ then E declines toward ε in direction $\delta_\varepsilon(x)$. If $E\le\varepsilon$ then Z declines in direction $\delta_\varepsilon(x)$ toward a minimum of Z in $\{x;\ E(x)\le\varepsilon\}$; at a minimum $D(x)=0$ so as Z declines $D(x)$ must tend to 0.

Let x_0 be any given starting point, let $E_0=E(x_0)$ and $D_0=D(x_0)$. From x_0 follow (2.3) with $\varepsilon=E_0/8$ until $E(x)<E_0/6$ and then follow (2.3) with $\varepsilon=E_0/4$ until $D(x(t))\le D_0/4$. Suppose that x_1 is a point reached by this two part trajectory. Then $E(x)\le E_0/4$ throughout the second part of this trajectory and so

$$E(x_1)\le E_0/4 \quad\text{and}\quad D(x_1)\le D_0/4.$$

Repeat the above two-part procedure to obtain a sequence of points $x_0,x_1,x_2,\ldots$, satisfying

$$E(x_n)\le E(x_{n-1})/4\le\ldots\le E(x_0)/4^n=E_0/4^n,$$

$$D(x_n) \le D(x_{n-1})/4 \le \ldots \le D(x_0)/4^n = D_0/4^n.$$

Thus $E(x_n) \to 0$ and $D(x_n) \to 0$ as $n \to \infty$. As $n \to \infty$, $E(x_n) \to 0$; and so x_n converges to the set $\{x; E(x) = 0\}$ of equilibria and $D(x_n) \to 0$ so the gradient of Z is increasingly opposite to the gradient of E.

It seems natural to measure the lack of weak local-optimality of x by the sum $E(x) + D(x)$. Then the lack of weak-optimality $E(x_n) + D(x_n) \to 0$ as $n \to \infty$ for our sequence above. It further seems natural to agree that x is *asymptotically weakly locally optimal* iff x is the limit of a sequence of points whose lack of weak local optimality tends to zero. With these agreements any limit point of our sequence is asymptotically weakly locally optimal.

4. THE BILEVEL METHOD WITH TWO EQUILIBRIUM FUNCTIONS

In this section we extend the bilevel procedure (outlined in sections 2 and 3) by splitting a single equilibrium function E into two equilibrium functions E_1 and E_2. Both these functions are positive away from equilibrium and they both must be zero at equilibrium. The advantage of splitting the equilibrium function E into E_1 and E_2 is that the descent direction becomes more constrained as will be seen below.

The direction -grad E_1 reduces the function E_1 most rapidly and the direction $-$grad E_2 reduces E_2 most rapidly. Let $C = \{\delta; \delta \cdot (\text{grad } E_1) \le 0 \text{ and } \delta \cdot (\text{grad } E_2) \le 0\}$ be the set of vectors which do not increase either of the functions E_1 or E_2. The set C is the intersection of two half-spaces and is therefore a cone. If we wish to reduce Z as much as possible without moving further away from equilibrium then it is natural to follow the projection of $-$grad Z onto the cone C. We therefore choose the direction

$$- \text{grad } E_1 - \text{grad } E_2 + \text{Proj}_C (- \text{grad } Z), \tag{4.1}$$

where $\text{Proj}_C (- \text{grad } Z)$ is the steepest descent direction for Z, $-$grad Z, projected onto the cone C. The first part of equation (4.1) uses the steepest descent directions, - grad E_1 and $-$grad E_2 to reduce E_1 and E_2; whereas the last part of (4.1) attempts to reduce Z, by projecting the steepest descent direction - grad Z onto the cone C of non-increasing E_1, E_2. The method ensures that reaching equilibrium still always gets priority. To introduce continuity, let ε be positive, let $E = \max(E_1, E_2)$ and revise direction (4.1) to

$$\max ((E/\varepsilon - 1),0) (- \text{grad } E_1 - \text{grad } E_2) + \max ((1 - E/\varepsilon),0)(- \text{grad } Z) + E/\varepsilon \text{ Proj}_c (-\text{grad } Z).$$

The same method as that outlined in section 3 can be used for the discrete process of steps towards the optimal solution, the only difference being that the first stage of the process stops only when both of the conditions $E_1 < \varepsilon$ and $E_2 < \varepsilon$ are satisfied.

5. SOME SIMPLE APPLICATIONS

The bilevel method described in sections 2 and 3 has been tested on two simple applications taken from Marcotte (1988). The first very simple example involves minimising the function

$$Z = x + 2y$$

subject to an "equilibrium function" E being zero where

$$E = (2 - x - y)^2 + (x - 1)_+^2 + y_-^2 + x_-^2,$$

and

$$(x)_+^2 = x^2 \text{ if } x > 0 \text{ and zero otherwise,}$$

$$(x)_-^2 = x^2 \text{ if } x < 0 \text{ and zero otherwise.}$$

The equilibrium function within a real traffic problem consists of constraints very similar to these. This simple collection of constraints was chosen merely to test the method using a problem already in the bilevel literature. The above bilevel problem has an optimal value $Z = 3$, and naturally the equilibrium function E should be zero at this optimal solution. Figure 1 shows the progress of the bilevel optimisation. The zig-zag nature of the progress towards the optimal point is caused by the fact that the method approaches equilibrium in stages, and at each stage the objective function Z is minimised within the constraint that E does not increase more than some specified amount. It can be seen from Figure 1 that the optimal solution is reached successfully even though the procedure involved a very basic step length at each stage of the algorithm.

The second application, also taken from Marcotte (1988), is a simple network problem. It is derived from a network which has one origin and one destination with two links between the two nodes. The flows along the links are represented by x_1, x_2; the capacities by y_1, y_2 and congestion functions are given by $f_1(x_1) = 2x_1 / (1 + y_1)$ and $f_2(x_2) = 8x_2 / (1 + y_2)$. We aim to minimise the following objective function representing total cost (congestion cost and construction cost)

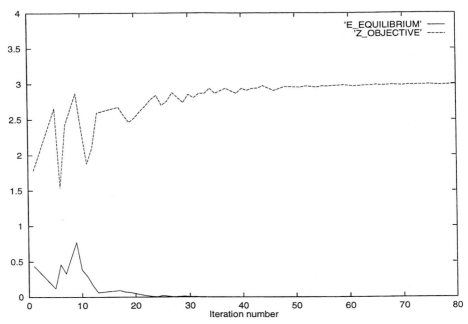

Figure 1. Convergence for first simple problem

$$Z = x_1\left[\frac{2x_1}{1+y_1}\right] + x_2\left[\frac{8x_2}{1+y_2}\right] + y_1 + y_2$$

subject to the equilibrium function E being zero where

$$E = \left\{\left[\frac{2x_1}{1+y_1} - \frac{8x_2}{1+y_2}\right]x_1\right\}_+^2 + \left\{\left[\frac{8x_2}{1+y_2} - \frac{2x_1}{1+y_1}\right]x_2\right\}_+^2 + \left(x_1 + x_2 - 1\right)^2 + x_{1_-}^2 + x_{2_-}^2 + y_{1_-}^2 + y_{2_-}^2 .$$

The first two terms of the above equilibrium function ensure that drivers follow their cheapest route, the middle term ensures that the demand is correct and the last terms are the constraints that all variables must be non-negative. The method outlined in sections 2 and 3 of this paper produced an optimal value for the objective function of $Z = 1.591$ compared with the optimal value quoted in Marcotte (1988) of $Z = 1.578$. Since this method converged to a value slightly greater than the true optimal value, the method outlined in section 4 was tested. This involved splitting the equilibrium function into the following two parts

$$E_1 = \left\{\left[\frac{2x_1}{1+y_1} - \frac{8x_2}{1+y_2}\right]x_1\right\}_+^2 + \left(x_1 + x_2 - 1\right)^2$$

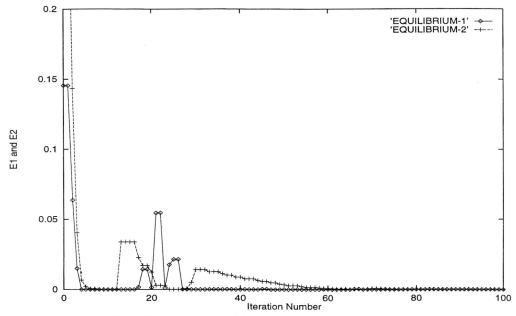

Figure 2. Convergence of the two equilibrium functions

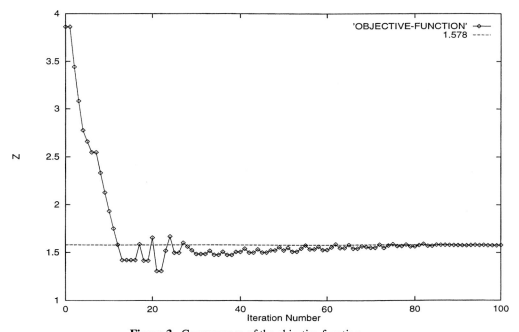

Figure 3. Convergence of the objective function

$$E_2 = \left\{ \left[\frac{8x_2}{1+y_2} - \frac{2x_1}{1+y_1} \right] x_2 \right\}_+^2 + x_{1-}^2 + x_{2-}^2 + y_{1-}^2 + y_{2-}^2 ,$$

with the same objective function Z as above.

Using this method the optimisation succeeded in reaching Marcotte's value, 1.578. Figure 2 displays the convergence of E_1 and E_2 to zero (i.e. equilibrium) and figure 3 shows the approach of Z to its optimal value.

The main difference between the two methods, the first outlined in section 2 and the second discussed in section 4, is that the direction chosen is more constrained when using the procedure in section 4. This is because the equilibrium descent direction is restricted to being within a cone whereas in the simpler method of section 2 the equilibrium descent direction is within an entire half-space. The particular problem above, for which both methods have been used, shows how this more constrained direction can be useful. In general, the method outlined in section 4 can be extended to cases where there are more than two equilibrium functions, and as the number of equilibrium functions increases so the descent direction will become even more constrained.

9. REFERENCES

Marcotte P (1988), "A note on a bilevel programming algorithm by Leblanc and Boyce", Transpn. Res. B, Vol 22B, No 3, pp 233-237.

Smith M J (1984a), "A descent algorithm for solving a variety of monotone equilibrium problems", Proceedings of the ninth International Symposium on Transportation and Traffic Theory, The Netherlands, VNU Science Press, Utrecht, 273-297.

Smith M J (1984b), "The stability of a dynamic model of traffic assignment - an application of a method by Lyapunov", Transportation Science, 18, 245-252.

Smith M J (1984c), "A descent method for solving monotone variational inequalities and monotone complementarity problems", Journal of Optimisation Theory and Applications, 44, 485-496.

Wardrop J G (1952), "Some theoretical aspects of road traffic research", Proceedings, Institution of Civil Engineers II, 235-278.

TRAFFIC CONTROL SYSTEM OPTIMISATION: A MULTIOBJECTIVE APPROACH

Tessa Sayers, Jessica Anderson and Michael Bell, Transport Operations Research Group, Department of Civil Engineering, University of Newcastle upon Tyne, UK

ABSTRACT

The increasing awareness of the importance of the wider objectives of traffic management and control has led to the work described in this paper. The aim of the study is to develop a flexible signal controller which may be configured so that it embodies the objectives appropriate for the situation in which it is to be used. This paper describes the optimisation of a prototype fuzzy logic signal controller with respect to several criteria simultaneously. Having demonstrated the controller's sensitivity to changes in its parameters, a multiobjective genetic algorithm (MOGA) optimisation technique is used to derive a family of solutions, each of which is optimal with respect to at least one of the criteria, whilst minimising the trade-off with respect to the other criteria.

1. INTRODUCTION

Outline

The first section of this paper describes the context in which this study is being carried out. There follows an introduction to fuzzy logic in traffic signal control and an outline of the prototype traffic signal controller used in this study. The use of Genetic Algorithms for optimisation problems is then described, followed by an outline of the evaluation of traffic controllers using microscopic simulation and an emissions model. Having introduced the main components of the experiment, the study and its outcome are described.

Background

The Urban Traffic Management and Control (UTMC) initiative, launched in 1997 by the Department of the Environment Transport and the Regions, together with the Traffic Director for London, recognises that the objectives of traffic management and control must address the problems resulting from increasing urban congestion (Routledge *et al* (1997)). Flexibility and inter-operability are seen as key features of modern UTMC systems, allowing them to "achieve a wider range of transport and environmental objectives". The advent of these new perspectives on urban traffic management and control has challenged the assumption that the main measure for judging the efficacy of a traffic signal control system is the extent to which it reduces vehicular delay and stops.

In parallel with the UTMS initiative, the Transport Operations Research Group (TORG) was awarded funding by the EPSRC for a two year project entitled *policy-sensitive traffic signal control*, starting March 1997. This paper describes work carried out to date in this project.

The first step was to carry out a Delphi study amongst the UK transport community with the aim of ascertaining the range of objectives that could pertain to various scenarios, such as an isolated intersection, or an urban network. Detailed results from the two rounds are given in Anderson *et al* (1998a), but the general outcome was that it was unrealistic to try to reach consensus on the objectives of traffic signal control and their relative priorities, since every installation will be faced by different constraints and requirements, and these may vary over time. The study concluded that one of the most important features of a traffic signal controller is flexibility.

2. FUZZY LOGIC IN TRAFFIC SIGNAL CONTROL

Background

The heart of a fuzzy logic control system is a set of rules (rulebase) which describes the relationship between the inputs and the output in qualitative "natural language" terms. As in a knowledge-based expert system, these rules provide an easily understood scheme for explaining the input/ output mapping. In contrast to expert systems, however, a fuzzy logic rulebase can be relatively simple and concise, without compromising the smoothness of the output, due to the mapping of the individual discrete input and output values onto overlapping user-defined fuzzy sets. These capture the significant categories of input and output values. The combination of fuzzy sets, linked by a rulebase, creates the appropriate input-output mapping, while avoiding unwanted "steps" in the output values caused by the simple use of thresholds in the input values. A thorough introduction to fuzzy control can be found in Lee (1990).

Since the seminal work of Pappis and Mamdani (1977) describing a fuzzy controller for traffic signals, there have been many interesting applications of fuzzy logic control to traffic signals. A fuller discussion of these can be found in Sayers (1997). Previous experience in TORG of developing a fuzzy logic traffic signal control system for an isolated intersection (Sayers *et al.*, 1996) demonstrated its potential to provide the required flexibility.

Competitive control

There are a number of signal control methods into which fuzzy logic can be usefully incorporated. The method chosen for this experiment was initially based on that of Lee *et al* (1995) and subsequently revised substantially. It employs a competitive technique based upon an urgency value for each signal group (i.e phase).The signal group's urgency represents the importance of giving it green time and is derived each second by means of several fuzzy control modules. The urgency values are then used to determine the best signal settings for the current traffic demand. A novel phase-based approach is taken which allows this assessment to be made every second, and acted upon within the safety constraints of the intersection. Even once a transition out of the current state is underway, the prevailing traffic situation continues to determine the goal of the transition. The input data are supplied by inductive loop detectors on the junction approaches (upstream and at the stopline), pedestrian detectors and public transport telegrams.

Optimisation of Fuzzy Logic Controllers

One criticism often levelled at the principle of fuzzy logic control is the subjectivity of the fuzzy set definitions and the rulebases which determine the system's performance. The derivation and optimisation of these components have been the subject of a number of studies, some examples of which are discussed in Sayers (1997).

There are two main candidates for optimisation in a classic fuzzy logic control system. These are the fuzzy set definitions (membership functions) for the input and/or output variables, and the rulebase, which constitutes the mapping between combinations of input variables and the output. These two components are related and changes to either (or both) can have a profound effect on the operation of the control system.

In the context of the fuzzy logic traffic signal controller described above, changes to the parameters of the fuzzy logic modules affect the mapping of input data to signal group urgency values, and thus alter the stage change decisions made each second based on those urgency values. It is therefore necessary to explore the effects of changing the controller parameters. Different parameter sets will result in controllers which operate very differently. For example, one parameter set might create a controller which minimises delay to pedestrians at the expense of increased vehicular delay, with concomitant increase in some emissions and decrease in others. Another parameter set might result in a controller which incurs medium pedestrian and vehicular delay without favouring either party.

The resulting optimisation problem has a very large search space (in the order of 10^{30} different combinations of parameters) with several attributes associated with each possible combination, including various measures of delay to vehicles and pedestrians, measurement of emissions of up to five pollutants, and measurements of public transport efficiency.

3. GENETIC ALGORITHMS FOR OPTIMISATION

Introduction

The most appropriate optimisation method for a given problem depends to a large extent on the nature of the search space, in terms of its size and complexity, and also on the nature of the desired result. For the problem under consideration, an enumerative method was not viable due to the number of possible solutions, outlined above. Since the optimisation problem in question has multiple attributes, a search technique which yields many good solutions is preferable to one which attempts to find a single "best" solution.

In this study the optimisation method of Genetic Algorithms (GAs) has been adopted. This method has its roots in a simulation of the natural selection process, working on a population of possible solutions in parallel, combining them to produce successive generations and eventually converging on a group of near-optimal solutions. The technique has a stochastic element in that it makes small random changes (mutations) and also in the generation of the initial population. However, it is also directed in that solutions which perform better (have a higher fitness) are more likely to produce offspring to take part in the next generation. A full explanation of GAs in all their diverse forms is found in Goldberg (1989).

Multiobjective Genetic Algorithms

In order to achieve the desired flexibility, the parameters of the signal controller must be optimised with respect to a number of diverse criteria. Traditionally, this kind of problem has been solved piecemeal by using a number of different weighting coefficients for each criterion, reflecting the relative importance of each criterion in determining the final solution. For each chosen set of coefficients, the weighted criteria are summed to give a single objective. These may then be optimised using an appropriate method, yielding a solution corresponding to each coefficient set. The recently developed Multiobjective Genetic Algorithm (MOGA), described by Fonseca and Fleming (1995), eliminates the need to use weighting coefficients by permitting a diverse range of optimal candidate solutions to be found. Each optimal solution reflects a different trade-off between the desired objectives. When implementing the controller in a particular context, the solution that performs best with respect to the desired objectives for that context may be chosen by the user from the optimal set.

The MOGA uses the Pareto ranking method to rank the solutions of each generation by the number of other solutions which dominate them. For one solution x to dominate another, y, the value for at least one attribute of x must be better than the value for the same attribute of y, with the value for every other attribute of x better than, or equal to its counterpart in y. The Pareto ranking method is then used to guide the process of selecting parents for the next generation. Normally, a GA will converge to a single optimal solution, but when the problem has multiple attributes, the GA must be modified to maintain the diversity of the population, which then converges to a surface (the Pareto-optimal front) rather than to a single point. This front represents the boundary between the feasible and the infeasible solutions. This is accomplished using the mechanism of *niching* which assigns a relatively lower fitness to good solutions in over-populated areas of the search space, thus reducing their chance of being selected as parents. This technique is described more fully in Horn *et al* (1994).

In the example of a signal control program whose parameters were to be optimised; if the conflicting criteria were reduction of delay to vehicles and to pedestrians, then the outcome of the MOGA should be a collection of solutions from which the appropriate solution could be chosen according to whether priority should be given by the signal controller to vehicles or to pedestrians.

4. EVALUATION OF A TRAFFIC SIGNAL CONTROLLER

Introduction

The evaluation of a vehicle-responsive traffic signal controller is not a simple matter, due to its dynamic and adaptive nature. The controller's response to the approaching traffic affects the subsequent flow of traffic which in turn affects the operation of the controller, and the control of the signal for each approach is determined not only by the traffic it controls, but also by the traffic on opposing approaches. These factors mean that the only practical way to test the controller is using microscopic simulation, which models a junction with a given topology and other parameters such as input flows, turning movements, traffic mix and desired speeds. The simulator used for this study is VISSIM (Fellendorf, 1996), a microscopic simulator operating in discrete time steps of 1s, and using a psycho-physical vehicle driver model. It is augmented by the MODEM model (available in the UK from TRL) which allows vehicle emissions to be estimated from the detailed vehicle-by-vehicle speed and acceleration profile, which can be part of the VISSIM output .

The Test-bed Intersection

In order to test the signal controller in a realistic environment, the four arm intersection described by Allsop (1992) was used, with the addition of two conflicting bus lanes. This intersection has 13 phases (including four for pedestrians), which can be sensibly combined in 16 different ways (16 possible stages). Upstream of the junction on each approach is a fixed-time signal controlled junction, which gives a more realistic platoon-like flow of vehicles approaching the junction than a random generation of arrivals at a given flow rate.

For the initial implementation of the MOGA, evaluation of each candidate solution is done using constant flow rates in a 15 minute simulation. More realistic flow profiles incorporating peaks will also be used in subsequent tests, and the resulting optimum controller parameters will be compared with those derived previously.

5. OPTIMISING THE CONTROLLER

Feasibility study

Before embarking on a MOGA which would be time-consuming to carry out due to the large number of simulation runs required in the evaluation stage, a feasibility study was carried out to determine the controller's sensitivity to changes in its membership function parameters. This involved evaluating a signal control program fifty times, each time using a different random set of membership function parameters. It was shown that altering these parameters could significantly affect the operation of the controller (Anderson *et al.*, 1998b).

Membership function parameters

In the optimisation of the membership functions, a number of constraints have been imposed in order to preserve the transparent and meaningful nature of the fuzzy control system. These have the effect of ensuring that:

- there are always the same number of fuzzy sets for each input value,
- the ordering of the fuzzy sets is unchanged (the set LOW always covers a range of lower values than the set HIGH),
- the sets maintain the same topology , although they may be skewed differently, and
- the point at which two sets overlap is always at degree of membership 0.5, thus ensuring that the degrees of membership of any input value always sum to 1.

A different set of membership function parameters for an input variable results in differently shaped fuzzy sets for that input variable. This results in a different urgency being assigned to the signal groups affected by that input variable which in turn will result in the signal controller making a different decision about how to set the signals during the simulation, thus affecting the controller's overall performance (judged by the various criteria) for better or worse.

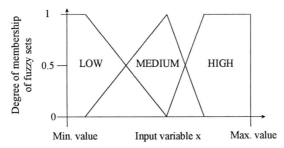

Figure 1. Illustration of typical fuzzy sets for input variable x

For example, the urgency of a pedestrian signal group that is currently red is determined by the number of seconds that have elapsed since the first pedestrian was detected waiting at the crossing and the number of seconds since the signal group was last green. There will be a number of rules such as "if *waiting time* is HIGH and *number of seconds since last green* is HIGH then *urgency* is VERY HIGH" which map the input data to a particular urgency. With different membership function parameters, the degree to which given inputs are considered high will alter, and thus alter the output of the rule. One set of parameters might have the effect that a pedestrian wait of 30 seconds is considered HIGH, whereas with a different parameter set, the same waiting time could be considered MEDIUM, with HIGH only reached when the pedestrian wait reaches 60 seconds. One would expect the pedestrians to have to wait for shorter times before getting green in the former case, all other things being equal.

Optimising the parameters using a MOGA

The complete optimisation process combines all of the processes described so far in a complex iterative loop whose end result is the set of Pareto-optimal fuzzy set parameters which govern the signal controller's operation. Figures 2 to 4 illustrate the processes at work within the MOGA, with processes represented by rectangles and data storage represented by ellipses. The arrows indicate the sequence of events and the passage of data. The term "solution" means one complete set of fuzzy membership function parameters for each of the input variables.

The event loop continues until either a fixed number of iterations (generations) has been completed, or until some other stopping criterion has been reached, such as the attainment of a minimum amount of improvement from one generation to the next. Figures 3 and 4 expand the two boxes describing the main actions within the loop.

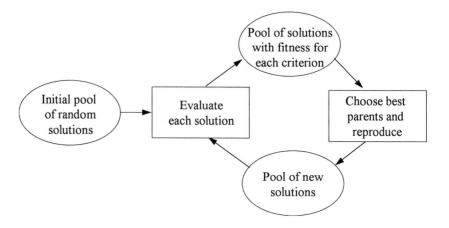

Figure 2. Sequence of events within the MOGA

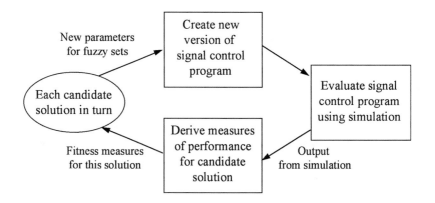

Figure 3. Process of evaluation of each candidate solution

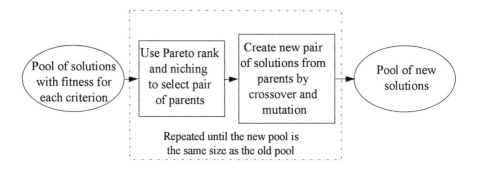

Figure 4. Process of creating successive generations in the MOGA

6. RESULTS AND CONCLUSIONS

Initial runs of the MOGA using only the two conflicting objectives of vehicle delay and pedestrian delay showed that the population converges toward the Pareto front where solutions are optimal with respect to the two objectives. Figure 5 shows the performance of the starting population (randomly initialised) and the population of the tenth generation, using a MOGA with a population of 30 solutions per generation.

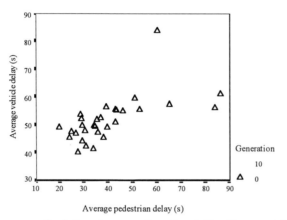

Figure 5. Performance of solutions in generations 0 and 10 with respect to two criteria

These preliminary results demonstrated the ability of this approach to identify solutions on the Pareto front for two objectives. Current work is concerned with the performance of the MOGA for nine criteria.

7. FURTHER WORK

The simulator will be encoded with varying traffic scenarios when evaluating the controller, in order to assess the ability of optimum solutions to deal consistently with them. The results will determine whether different controller parameters are required for optimum control under different traffic conditions. Following the optimisation of the controller parameters for an isolated intersection, the control program will be extended to operate in a network environment, providing co-ordination between distributed control modules. It will then be tested further by simulating a SCOOT-controlled network in Nottingham, for which data are available via the EPSRC Instrumented City facility. VISSIM will be calibrated and validated for this network using floating car data. A period for which the vehicle flows and signal timings are available will be reproduced in simulation, and the performance of SCOOT will be compared with that of the optimised fuzzy logic controller operating with different priority settings.

REFERENCES

Allsop R. E. (1992). Optimisation of signal-controlled road junctions. In *Mathematics in Transpor Planning and Control*, (J.D.Griffiths ed.), Clarendon Press, Oxford.

Anderson J. M., T. M. Sayers, and M. G. H. Bell (1998a). The Objectives of Traffic Signal Control *Traffic Engineering and Control*, Vol. 39, No 3, 167-170.

Anderson J. M., T. M. Sayers, and M. G. H. Bell (1998b). Optimisation of a Fuzzy Logic Traffi Signal Controller by a Multiobjective Genetic Algorithm. *IEE 9th International Conferenc on Road Transport Information and Control*, London, April.

Fellendorf M. (1996), VISSIM for traffic signal optimisation. *Traffic Technology International '96* 190-192.

Fonseca C. M. and P. J. Fleming (1995). An Overview of Evolutionary Algorithms in Multiobjectiv Optimization. *Evolutionary Computation*, Vol. 3, 1-16.

Goldberg D. E. (1989). *Genetic Algorithms in Search, Optimization and Machine Learning* Addison-Wesley Publishing Co. Inc., Reading, Massachusetts, USA.

Horn J., N. Nafpliotis, and D. E. Goldberg (1994). A Niched Pareto Genetic Algorithm fo Multiobjective Optimization *Proceedings of the First IEEE Conference on Evolutionar Computation, IEEE World Congress on Computational Intelligence*, Vol. 1, 82-87.

Lee C. C. (1990). Fuzzy Logic in Control Systems: Fuzzy Logic Controller - Parts I and II. *IEEI Transactions on Systems, Man, and Cybernetics*, Vol. 20, No. 2, 404-435.

Lee J. H., K. M. Lee and H. Leekwang (1995). Fuzzy Controller for Intersection Group. *Int IEEE/IAS Conference on Industrial Automation and Control*, Taipei, Taiwan, 376-382.

Pappis C. P. and E. H. Mamdani (1977). A Fuzzy Logic Controller for a Traffic Junction. *IEEI Transactions on Systems, Man, and Cybernetics*, SMC-7, 707-717.

Routledge I., S. Kemp and B. Radia (1996). UTMC: The way forward for Urban Traffic Control *Traffic Engineering and Control*, Vol. 37, 618-623.

Sayers T. M. (1997). Fuzzy Logic In Traffic Responsive Signal Control. *Proceedings 8th IFAC IFIP / IFORS Symposium on Transportation Systems*, Chania, Crete, June, 714-718.

Sayers T. M., M. G. H. Bell, Th. Mieden., and F. Busch (1996). Traffic responsive signal contro using fuzzy logic - a practical modular approach. *Proceedings EUFIT '96*, Aachen Germany, September 2-4, 2159-2163.

ACKNOWLEDGEMENTS

The authors would like to thank the UK EPSRC for funding the work under GR/L07833, Luka Kautzsch (PTV) for advice on VISSIM, Dr Tim Barlow (TRL) for help with MODEM, and D Andrew Hunter (University of Sunderland, England) for providing the SUGAL Genetic Algorithm package.

Fuzzy Logic Application to Public Transport Priorities at Signalized Intersections

Jarkko Niittymäki, Helsinki University of Technology, Transportation Engineering, P.O.Box. 2100, FIN-02015, HUT, FINLAND

Abstract

Traffic signal control is one of the oldest application areas of fuzzy sets in transportation. In general, fuzzy control is found to be superior in complex problems with multi-objective decisions. In traffic signal control, several traffic flows compete for the same time and space, and different priorities are often set to different traffic flows or vehicle groups

The public transport priorities are a very important part of the effective traffic signal control. Normally, the public transport priorities are programmed by using special algorithms, which are tailor-made for each intersection. The experiences have proved that this kind of algorithms can be very effective if some compensation algorithms and the traffic-actuated control mode are used. We believe that using the fuzzified public transport priority algorithms, the measures of effectiveness of traffic signal control can be even better. In this paper, our fuzzy control algorithm of the public transport priorities will be presented.

1. Introduction

Fuzzy set theory and fuzzy logic provide a systematic way for a handling qualitative information in a formal way. The fuzzy set theory opens new opportunities for solving difficult control process like traffic signal control with many approaches, multi-objective goals and different priorities. In addition, the simplicity of implementing fuzzy systems reduces design complexity.

Traffic signal control is one of the oldest application areas of fuzzy sets in transportation. The published applications are mainly theoretical but there is active research going on in this area, and the first real installations have been presented in 1990's. The public transport priorities are a very important part of the effective traffic signal control. Normally, the public transport priorities are programmed using special algorithms, tailor-made for each intersection. Experience has proved that these kinds of algorithms can be very effective if some compensation algorithms and the traffic-actuated control mode are used. We believe that using the fuzzy public transport priority algorithms with the fuzzy signal control mode, the measures of effectiveness of traffic signal control can be even better.

In this paper, our suggested fuzzy control rules for the public transport priorities (timing and selection of priority function) will be presented, and the structure of the fuzzy signal control algorithms with the infrastructure solution will be discussed. The discussed priority functions are phase extension and phase recall for two-stage control, and phase extension, phase recall and extra phase for multi-phase control. The testing of the fuzzy public transport priorities will be done later using the simulation and the field tests.

2. CONVENTIONAL PRIORITIES FOR PUBLIC TRANSPORT AND DETECTION

2.1 Detection

The ideal bus detection system must be able to distinguish single buses and their line numbers. It should be possible to use the bus detection equipment in other public transport telematics such as passenger information systems or fleet control. However, the number of roadside detectors should be as small as possible to avoid high maintenance costs. There are basically two different ways for bus-detection; using roadside equipment or on-board computers. Inductive loops, microwave antennas and infrared communication systems are the most common detection types.

The priority function of trams is based on a two-detector system - the request detector and the exit detector. The request detector normally starts the priority call. It is usually located 150 - 200 metres before the stopline. The use of the exit detector in the stop line is very useful - it indicates that the tram has passed the intersection and the priority extension can be finished. This kind of system has some advantages. The exit detector indicates that the tram has passed the intersection. It is a big advantage during the peak hours because of the large variations in travel time. It is also quite easy to check that both detectors are working because they should have the same traffic volumes.

The radio transmission communication concept will be the new detection method for public transport priorities. With this method, buses are instrumented with an onboard computer and a speedometer. The computer sends messages to the roadside antennas located at intersections. This concept works in the same way as in the tram priority detection.

2.2 Control Algorithm

Public transport priorities are tailor-made using some special algorithms. The most important of these algorithms are phase extension, phase recall, extra phase and rapid cycle. A public transport priority is given when a public transport vehicle is detected.

Phase extension is the function which continues the green phase if the public transport vehicle is approaching the intersection. The priority lasts as long as the public transport vehicle has passed the intersection or to some maximum value programmed to the controller. If the exit detector at the stop line of the bus approach is used, the extension can be stopped exactly at the right moment. Otherwise, some reprogrammed settings for the extension period during peak and off-peak should be used.

Phase recall is the function which cancels the conflicting green phases and starts the green phase of the public transport approach. This function should pay attention to some minimum green restrictions to avoid too short green times. The minimum green time should be traffic- actuated. If there is traffic demand in those approaches where the green time is to be cancelled, the minimum green time should always be longer than normal minimum green time to avoid excessive delays of those approaches..

Extra phase is the function which is used only multi-phasing intersections. If the normal order is A-B-C (A is the phase of the public transport), then the priority call during B forced the phase sequence to an extra phase. After that phase the order is A-B-A-C. Extra phase is very effective, especially when the extra phase is very short (4 - 6 s). The extra phase-function can always be used in isolated signals. In co-ordinated signals, especially with shorter cycles, *rapid cycle-*function is more useful. It keeps the length of all conflicting phases as short as possible until the priority phase has green again.

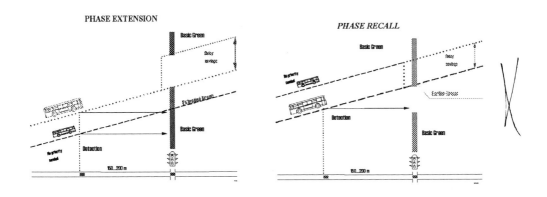

Figure 1. Public transport priority functions (Sane, 1995).

3. FUZZY APPROACH FOR TRAFFIC SIGNAL CONTROL

3.1 Traffic Signals as a Control System

Traffic signal control is used to maximize the efficiency of existing traffic systems without new road construction, to maintain safe traffic flows, to minimize total traffic delays and individual vehicle stops, and to reduce environmental impacts. Several dynamic signal control systems have been developed which calculate the signal timing parameters, cycle lengths, offsets and splits, on a real-time basis, in response to changing traffic patterns. However, most do not resolve the problem of accuracy in modeling dynamic traffic flows, since they can only provide a rough estimation on a real-time base, which is heavily dependent on the initial settings. In addition, the signal control of these algorithms cannot respond quickly enough to traffic changes to match abnormal traffic changes (Zhou, et al. 1997.). By providing temporal separation of rights of way to approaching flows, traffic signals exert a profound influence on the efficiency of traffic flow. They can operate to the advantage or disadvantage of the vehicles, pedestrians or public transport vehicles, depending on how the rights of ways are allocated.

In practice, uniformity of control is the principle followed in signal control for traffic safety reasons. This consideration sets limitations to the cycle time and sequence arrangements. Hence, traffic signal control in practice is based on tailor-made solutions and adjustments made by the traffic planners. For good results, an experienced planner and fine-tuning in the field is needed, because the human has the ability to see many situations. In traffic signal control, several traffic flows compete for the same time and space, and different priorities are often set for different traffic flows or vehicle groups. These features are typical features of fuzzy control.

3.2 Fuzzy Traffic Signal System

Fuzzy logic allows linguistic and inexact traffic data to be manipulated as a useful tool in designing signal timing plans. The base idea of fuzzy (signal) control is to model the control based on human expert knowledge, rather than the modeling of process itself. Fuzzy logic has the ability to comprehend linguistic instructions and to generate control strategies based on a priori verbal communication. The purpose of control is to influence the behavior of a system by changing input or inputs to that system according to a rule or set of rules that model how the system operates.

The knowledge base comprises a rule base, which characterizes the control policy and goals. The linguistic rules are the way that fuzzy control models the knowledge. A typical form of the linguistic rules is

Rule 1	If x is A_1,	then f(x) is B_1
Rule 2	If x is A_2,	then f(x) is B_2
... ...	...	
Rule N	If x is A_N,	then f(x) is B_N,

where x and f(x) are independent and dependent variables, and A_i and B_i are linguistic constants. These rules are referred to as if-then-rules because of their form. An if-clause is referred to as an antecedent (premise) and a then-clause a consequence.

Fuzzy logic based controllers are designed to capture the key factors for controlling a process without requiring many detailed mathematical formulas. Due to this fact, they have many advantages in real time applications. The controllers have a simple computational structure, since they do not require many numerical calculations. The "if-then-else" logic of their inference rules does not require much computational time. Also, the controllers can operate on a large range of inputs, since different sets of control rules can be applied to them. If the system related knowledge is represented by simple if-then-else fuzzy rules, a fuzzy-based controller can control the system with efficiency and ease. The control process for fuzzy signal control is shown in *Figure 2*. The ordinary fuzzy controller consists of different parts; a fuzzification part, a knowledge base, a database, a decision making part, and a defuzzification part.

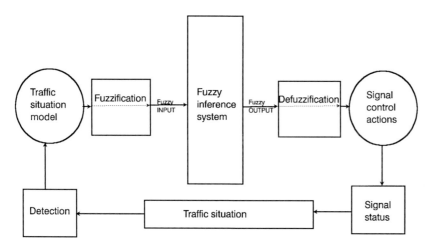

Figure 2.   Traffic signal control process (Niittymäki, 1998).

The decision-making logic processes fuzzy input data and rules to infer fuzzy control actions employing fuzzy implication and the rules of inference in fuzzy logic. For a multi-input multi-output system with n fuzzy rules derived by using the sentence connective 'and', the consequent y_k can be obtained by composing the antecedent x_k and the rules R_i

$$y_k = (x_k \circ R_1) \text{ or } (x_k \circ R_2) \text{ or } ...(x_k \circ R_n)$$
$$= x_k \circ (R_1 \cup R_2 \cup ...R_n)$$
$$= x_k \circ R$$

where R is called the system transfer relation. R is used to calculate the fuzzy output y_k from the fuzzy input x

$$Y = X^\circ R \Rightarrow \mu_V(Y) = \max\{\min\{\mu_U(X), \mu_R(X, Y)\}\}$$

where X and Y are the input and output vectors in fuzzy sets U and V, respectively. The output is still fuzzy (value of membership function), and some defuzzification method (COG, COA, min-max, MOM) will be needed.

4. FUZZY RULE BASE FOR PUBLIC TRANSPORT PRIORITIES

4.1 Motivation

The main reason why fuzzy set theory is a suitable approach to traffic signal control is the nature of uncertainties in signal control; decisions are made based on imprecise information, the consequences of decisions are not well-known, and the objectives have no clear priorities. It has been known through various experiments and applications that fuzzy control is well suited when the control involves these kinds of uncertainties and human perception, like traffic signal control.

In practice, the use of traditional optimization methods (like Miller, 1968) can be problematic or too complex to develop. The aim of fuzzy logic based control is to soften the decision making process by accepting human-like acquisition of information and executing soft decision rules. Such control will be robust and adaptive in terms of handling various objectives at the same time, and choosing the parameters should be rather simple (Niittymäki and Kikuchi, 1998).

In the case of public transport priorities, there are reasons to believe that the fuzzy public transport priorities can be better than the traditional binary-logic priorities:

- Public transport priorities add complexity of control policy, and the control policy has at least one additional objective.
- Consequences of fired public transport priority function are not known.
- Binary-logic calls public transport priority regardless of traffic situation.
- There can be large variations in travel time (uncertainty), especially during the peak hours. Travel time can be even longer than the maximum phase extension time.
- Infrastructure (simple detector configuration with traffic situation modeling) and control rules (not many) can be quite simple.
- The rule base can easily be modified for all kinds of isolated intersections, and the rule structure can be extended to coordinated signals.

4.2 Structure of Fuzzy Rule Base

The preliminary plan is that traffic signals will work using the multilevel fuzzy decision making algorithms (Niittymäki and Pursula, 1997). In general, the fuzzy rules work at three levels; traffic situation level, phase and sequence level, and green ending or extension level. The extension level can also be a multi-objective. The fuzzy public transport priority function has to be connected to the green extension level and the phase and sequence levels.

The basic idea of fuzzy approach to public transport priorities is that PT-requesting (the first approach detection, PT(time) > 0) starts rule combinations. In this case, PT is the general term for public transport. PT(time) is the most important fuzzy variable for time of public transport requesting, and it means the time a public transport vehicle remains between the priority detectors or first detector and stop line. It starts while requested in call detector and it stops (PT(time)=zero) while requested in the exit detector. If two or more PT-vehicles are coming, the fuzzy variable is the smaller value.

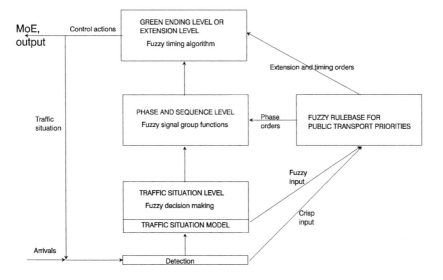

Figure 3. Multilevel fuzzy control algorithm with public transport priorities.

It can be noted that the fuzzy public transport priority rule base can also be implemented as a part of the basic rules, but this kind of structural rule base is easy to develop and to use.

4.3 Fuzzy Rule Base for Public Transport Priority

The main goals of the fuzzy rule base of public transport priority are

- to give a correct priority function as a function of request moment,

-    to make a correct priority decision, based on current traffic situation at the intersection.

In our suggested application, a request-exit detector-system is used and two detectors are located per approach. The locations of the request and the first detector define the used membership functions. The recommended distances are 100 – 150 meters. The public transport detection (crisp input) and the use of the traffic situation model give an input to fuzzy algorithms. The recommended priority functions are phase extension and phase recall for two-phase control, and phase extension, phase recall, rapid cycle and extra phase for multi-phase control. The two-phase control is a simple control case because the conclusion of the rule base is either to continue current phase or terminate it. If the public transport request is detected in the approach of a green signal group, the phase extension rules evaluate the current traffic situation based on two fuzzy parameters

- detection time of public transport (PT(time); zero, short, medium, long)
- weight of red signal group (W(red); short, medium, long).

The W(red) or W(next) is the fuzzy factor for the importance of the next phase, for example the estimated green time need in next phase or just the number of stopped vehicles (Q = queue). The rules are typical if-then-rules, for example

If PT(time) is zero and W(red) is short then use basic rules or

…

if PT(time) is medium and W(red) is medium then extend phase or

…

if PT(time) is long and W(red) is high then use basic rules.

PHASE EXTENSION RULES

W(red)	PT(time)			
	zero	short	medium	long
short	basic	extend	extend	basic
medium	basic	extend	extend	basic
high	basic	basic	basic	basic

If the public transport request is detected in the approach of a red signal group, the phase recall rules evaluate the current traffic situation, based on two fuzzy parameters

- detection time of public transport (PT(time); zero, short, medium, long)
- weight of green signal group (W(green); short, medium, long).

The W(green) or W(next) is the fuzzy factor for the importance of the next phase, for example, in the estimated green time needed current phase or just the number of approaching vehicles (A = arrivals). The phase recall order means that the green signal group is ready to be terminated. For example,

if PT(time) is zero and W(green) is short then use basic rules or

...

if PT(time) is medium and W(green) is medium then recall phase or

...

if PT(time) is long and W(green) is high then use basic rules.

PHASE RECALL RULES

		PT(time)			
		zero	short	medium	long
	short	basic	recall	recall	recall
W(green)	medium	basic	recall	recall	basic
	high	basic	basic	recall	basic

The multi-phase control is more complicated because there exists the opportunity to affect the phase and sequence order too. The decision rules for priority algorithms are based on the moment (phase) of public transport detection (crisp) and the fuzzy parameter of next phase (W(next)).

MULTI-PHASE CONTROL RULES

		PHASE WHILE PT-REQUEST			
		A	B	C	(crisp input)
	low	extension	extra phase	recall	
W(next)	medium	extension	rapid cycle	recall	
	high	extra phase	rapid cycle	basic	

If the selected control algorithm is phase extension or phase recall, then we can evaluate the traffic situation using the phase extension and phase recall rules. If extra phase or rapid cycle as in selected, then the control algorithm works as in the conventional control.

5. DISCUSSION

This paper proposes the structure and the preliminary rule base for fuzzy public transport priorities. The membership functions have to define each case individually, and the final definition of membership functions can be done using the neural networks, genetic algorithms or some other method. The testing of fuzzy rule base will be done using the simulation, and if the results indicate better effectiveness of traffic signal control the real installation will be done in Helsinki. We believe that the fuzzy control provides a compromise between two objectives; fast-moving vehicle traffic and fast-moving public transport traffic.

Because the decision rules are fuzzy, we have to use some defuzzification method to achieve a crisp signal control action. The most common methods are the min-max, the mean of maximum, and the center of gravity or area methods. Brase and Rutherford (1978) have discussed more the features of defuzzification methods. The final defuzzification method will be based on simulation experiences. If the output is crisp, then we can use min-max-method, and the rule with the highest value of membership function will be selected. In reality, we have some kind of uncertainty concerning the right decision or output. This uncertainty means that if the rule output is a fuzzy number then some other defuzzification method should be used.

6. CONCLUSIONS

This paper specifies the proposed fuzzy control algorithm for public transport priorities at the isolated intersections. The algorithm has not been tested yet, but it will be done using the HUTSIM-simulation model. The previous studies have indicated that fuzzy control is well suited to traffic signal control, and the results have shown even better performance than traditional vehicle-actuated control. The benefits are mostly based on the adaptability of fuzzy control. Fuzzy control has the ability to fire many soft rules simultaneously, and to offer a compromise between objectives and approaches. The linguistic rules using fuzzy control can be based on the definition of control and its goals.

The benefits of fuzzy public transport priorities are based on detection of the current traffic situation. The momentary rush-peaks and the changes of traffic flow can be noticed by traffic situation modeling, which is the input for the fuzzy rule base. In this fuzzy way, the need for priority call and the selection of priority function can be handled in the systematic way.

ACKNOWLEDGEMENTS

Part of this research was performed with the support of the Technology Development Centre of Finland (TEKES) as a part of FUSICO-research project mostly financed by the Academy of Finland.

REFERENCES

Brase M. and Rutherford D. (1978). *Fuzzy Relations in a Control Settings.* Kybernetics, vol.7, no.3, pp. 185 - 188.

Kim S. (1994*). Applications of Petri Networks and Fuzzu Logic to Advanced Traffic Management Systems.* Ph.D.-thesis, Polytechnic University, USA. 139 p.

Niittymäki J. (1998). *Isolated Traffic Signals – Vehicle Dynamics and Fuzzy Control.* Helsinki University of Technology, Transportation Engineering. 128 p.

Niittymäki J. and Kikuchi S. (1998). *Application of Fuzzy Logic to the Control of a Pedestrian Crossing Signal.* Paper presented a the 1998 Transportation Research Board Annual Meeting, and accepted for publication in Transportation Research Record. 15 p.

Niittymäki J. and Pursula M. (1997). *Signal-Group Control Using Fuzzy Logic.* Paper presented at the 9th Mini-EURO Conference, Fuzzy Sets in Traffic and Transport Systems, EURO Working Group on Transportation. Budva, Yugoslavia. September 15 – 19, 1997. 17 p.

Sane K. (1995). *Public Transport Priorities in Traffic Signals – Theory and Practise from the City of Helsinki.* Paper presented in European Workshop (DG XVIII), 16.3.95, Barcelona, Spain. 9 p.

Zhou W-W., Wu J., Lee A, Fu L. and Miska E. (1997). *Fuzzy flows.* ITS: Intelligent transport systems, May/June 1997. pp. 43 – 45.

Zimmermann H.-J. (1996). *Fuzzy Set Theory and Its Applications.* Kluwer Academic Publishers, Boston, Dordrect, London. 435 p.

BI-LEVEL FORMULATION FOR EQUILIBRIUM TRAFFIC FLOW AND SIGNAL SETTINGS

Suh-Wen Chiou, University of London Centre for Transport Studies, University College London, UK

ABSTRACT

A bi-level programming approach has been used to tackle an area traffic control optimisation problem subject to user equilibrium traffic flows. In the upper level problem, the signal timing plan for coordinated fixed time control has been defined. In the lower level problem, user equilibrium traffic assignment obeying Wardrop's first principle has been formulated as a variational inequality problem. Mathematical expressions for various components of the performance index in the upper level problem and the average delay in the lower level problem have been derived and reported (Chiou 1997a). A mixed search procedure has been proposed as the solution method to the bi-level problem and a range of numerical tests have been carried out (Chiou 1997b, 1998a,b). In this paper, further numerical tests are performed on Allsop and Charlesworth's (1977) road network in which various traffic loads are taken into account. Effectiveness in terms of the robustness and reliability of the mixed search procedure in congested and uncongested road networks is thus investigated further. Comparisons of the performance index resulting from the mixed search procedure and that of mutually consistent TRANSYT- optimal signal settings and traffic flows are made for the congested road network.

1. INTRODUCTION

The combined problem for traffic equilibrium flow and optimisation of signal settings has long been recognized an important problem. A number of solution methods to this combined problem have been discussed and good results have been reported in a medium sized example road network. For example, a mutually consistent calculation was carried out by Allsop and Charlesworth (1977), which alternately solved the traffic signal settings problem with the current link flow and calculated user equilibrium flows for the current signal settings until an intuitively expected convergence was achieved. On the other hand, a bi-level formulation for the combined problem has been widely used (Heydecker and Khoo 1990, Yang and Yagar 1995). In this bi-level formulation the optimisation of traffic signal settings is regarded as the upper level problem whilst the user equilibrium traffic assignment is regarded as a function of the signal settings and therefore it can be dealt with as the lower level problem. It has been recognized that although the link flows must be in equilibrium for the resulting signal settings, the signal settings will not in general be optimal for the resulting link flows if the latter is regarded as fixed. Furthermore, application of sensitivity analysis for user equilibrium flow was introduced by Tobin and Friesz (1988) and subsequent applications of the sensitivity analysis to solving the bi-level problems of interest in this aspect of transportation have been investigated extensively (Yang 1997).

In this paper, we adopt the bi-level programming to formulate this combined problem and application of sensitivity analysis for user equilibrium traffic flow to obtain relevant derivatives for the objective function with respect to the changes of the signal settings. In the upper level problem, traffic signal settings are defined by the common cycle time, and the starts and durations of all greens. Also the performance index is defined as the weighted sum of the estimated rate of delay and number of stops per unit time for all traffic streams, as evaluated by the traffic model in TRANSYT (Vincent *et al.* 1980). In the lower level problem, a general expression for user equilibrium traffic assignment is formulated as a variational inequality in which the separable link travel time is defined as the sum of undelayed travel time on the link and average delay incurred by traffic at the downstream signal-controlled junction estimated as a function of flow on the link by the TRANSYT traffic model. This adopts expressions from previous work (Wong 1995) and newly derived expressions from Chiou (1997a). The separable equilibrium traffic assignment is solved by the convex combination method. Because of the non-convexity of the objective function, only local solutions can be found. A mixed search procedure is proposed in which a global search heuristic is particularly included for achieving better local solutions. In the next section, the bi-level formulation for this problem is given. The solution method for the bi-level formulation is discussed in section 3. Numerical results implemented on congested and uncongested example road networks are discussed in section 4. Conclusions and discussions for this paper are given in section 5.

2. PROBLEM FORMULATION

A bi-level problem for optimisation of traffic signal settings subject to user equilibrium flow is to

$$\underset{\psi \in S}{Minimise} \quad P = P_0(\psi, q) \tag{2.1}$$

$$subject\ to \quad q = q^*(\psi)$$

In problem (2.1) the total travel cost P can be minimised via function P_0 in terms of the signal settings ψ and the equilibrium flow q. The signal settings ψ are determined in the upper level problem whilst the equilibrium flow q are determined in the lower level problem via function $q^*(\psi)$ which is in terms of the signal settings ψ, where S is the feasible set of the signal settings.

3. SOLUTION METHOD

For any feasible signal settings, the neighboring local optimum for bi-level problem (2.1) can be found by a single level optimisation problem in which the gradient projection method is used to determine the descent direction at each iteration and the associated length of move can be decided by a bisection method along this feasible direction. Let A and B be respectively the coefficient matrix and constant vector for linear constraints on the signal settings ψ, the bi-level problem (2.1) can be re-expressed as to

$$\underset{\psi}{Minimise} \quad P = P_1(\psi) \tag{3.1}$$

$$subject\ to \quad A\psi^T \le B$$

where superscript T is the matrix transpose operator.

3.1 Gradient projection method

For any feasible signal settings ψ^0, let $A^T = [A_b^T, A_{nb}^T]$ and $B^T = [B_b^T, B_{nb}^T]$ decomposed into the binding and non-binding coefficient matrices and constant vectors. Following the results of gradient projection method (see Luenberger, 330-337) a feasible descent direction d for problem (3.1) at ψ^0 can be decided as follows. If the projection matrix $g = I - M^T(MM^T)^{-1}M$ such that $g\nabla_\psi P_1(\psi^0)^T \ne 0$ where I is the identity matrix and $M = A_b$

$$d^T = -g\nabla_\psi P_1(\psi^0)^T \tag{3.2}$$

and if $g\nabla_\psi P_1(\psi^0)^T = 0$ then the descent direction d can be decided as

$$d^T = -\overline{g}\nabla_\psi P_1(\psi^0)^T \tag{3.3}$$

where $\overline{g} = I - \overline{M}^T(\overline{M}\,\overline{M}^T)^{-1}\overline{M}$ and $\overline{M} = \overline{A_b}$ where $\overline{A_b}$ is obtained from A_b by deleting the rows of A_b corresponding to negative Lagrange multipliers at ψ^0, and $\nabla_\psi P_1(\psi^0)$ is the gradient at ψ^0 of the objective function in problem (3.1).

3.2 Determination of move length

Let $C = B_{nb} - A_{nb}(\psi^0)^T$ and $D = A_{nb}d^T$, the determination of optimal choice α_{opt} along d can be regarded as a one-dimensional search problem.

$$\underset{0 \le \alpha \le \alpha_{max}}{Minimise} \quad Z_1(\alpha) = P_1(\psi^0 + \alpha d) \tag{3.4}$$

where $\alpha_{max} = \infty$ if $D \le 0$ or $\alpha_{max} = Min\{\dfrac{C_i}{D_i} : D_i > 0\}$ otherwise; and can be solved by the bisection method.

3.3 Global search heuristic for offsets

In problem (3.1) only local optima can be found, and the fact that direction d will in general include changes in durations of green means that α_{max} is in practice quite small; however, because there is no practical constraint for offsets, an unconstrained optimisation problem (3.5) for offsets can be formulated and a global search for optimal step length along the steepest descent direction is carried out by the uniform search method so that if a better local optimum outside the neighborhood of current signal settings ψ^0 can be located in this direction it will be found. Although the author knows of no way in which information local to ψ^0 can be used to identify a direction in which a better local optimum is particularly likely to be found, the steepest descent direction for changes in offsets from ψ^0 is used heuristically to avoid taking a direction that is initially unnecessarily unfavorable. The relevant search interval is equally divided into many sub-intervals where the objective function is evaluated and the optimal step length is determined by which of these values of the objective function is the least. During this uniform search a re-assignment process for traffic flows needs to be carried out at each sub-interval as the objective function is evaluated. Let $\varsigma^0, \theta^0, \phi^0$ be respectively current signal settings for common cycle time, starts and durations of greens, and $\psi_1 = [\varsigma^0, \theta^0 + \Theta, \phi^0]$, and $\Theta = [\Theta_1, \ldots, \Theta_N]$ where N is the number of signal-controlled junctions in the road network and for each junction on Θ_m is the vector $\theta_m[1, \ldots, 1]$ of order G_m which denotes the number of signal groups for junction m. Then in ψ_1 each element θ_{jm} for particular signal group j at junction m can be expressed as

$$\theta_{jm} = \theta_{jm}^0 + \Theta_m \ , \ 1 \le m \le N$$

The optimal choice of offsets in direction $d = [0, \Theta, 0]$ is to

$$\underset{0 \le \alpha \le \alpha_{max}}{\text{Minimise}} \qquad Z_2(\alpha) = P_1(\psi_1 + \alpha d) \tag{3.5}$$

where $\alpha_{max} = \underset{1 \le m \le N}{Max} \{|\Theta_m|^{-1}\}$

3.4 Application of sensitivity analysis

The calculation for gradient $\nabla_\psi P(\psi^0)$ in equation (3.2) taking account of consequential variation in q can be written as

$$\nabla_\psi P = \nabla_\psi P_1 + \nabla_q P_1 \nabla_\psi q^T \tag{3.6}$$

Detailed derivatives for use in equation (3.6) have been derived and used numerically in Chi (1997a,b; 1998a,b). For the derivatives of equilibrium flow, the equation adopted from Tobin and Friesz (1988) is

$$\nabla_\psi q^T = -\delta B \delta^T \nabla_\psi c^T \tag{3.7}$$

where δ, Δ respectively denotes the link-path and OD-path incidence matrices and c represents the link travel time functions and B takes the following form

$$B = (\delta^T \nabla_q c \delta)^{-1} (I - \Delta^T (\Delta (\delta^T \nabla_q c \delta)^{-1} \Delta^T)^{-1} \Delta (\delta^T \nabla_q c \delta)^{-1})$$

3.5 Mixed Search Procedure

3.5.1 Basic strategies

For current signal settings, a practical way searching for a better solution for problem (3.1) is set out in the following three type of steps.

Type I. Use of the derivatives in equations (3.6-3.7), the feasible descent direction is decided by equations (3.2-3.3) and the optimal move length is decided by equation (3.4).

Type II. For specified common cycle time, use of the derivatives in equations (3.6-3.7), the feasible descent direction is decided by equations (3.2-3.3) and the optimal move length is decided by equation (3.4).

Type III. For specified common cycle time and duration of greens for all signal groups, an unconstrained optimisation problem for the starts of green for all signal groups at each junction is formulated in (3.5) and a global search for the optimal step length along the steepest descent direction is carried out as discussed in Section (3.3).

The rationale for the mixed search procedure starting with arbitrary initial signal settings is to find the way to locate a good point in the neighborhood of current signal settings by using the steps of type I or II, and particularly with help of the steps of type III to move to another part of the feasible region which contains better local optimal points then use the steps of type I or II again to locate these local better points. This search procedure will continue by using various sequences of alternate steps of type I or II then type III, and type I or II again and so on until a predetermined threshold is satisfied.

3.5.2 Implementation heuristic.

An implementation heuristic used in this paper for the mixed search procedure is set out as follows.

Step 1. Use *type III* to determine a good initial solution for subsequent searches.

Step 2. Use *type I* to determine the appropriate common cycle time for the whole network and the corresponding duration of green for each signal group.

Step 3. Use *type III* to localize the good solution for subsequent searches again.

Step 4. Use *type II* to determine the optimal duration of green for each signal group in the network by which the values of the performance index is minimised.

Step 5. Use *type III* to localize the optimal signal settings by carrying out a global search for the offsets, i.e. the starts of greens for each junction.

Step 6. Use *type II* again to fine-tuning the resulting signal settings until the difference of the values of the performance index between successive iterations is negligible.

4. NUMERICAL EXAMPLE

An example network is illustrated based upon the one used by Allsop and Charlesworth (1977). Configurations of the test road network are given in Figure 4.1. Base travel demands for each pair of origin and destination are given in Table 4.2. This numerical test includes 22 pairs of trip-ends, 23 links, 40 feasible routes and 18 signal groups at 6 signal-controlled junctions. Two arbitrary distinct sets of initial signal settings are specified, each of which is given in Table 4.3. Two methods of calculations are implemented on the congested and uncongested road networks: the mixed search procedure, and the mutually consistent calculations used by Allsop & Charlesworth (1977). Good results in terms of the effectiveness and robustness for the mixed search when implemented on the uncongested road network have been reported (Chiou 1998 a,b). In this paper, further numerical tests are made on the congested road network by increasing the travel demand to 110% and 120% of the base demand. Results for the mixed search and mutually consistent calculation with the various traffic loads are shown in the Tables 4.4 and 4.5.

Table 4.2 Travel demand for Allsop & Charlesworth's network in veh/h

O/D	A	B	D	E	F	totals
A	-	250	700	30	200	1180
C	40	20	200	130	900	1290
D	400	250	-	50	100	800
E	300	130	30	-	20	480
G	550	450	170	60	20	1250
totals	1290	1100	1100	270	1240	5000

Table 4.3 Two sets of initial starts of greens for Allsop & Charlesworth's network in seconds

junction	cycle time	start of green group 1	group 2	group 3	group 4	group 5	cycle time	start of green group 1	group 2	group 3	group 4	group 5
1	45	0.0	22.5	0.0			90	0.0	45.0	0.0		
2	45	0.0	22.5				90	0.0	45.0			
3	45	0.0	22.5				90	0.0	45.0			
4	45	0.0	15.0	30.0	0.0	30.0	90	0.0	30.0	60.0	0.0	60.0
5	45	0.0	15.0	30.0	0.0		90	0.0	30.0	60.0	0.0	
6	45	0.0	22.5				90	0.0	45.0			

As is seen in Table 4.4, for the two sets of initial signal settings in the base travel demand the mixed search procedure improves the system performance by 86% and 91% and the difference between the resulting values is less than 1.5%; for 110% travel demand the system performance is improved by 86% and 90% and the difference between the resulting values is less than 0.5%; and for 120% travel demand the system performance is improved by 89% and 91% and the

Table 4.4 Results for mixed search procedure and mutually consistent calculations

iteration number	First set		signal settings		Second set		signal settings	
	mixed search		MC calculation		mixed search		MC calculation	
	cycle time	performance index (veh-h/h)	cycle time	performance index (veh-h/h)	cycle time	performance index (veh-h/h)	cycle time	performance index (veh-h/h)
Base travel demand								
1	45	540.08	45	540.08	90	822.37	90	822.37
2	45	537.64	100	92.73	90	821.15	110	216.59
3	97	155.64	100	94.48	108	94.03	110	187.96
4	97	152.42	100	91.96	108	93.28	110	206.65
5	97	73.35	100	92.24	108	72.24	110	247.60
6	97	73.31	100	92.24	108	72.22	110	157.21
7	97	73.23			108	72.20	110	121.64
8							110	121.64
110% travel demand								
1	45	611.63	45	611.63	90	920.42	90	920.42
2	45	609.70	90	117.13	90	918.24	125	261.82
3	115	224.33	90	121.15	116	239.51	125	245.80
4	115	221.23	90	116.04	116	237.02	125	197.78
5	115	88.12	90	116.04	116	89.18	125	165.88
6	115	87.78	90	119.21	116	88.53	125	193.45
7	115	87.75	90	119.29	116	88.18	125	190.24
8			90	119.29			125	190.24
120% travel demand								
1	45	893.46	45	893.46	90	1029.63	90	1029.63
2	45	892.78	115	162.10	90	1028.47	135	427.07
3	137	317.28	115	162.34	129	430.65	135	495.49
4	137	315.41	115	169.54	129	427.09	135	466.61
5	137	95.45	115	166.26	129	97.39	135	395.98
6	137	94.98	115	166.12	129	95.69	135	405.27
7	137	94.75	115	165.58	129	95.45	135	342.79
8			115	167.23			135	342.30
9			115	167.23			135	316.71
10							135	306.65
11							135	306.65

Table 4.5 Results for rerun mutually consistent calculations

iteration number	First set		signal settings				Second set		signal settings			
	base	demand	110%	demand	120%	demand	base	demand	110%	demand	120%	demand
	cycle time	PI (in veh-h/h)	cycle time	PI (in veh-h/h)	cycle time	PI (in veh-h/h)	cycle time	PI (in veh-h/h)	cycle time	PI (in veh-h/h)	cycle time	PI (in veh-h/h)
1	97	73.23	115	87.75	137	94.75	108	72.20	116	88.18	129	95.45
2	97	90.63	115	118.77	137	174.91	108	100.85	116	141.64	129	192.08
3	97	89.98	115	118.69	137	188.89	108	100.40	116	133.33	129	217.66
4	97	89.98	115	119.45	137	211.27	108	96.51	116	132.57	129	243.96
5	97	92.06	115	129.94	137	213.23	108	96.85	116	132.22	129	251.98
6	97	92.06	115	126.14	137	217.84	108	96.85	116	130.10	129	262.18
7			115	126.97	137	247.78			116	129.70	129	315.58
8			115	126.97	137	245.03			116	129.70	129	320.91
9					137	245.03					129	322.53
10											129	322.53

difference between the resulting values is less than 1%. Furthermore, the final signal settings for the three cases above are appreciably different for the two sets of initial signal settings.

Results for the mutually consistent signal settings and equilibrium link flows are also shown in Table 4.4. For the base travel demand the system performance is improved by 83% and 85% but

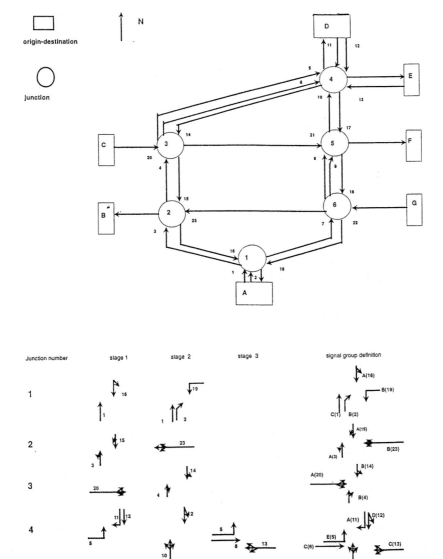

Figure 4.1 Cofigurations for Allsop & Charlesworth's (1977) road network

the difference between the resulting values is greater than 30%, for 110% travel demand the system performance is improved by 80% and 79% but the difference between the resulting values is nearly as high as 60%, and for 120% travel demand the system performance is improved by 81% and 70% but the difference between the resulting values is over 80%.

Comparisons for the two solution methods in terms of the improvements for system performance: for base travel demand the mixed search procedure improves on the mutually consistent calculation by 20% and 40%; for 110% travel demand the mixed search procedure improves on the mutually consistent calculation by 26% and 54%; and for 120% travel demand the mixed search procedure improves on the mutually consistent calculation by 43% and 69%. Furthermore, a rerun mutually consistent calculation carried out starting from the equilibrium flows given by the mixed search procedure is shown in Table 4.5. Again it shows that the non-optimal mutually consistent calculation performed worse than did the mixed search procedure over all the cases discussed above.

The computation efforts for the mixed search procedure performed on PC 486SX 25/33 Zenith machine, each iteration for this numerical example was performed in less than 30 seconds CPU time in the C++ integrated development environment. Total computation efforts for complete run of the mixed search procedure did not exceed 10 minutes of CPU tine on that machine.

5. CONCLUSIONS AND DISCUSSIONS

In this paper, we showed a bi-level formulation for the combined problem of the signal settings and equilibrium flow and proposed the mixed search procedure to solve this optimisation problem. Following earlier tests on an uncongested road network, in this paper, we went on to implement this mixed search procedure on the same network under congested conditions, and good results again showed the reliability and robustness of the proposed method on various increasing traffic loads in comparison with the other alternative.

Concerning further extensions of this bi-level formulation, a detailed examination in the numerical implementations of the differences of the results produced by these methods on a link by link basis needs to be undertaken. Furthermore, Heydecker and Khoo, Yang and Yagar respectively presented optimisation based algorithms to solve the bi-level formulation problems; however, so far no empirical studies have been reported about the efficiency and robustness of these optimisation algorithms. Therefore, the work reported here will continue to investigate these interesting topics of the optimisation algorithms especially as they are implemented on various patterns of example networks.

ACKNOWLEDGMENTS

The author would like to express her deep gratitude to Professor Allsop and Dr. Heydecker for their invaluable discussions on this work. The work reported here was sponsored by the scholarship of Taiwan government.

REFERENCES

Allsop, R.E. and J. Charlesworth (1977). Traffic in a signal-controlled road network: an example of different signal timings inducing different routeings. *Traffic Engineering Control*, **18**, 262-264.

Chiou, S-W. (1997a). Derivatives in an area traffic control problem subject to user equilibrium traffic assignment. *Universities Transport Study Group 29th annual conference*, Bournemouth University, 1997. (Unpublished)

Chiou, S-W. (1997b). Optimisation of area traffic control subject to user equilibrium traffic assignment. 25th European Transport Forum, *Proceedings of Seminar F, Transportation Planning Methods: Volume II, PTRC*, London, 53-64.

Chiou, S-W. (1998a). Optimisation of traffic signal timings for user equilibrium flow, *Proceedings of the Third International Symposium on Highway Capacity*, Copenhagen, Denmark, June 22-27, 1998.

Chiou, S-W. (1998b). Area traffic control optimisation for equilibrium network flow, *Proceedings of the 8th World Conference on Transportation Research*, Antwerp, Belgium, July 12-17, 1998.

Heydecker, B.G. and Khoo, T.K. (1990). The equilibrium network design. *Proceedings of AIRO '90 Conference on Models and Methods for Decision Support,* Sorrento, 587-602.

Luenberger, D.G. (1989) *Linear and Nonlinear Programming*. 2nd edition, Addison-Wesley, Massachusetts.

Tobin, R.L. and Friesz, T.L. (1988). Sensitivity analysis for equilibrium network flow. *Transportation Science*, **22**(4), 242-250.

Vincent, R.A., Mitchell, A.I. and Robertson, D.I. (1980). User guide to TRANSYT Version 8, *TRRL Report* **LR888**, Transport and Road Research Laboratory, Crowthorne.

Wong, S.C. (1995) Derivatives of the performance index for the traffic model from TRANSYT. *Transportation Research* **29B**(5), 303-327.

Wong, S.C. (1996) Group-based optimisation of signal timings using the TRANSYT traffic model, *Transportation Research* **30 B**(3), 217-244.

Yang, H and Yagar, S. (1995). Traffic assignment and signal control in saturated road networks. *Transportation Research* **29B**(4), 231-242.

Yang, H. (1997). Sensitivity analysis for the elastic-demand network equilibrium problem with applications, *Transportation Research*, **31B**, 55-70.

HEADWAY-BASED SELECTIVE PRIORITY TO BUSES

Fraser McLeod, Transportation Research Group, Department of Civil and Environmental Engineering, University of Southampton, UK

ABSTRACT

This paper addresses the problem of bunching of buses and how it may be counter-acted. An algorithm is presented for providing selective priority to buses at traffic signals according to their headways, the highest levels of priority being given to those buses with the highest headways, i.e. those buses which are running late or falling behind the bus in front.

Alternative selective priority strategies are evaluated in terms of their effects on bus journey time regularity, bus delay and general traffic delay. A simulation model SPLIT (Selective Priority for Late buses Implemented at Traffic signals) has been developed to investigate the performance of different priority strategies. The paper describes the details of the model, including bus stop dwell times and overlapping bus services, and compares results obtained from the model for a number of different priority strategies.

This work was driven by the keen interest in bus priority applications in London within the EC DGVII project INCOME.

1. INTRODUCTION

The old cliché of three buses coming along at the same time is a real problem challenging bus operators. Bunching of buses is undesirable since passengers have to wait longer, on average, at bus stops (as shown in Section 2). These delays could cause passenger dissatisfaction with public transport and, possibly, lead to a mode shift to private travel.

The problem of bus bunching comes about for a variety of reasons including

- buses not setting off on time,
- late buses being delayed further at bus stops as they tend to have more passengers to pick up,
- the platooning effect of traffic signals,

- 'scratching' - the practice of a bus driver following closely behind the bus in front to do less work (picking up passengers, checking tickets, taking fares etc.) .

Potential methods for improving bus service regularity include
- automatic, pre-paid ticketing,
- delaying the 'front' bus when the headway becomes too big,
- selective priority for buses at traffic signals.

The option of automatic ticketing, where tickets are bought before travelling and, optionally, stamped by an on-bus machine, rather than checked by the driver, would be of great benefit to improving bus service regularity. This option is widely used in other European countries but has not been favoured in the United Kingdom.

Deliberately delaying the bus in front (e.g. by radio request to the driver to slow down) would not only be unpopular with passengers on the delayed bus, but is contrary to another important objective of bus operators, that of reducing operating costs and, *ipso facto*, reducing journey times.

This paper is concerned with selective bus priority at traffic signals, that is, giving varying levels of priority to buses. By giving higher levels of priority to buses which are running late it is hoped that bus service regularity will be improved. Alternative strategies for assigning priority are described in Section 3.

A simulation model was developed to aid evaluation of alternative priority strategies in terms of their impacts on bus headway regularity, bus journey time and delay to general traffic. A brief description of the model is given in Section 4 and results of the evaluation in Section 5.

2. THEORETICAL BACKGROUND

One of the objectives in this paper is to minimise the average waiting time of passengers at bus stops. In this section the relationship between average passenger waiting time and bus headway is derived.

If t_N and t_{N+1} are the arrival times of two consecutive buses at a bus stop then, assuming that the arrival of passengers at the bus stop is uniformly distributed, the expected average bus passenger waiting time, W, is given by

$$W = \int_{t_N}^{t_{N+1}} (t_{N+1} - x)\, dx \, / \int_{t_N}^{t_{N+1}} 1\, dx \quad = \tfrac{1}{2}\,(t_{N+1} - t_N)^2 / (t_{N+1} - t_N) \; = \tfrac{1}{2}\, H^2 / H \quad (= H/2) \quad (1)$$

where H is the headway between the two buses, i.e. $H = t_{N+1} - t_N$

For a series of buses with headways H_i , the above formula generalises to

$$W = \tfrac{1}{2} \sum_i H_i^2 / \sum_i H_i \qquad\qquad (2)$$

Since the denominator in the above formula, the sum of the headways, is a constant term, equalling the time period between the first and last buses, minimising the expected average waiting time is equivalent to minimising the sum of the squared headways. This, in turn, is equivalent to minimising the variance (or standard deviation) of bus headway. The minimum variance occurs when all the headways are the same, i.e. the buses are equally spaced. This motivates the desire to improve headway regularity.

3. METHODOLOGY

In this section some of the practical aspects of providing selective bus priority at traffic signals are described:

Priority levels
The two fundamental ways in which a bus may receive priority at a traffic signal are:
- extensions - where the bus is detected on a green aspect and the green is held until the bus clears the signals
- recalls - where the bus is detected on red and the green signal is 'hurried', i.e. intervening signal stages are shortened to give green to the bus more quickly

Extensions tend to bring about more benefit for buses than recalls as the delay associated with buses which are eligible for an extension but 'just miss' the green light is greater than that for buses which are eligible for a recall which could arrive at a traffic signal during the middle, or towards the end of the red period. In addition, extensions cause less disruption to other traffic than recalls since there is less interference with traffic signal settings. If too many recalls are awarded the disruption can be detrimental to buses as well as to other traffic.

In practice, therefore, it is advisable to give preference to extensions over recalls, where there are conflicting demands from two or more buses at a set of lights, and to constrain the number of recalls which are awarded.

Different levels of priority can be defined in terms of whether or not extensions and/or recalls are awarded and by degree of saturation constraints, whereby priority is only granted if there is spare capacity at the junction, defined by the specified degree of saturation threshold value. For example, the following priority levels can be defined:

Priority level	Description
0	No priority
1	Extensions only
2	Extensions and recalls (constrained by degree of saturation)
3	Extensions and recalls (unconstrained)

Priority benefits

Figure 1 shows the impact of different priority levels, in terms of delay savings to buses and delays incurred by general traffic, for varying levels of bus flow. Figure 1 was derived from field studies undertaken in Camden and Edgware Road in London since 1994 (Hounsell *et al*, 1996). Figure 1 indicates the amount of time saving which can be expected for a bus at a traffic signal operating priority. It should be pointed out that this is an average time saving over junctions: at highly saturated junctions there may be little or no benefit gained by priority whereas higher benefits would be expected at junctions having plenty of spare capacity.

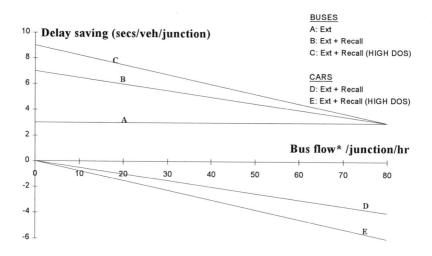

Figure 1. Relationships for delay savings

Notes on Figure 1

* Bus flow refers to buses given priority

EXT = extension, DOS = degree of saturation

Lines B and D correspond to a DOS constraint of 95% threshold for recalls

Lines C and E correspond to unconstrained recalls

Bus headways

Time gaps between buses, i.e. headways, have to be measured on-street. Automatic vehicle location (AVL) systems, such as COUNTDOWN in London (Smith *et al*, 1994) and STOPWATCH in Southampton (Wren, 1996), which are normally used to provide information to passengers at bus stops, can be used to obtain the necessary bus location information from which headways can be calculated.

AVL systems tend to use radio communications to poll buses for their position at regular time intervals (for example, every 30 seconds). The calculated bus headway, therefore, could be 'out of date' by 30 seconds, say. Where buses have become closely spaced it is possible that in 30 seconds one bus may have overtaken another. The possibility of errors in calculated headway and, as an extreme example of this, overtaking, should be considered when assigning priority levels to buses based on headways.

Priority strategies

A priority level has to be assigned to each bus according to its current headway to the bus in front. Different priority strategies will have different impacts on bus service regularity, bus journey time and delay to general traffic. In the results shown in Section 4 these performance indicators are translated into monetary costs in order to compare them.

The preferred priority strategy will be dependent on policy objectives. In this paper the main objective was to improve bus service regularity. A heuristic algorithm, based on the ratio of the actual bus headway to the expected average bus headway, was developed to select priority levels with this objective in mind. This *headway algorithm* consists of two steps:

1. For each bus calculate the *headway ratio* R as

$$R = actual\ headway\ /\ nominal\ headway \tag{3}$$

where the *nominal headway* is the expected average bus headway for the bus service.

2. Compare R with pre-defined threshold ratios R_1, R_2, R_3 to assign priority level:

    If $R \geq R_3$ then   *priority level* = 3

    else if $R \geq R_2$ then   *priority level* = 2

    else if $R \geq R_1$ then   *priority level* = 1

    else   *priority level* = 0

where $R_3 \geq R_2 \geq R_1$

By modifying the values taken by R_1, R_2, R_3 and defining the meanings of the different priority levels it was possible to 'fine-tune' the priority system to improve results. The combinations which were tested are shown in Table 1: logic 1 provides no priority and is the base case against which results are compared in Section 5; logics 2 and 3 are *non-selective* priority strategies, providing the same level of priority for all buses, which will benefit bus journey times but will not improve headway regularity; logics 4 to 9 are all *selective* strategies which use the headway ratio (R) of a bus to decide its level of priority. The selective strategies are expected to improve both bus journey time and headway regularity.

Table 1. Alternative priority strategies tested

Logic no.	Description
1	No priority for any bus
2	Extensions only for all buses
3	Extensions +R_{HIGH} for all buses
4	Extensions +R_{HIGH} for buses with $R \geq 1.0$
5	Extensions +R_{HIGH} for buses with $R \geq 1.25$
6	Extensions +R_{HIGH} for buses with $R \geq 1.5$
7	Extensions +R_{HIGH} for buses with $R \geq 1.75$
8	Extensions +R_{HIGH} for buses with $R \geq 1.0$ Extensions only for buses with $R \leq 1.0$
9	Extensions +R_{HIGH} for buses with $R \geq 1.5$, Extensions +R_{NORMAL} for buses with $1.5 \geq R \geq 1.0$ Extensions only for buses with $R \leq 1.0$

Notes on Table 1

R_{HIGH} = recalls using a high degree of saturation threshold

R_{NORMAL} = recalls using the 'normal' degree of saturation threshold (95%)

Testing and evaluation of the priority strategies shown in Table 1 were undertaken with the aid of a simulation model developed for the study. This simulation is described in Section 4.

4. SIMULATION MODEL

The headway algorithm described in Section 3 was incorporated into a simulation model. The model was built using survey data from Uxbridge Road in West London. The features of the model included:

- *Network* - Linear, containing 29 bus stops and 14 signalised junctions.

- *Buses* - Bus flows, starting headways and occupancies derived from survey data. Four different bus routes, displaying a range of different headway and stopping characteristics and overlapping on parts of the route.

- *Passengers* - Boarding and alighting passengers at bus stops calculated as

$$no.\ of\ boarders = boarding\ rate * bus\ headway \qquad (4)$$
$$no.\ of\ alighters = alighting\ rate * bus\ headway \qquad (5)$$

  where boarding and alighting rates were derived from survey data.

- *Bus stops* - Bus dwell times at bus stops calculated using formulae derived from a paper by York (1993). For example, for one-person operated, two-door buses:

$$dwell\ time\ (secs) = max\ (6.44+1.42*no.\ of\ alighters,\ \ 8.26+4*no.\ of\ boarders)\ \ (6)$$

  where max (x, y) means the larger value of x and y.

The effect of dwell time on the bunching of buses is illustrated in Figure 2, where the progress of 14 buses are shown, relative to the front bus, as they travel through the network, stopping at all 29 bus stops. The vertical distance between points represents the headway between buses. The points on the y-axis show the headway distribution near the start of the route (at bus stop no. 2). It can be seen from Figure 2 that, at bus stops further downstream, buses are tending to form bunches.

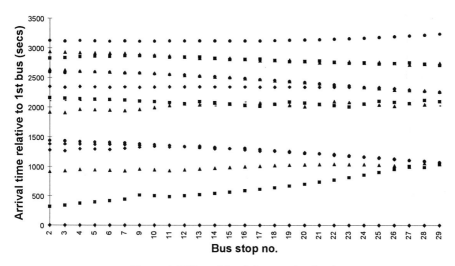

Figure 2. Effect of dwell times on bus headways

Note on Figure 2
Each symbol represents an individual bus (route 207).

- *Priority delay savings* - the delay saving accrued by a bus at a traffic signal taken from the data shown in Figure 1.

5. RESULTS

The nine different priority strategies, shown in Table 1, were tested using the simulation model, described in Section 4. Table 2 shows some results from the simulation runs on the effects of the different priority strategies on average bus journey time and on average passenger waiting time for two different bus routes: route 207 is a high frequency service (scheduled headway of 4½ minutes), stopping at every bus stop; route 607 is an 'express' service, stopping at 8 of the 29 bus stops and has a scheduled headway of 13 minutes.

Table 2. Effect of headway algorithm on journey time and average passenger waiting time

	Bus route 207		Bus route 607	
Logic no.	Journey time secs (saving)	Average passenge waiting time secs (saving)	Journey time secs (saving)	Average passenge waiting time secs (saving)
1	1378 (-)	236 (-)	1013 (-)	474 (-)
2	1336 (3.0%)	236 (0%)	971 (4.1%)	474 (0%)
3	1305 (5.3%)	236 (0%)	940 (7.2%)	474 (0%)
4	1331 (3.4%)	221 (6.4%)	959 (5.3%)	461 (2.7%)
5	1338 (2.9%)	218 (7.6%)	989 (2.4%)	463 (2.3%)
6	1345 (2.4%)	219 (7.2%)	997 (1.6%)	469 (1.1%)
7	1350 (2.0%)	218 (7.6%)	1004 (0.9%)	470 (0.8%)
8	1309 (5.0%)	227 (3.8%)	939 (7.3%)	467 (1.5%)
9	1312 (4.8%)	226 (4.2%)	946 (6.6%)	468 (1.3%)

Notes on Table 2
See Table 1 for meaning of logic numbers
Percentage savings are calculated relative to logic 1 (no priority)

From Table 2 it can be observed that average passenger waiting time is reduced by up to 18 seconds (7.6%) for the stopping service (bus route 207) and by up to 13 seconds (2.7%) for the express service (bus route 607). Higher benefits are gained by the stopping service since the greater number of stops disrupts the headway distribution to a greater extent and there is more scope for improvement, therefore.

Priority logics 8 and 9 give two different levels of priority, according to bus headway, and provide a good 'compromise' solution in terms of savings in both journey time and passenger waiting time.

Table 3 gives an economic assessment of the priority strategies using the following cost equations (using rounded figures from HEN2 (1996)):

passenger waiting cost = £6 (total passenger waiting time) /hr   (7)

bus travel cost = £12 (total bus travel time) /hr + £3 (total passenger travel time) /hr   (8)

other traffic disbenefit = £8 (flow) (no. of signals) (additional delay per signal) /hr   (9)

where a traffic flow of 1000 vehicles per hour was assumed. Note that bus passenger costs are totalled over all passengers; the cost of £8 per vehicle is based on average vehicle occupancy.

Table 3. Cost savings

Logic no.	Saving in travel cost (£/hr)	Saving in waiting cost (£/hr)	Additional cost to other traffic (£/hr)	Overall saving (All vehicles) (£/hr)	Overall saving (Buses only) (£/hr)
1	-	-	0.0	0.0	0.0
2	33.7	0.0	0.0	33.7	33.7
3	56.9	0.0	121.3	-64.4	56.9
4	41.4	35.7	53.7	23.4	77.1
5	31.0	39.5	40.7	29.8	70.5
6	28.5	37.9	30.3	36.1	66.4
7	23.3	37.7	24.3	36.7	61.0
8	53.0	21.9	55.0	19.9	74.9
9	50.6	22.0	46.7	25.9	72.6

Note on Table 3

See Table 1 for meaning of logic numbers

From Table 3 it can be observed that:

- Overall savings for buses using selective priority strategies are, typically, at least twice those available from acceptable non-selective strategies (acceptable in their impact on non-priority traffic).
- Considering results for buses only, waiting cost savings are maximised for strategies giving the highest differential in priority between "late" buses and other buses (i.e. Strategies 4 to 7). However, the other buses in these strategies lose significant travel cost savings through not being awarded priority. Two general strategies produce the best result for buses:
  1. Extensions and recalls for all late buses (Strategy 4)
  2. As 1, but where other buses are awarded extensions only (Strategy 8).
- Considering buses and other vehicles combined, the best overall performances are achieved by the 'moderate' strategies which either give a low level of priority, i.e. extensions only (Strategy 3), or give a high level of priority to relatively few buses by setting a high threshold value for the headway ratio (Strategies 6 and 7).

6. CONCLUSIONS

The paper has shown that bus headway regularity can be improved by providing selective bus priority at traffic signals according to bus running headway. Average passenger waiting times at bus stops were shown, by means of a simulation model of Uxbridge Road, West London, to be reduced by up to 7.6%.

Alternative priority strategies were proposed and evaluated according to criteria of bus journey time, passenger waiting time and impacts on other traffic. The optimum strategy will be dependent, in practice, on policy objectives. The best strategies for improving bus headway regularity are those which give a high level of priority to buses which are running late and no priority to other buses; the best strategies for improving bus journey time are those which give a high level of priority to all buses but these tend to bring high disbenefits to other traffic; the best overall strategies, in terms of combined passenger costs, are those which give a high level of priority to buses which are running late and a lower level of priority, e.g. extensions only, to other buses.

REFERENCES

HEN2 (1996) Highway Economics Note, Section 2, *Design Manual for Roads and Bridges*, **13**, Economic Assessment of Road Schemes, September 1996, Highways Agency, HMSO.

Hounsell, N.B., McLeod, F.N., Bretherton, R.D. and Bowen, G.T. (1996) PROMPT: Field Trial and Simulation Results of Bus Priority in SCOOT. *8th International Conference on Road Traffic Monitoring and Control, Conference Publication No. 422*, pp 95-99, *IEE*, London, 23-25 April 1996.

Smith, R., Atkins, S. and Sheldon, R. (1994) London Transport Buses: ATT in Action and the London COUNTDOWN Route 18 Project. *Proceedings of the First World Congress on Applications of Transport Telematics and Intelligent Vehicle-Highway Systems*, Paris, Nov/Dec, 1994, pp 3048-3055.

Wren, A. (1996) ROMANSE - Information Dissemination. *Proceedings of the Third World Congress on Applications of Transport Telematics and Intelligent Vehicle-Highway Systems*, Orlando, Florida, U.S.A., October 14-18 1996, pp 4688-4696.

York, I.O. (1993) Factors affecting bus-stop times *TRL Project Report 2, T1/25*.

A Study Into Urban Roadworks With Shuttle-Lane Operation

Nashwan A. Samoail and Saad Yousif, Department of Civil and Environmental Engineering, University of Salford, UK

Abstract

In urban areas where roadworks are required, single lane shuttle operation is applied, especially where there is limited road space. There are operational problems relating to the site such as site geometry, visibility, length of roadworks zone, position of signs with other traffic control devices and signal timing. Other problems are mainly related to drivers' behaviour and their compliance with traffic controls on site.

The reduced road width caused by the works will interrupt the free flow of traffic and it can also add to the risks to road users. In addition, shuttle operation may introduce long queues and increase delays especially during peak periods.

There is a need to identify those parameters and behaviours which might influence traffic performance in terms of safety and capacity. An investigation of four roadworks sites in urban roadworks within the Greater Manchester area was undertaken for this purpose. Parameters included in the examination were position of the STOP sign, signal timing, weather conditions, time headway, vehicle speed and percentages of heavy goods vehicles (HGV) in the traffic stream. Statistical analysis and comparisons between sites were conducted. Other factors related to the operation of the shuttle-lane were provided based on site observations.

1. Introduction

Maintenance works on single carriageway roads in urban areas are normally carried out by closing one lane in one direction leaving the other lane for use by both directions. This type of operation is referred to as single lane shuttle operation. This one-way traffic operation requires an 'all-red' interval of sufficient duration for traffic to clear the shuttle-lane at the minimum operating speed in the work area (OECD, 1989).

During shuttle operation under congested conditions, severe delays and long queues may reach to the point of causing blockage to side roads in the vicinity of the roadworks site due to reduction in road capacity. Drivers' behaviour towards, for example, non-compliance with traffic signs and signals may be increased by road congestion during temporary roadworks which may lead drivers to taking unnecessary risks. This in turn could have adverse effects on both capacity and safety.

Choosing the type of control for shuttle operation depends on many factors such as traffic flow, length of controlled area, visibility through the work area and duration of work (Freeman Fox and Associates, 1973). The capacity (maximum two-way flow) of the shuttle-lane controlled by temporary traffic signal depends on site length. Capacity can reach 1600 veh/hr for site lengths of up to 50 metres and about 1100 veh/hr for 300 metres site lengths (Summersgill, 1981).

This paper aims to examine the main parameters which could influence shuttle operation in terms of maximising capacity and improving safety. Figure 1 shows a typical site layout with the locations of the main traffic control devices as normally used in practice.

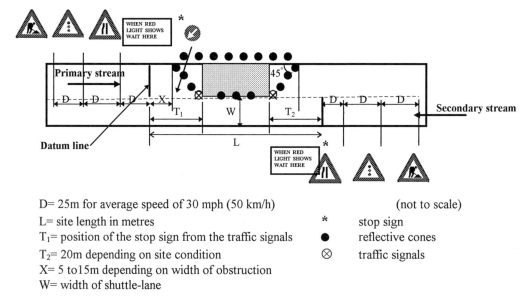

D= 25m for average speed of 30 mph (50 km/h)　　　　　　　　　(not to scale)
L= site length in metres　　　　　　　　　　　*　　　　stop sign
T_1= position of the stop sign from the traffic signals　　●　　　reflective cones
T_2= 20m depending on site condition　　　　　　⊗　　　traffic signals
X= 5 to15m depending on width of obstruction
W= width of shuttle-lane

Figure 1. Typical site layout and location of signs

Field surveys from different sites on urban roadworks were carried out to examine the following (for further details on site selection and data collection see Samoail, 1997):

⇒ time headways for both primary and secondary streams,

⇒ perception time of drivers on crossing the stop line after being in a queue and time required for queue dispersion, and

⇒ observations of driver's non-compliance with traffic signs and signals.

2. DATA COLLECTION METHODOLOGY

Data were collected for at least two hours from four roadworks sites in urban areas as shown in Table 1. This was done during day light for both peak and off-peak periods. Traffic counts, time headways and signal timings were recorded using video cameras from these sites which were controlled by temporary traffic signals. Half of these signals were vehicle actuated (VA) and the other half were fixed timing (FT). The data were then abstracted from video playbacks using an event recorder. Average speed data were obtained manually by measuring the time taken by a sample of vehicles to travel between the two temporary traffic signals using a stopwatch. Queue lengths, physical characteristics and drivers behaviour were observed on sites.

Table 1. Summary of the physical characteristics for the four sites

Parameters	Site 1		Site 2		Site 3		Site 4	
	P	S	P	S	P	S	P	S
Date of Survey	Sunday 28/2/96		Friday 26/7/96		Monday 16/9/96		Thursday 24/4/97	
Period of Survey (Hours)	13:35 - 15:45		15:18 - 18:13		11:10 - 14:10		8:32 - 10:47	
Weather	Snowy		Cloudy		Sunny		Sunny	
Site Length (metre)	72.00		33.80		137.20		46.00	
Width of Hazard (metre)	4.30	4.30	5.30	5.30	3.40	3.40	5.20	5.20
W (metre)	3.90	3.90	5.30	5.30	3.60	3.60	5.20	5.20
T_1 (metre)	12.00		10.00		16.70		26.00	

Note: P = primary stream, S = secondary stream

3. ANALYSIS OF DATA

Time Headway

Time headway may be defined as the time difference between the arrival of successive vehicles at a reference point. Time headway data were obtained by recording the time between two successive vehicles in the same direction on touching the datum line. The time headway data was compiled for each direction, primary and secondary stream, separately. This allowed comparisons between both distributions and their respective means.

Headway Distribution. A maximum headway of 5 seconds was chosen for the analysis of the saturation flow for both primary and secondary streams for all sites. This was to ensure that vehicles were travelling in platoons. The total number of observed headways was 6317. Figure 2 shows typical headway distributions for the primary and secondary streams. Both are positively skewed. The secondary streams have higher percentage of time headways of less than 2 seconds compared with that of primary streams for all the sites (for more details see Samoail, 1997).

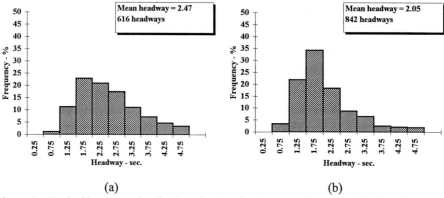

(a)   (b)

Figure 2. Typical headway distributions for time headways ≤ 5 seconds, (a) for primary stream and (b) for secondary stream

<u>Average Mean Time Headways</u>. Average mean headways between primary and secondary streams were compared to give an indication on their highest equivalent saturation flows. Mean time headways for each cycle were first calculated then an average value of these headways were obtained for all cycles for each site. The average mean headway for all the cycles for each direction was based on excluding the first vehicle from both directions to rule out the effect of site length. Results showed that the highest observed flows for the secondary stream were always greater than that of the primary stream for all sites as shown in Table 2.

Table 2. Typical comparison of μ_p and μ_s

		n	$\overline{x}$	σ^2	Z-Value	Critical	Saturation Flow (pcu/hr)
Site 1	P	88	2.58	0.2756			1410
	S	88	2.05	0.0748	8.44	1.96	1780
Site 2	P	129	2.42	0.2100			1630
	S	129	2.25	0.1595	3.39	1.96	1740
Site 3	P	108	2.73	0.3210			1400
	S	102	2.65	0.6250	0.84	1.96	1470
Site 4	P	101	2.13	0.0635			1750
	S	98	2.08	0.0655	1.38	1.96	1840

Note: P = primary stream, S = secondary stream, n = sample size (number of cycles during the survey); $\overline{x}$ = Average sample mean; σ^2 = sample estimate of population variance; Z = standardised normal distribution; μ_p = population average mean headway for primary stream; μ_s = population average mean headway for secondary stream.

Tests of significance were carried out, initially on the hypothesis that the average mean headway for all cycles in the primary stream did not differ from that in the secondary stream for all sites. Examination of primary and secondary streams average mean time headways at the 5% level of significance showed that the calculated values of Z (as shown in Table 2) were higher than the critical ones for Sites 1 and 2. The results could be attributed to the effect of the physical characteristics and geometric layout and the differences in mean speed and acceleration. Vehicles travelling in the primary streams need more manoeuvring time to clear the taper (obstructed work) than those in the secondary stream. Also bends or curvatures in the path might result in increased average gaps between vehicles. The measured average speed through the shuttle-lane for the secondary stream was higher than that of the primary stream for all the sites. This tends to shorten the time gaps between vehicles which may result in a higher overall capacity.

Headways Between Sites. Another tests of significance were carried out to compare the average mean headway between different sites for both primary and secondary streams. Table 3 shows a typical comparison, for example, between Sites 1 and 4. A test of significance at the 5% level indicated that primary average mean time headway for Site 1 does differ from that of Site 4.

Table 3. Typical comparison of μ_{p1} and μ_{p4} between Sites 1 and 4

n_{p1}	n_{p4}	$\overline{x}_{p1}$	$\overline{x}_{p4}$	σ_{p1}^{2}	σ_{p4}^{2}	Z	Critical
88	101	2.58	2.13	0.2756	0.0635	7.34	1.96

Note: $\%HGV_{p1} = 2.22$, $\%HGV_{p4} = 4.05$

The differences between the two averages could be attributed to weather conditions, position of the warning stop sign from the signals (which is more than double the distance for Site 4 compared with that of Site 1) and the percentages of HGV's. To examine which of these elements had a major impact on the difference between the two averages, another test was carried out by taking two samples (each sample with {n} number of cycles) from both sites with similar or identical percentages of HGV's. Although the sample sizes were small, a test of significance indicated that the primary averages mean time headway for Site 1 did differ significantly from that of Site 4 at the 5% level of significance, as shown in Table 4. This could be attributed to the position of the warning stop signs at the primary stream (T_1). If T_1 is too small, it may create manoeuvring difficulties for vehicles in the primary streams (especially for HGVs) before entering into the shuttle-lane. However, a longer distance may reduce the effect of the tapered section. This may bring the operational conditions of the primary streams to be similar to those of the secondary streams. This needs further examination to optimise the most appropriate positioning which can provide higher capacity.

Table 4. Typical comparison of μ_{p1} and μ_{p4} between Sites 1 and 4

n_{p1}	n_{p4}	$\overline{x}_{p1}$	$\overline{x}_{p4}$	σ_{p1}^{2}	σ_{p4}^{2}	t	Critical
13	21	2.37	2.19	0.0493	0.0504	2.24	2.06

Note: $\%HGV_{p1} = 9.86$, $\%HGV_{p4} = 9.89$

Other Parameters. Another parameter which might influence driver's behaviour when entering the shuttle-lane is road surface conditions. On a wet surface, observations show a general increase in the time headway between vehicles which indicates that drivers enter the shuttle-lane more cautiously. This may cause a general decrease in the capacity of the shuttle-lane.

Perception Time

The effect of the time headways between vehicles in the primary stream on crossing the stop line after queuing at the traffic signals was also examined for the four sites. The number of vehicles passing through in each cycle depends to a large extent on these headways. Average values for these headways as obtained from the four roadwork sites are shown in Table 5. For comparison, typical values as reported by Briggs (1997) for signalised intersections are also included.

Table 5. Average time headways on moving off from traffic lights

Average time headway (sec)	1 Site 1	2 Site 2	3 Site 3	4 Site 4	5 Average of the four roadwork sites	6 Typical values at signalised intersections (Briggs, 1977)
for 1st vehicle	2.81	2.82	3.44	2.34	2.85	3.8
between 1st and 2nd vehicle	3.29	2.73	3.09	2.45	2.88	2.56
between 2nd and 3rd vehicle	2.69	2.52	2.57	2.43	2.55	2.25
between 3rd and 4th vehicle	2.28	2.29	2.54	2.29	2.35	2.10
between 4th and 5th vehicle	2.35	2.21	2.39	2.17	2.28	1.98
between 5th and 6th vehicle	2.35	2.21	2.39	2.05	2.25	1.93
Total	15.77	14.78	16.42	13.73	15.16	14.62

In general, there are differences in the average time headways between the four sites (Column 5) and those obtained at intersections (Column 6). The results showed that for the 6th vehicle in the queue these average headways gradually reduced to 2.25 seconds (Column 5) at roadwork sites compared with 1.93 seconds (Column 6) at traffic signals. The total average headway for all six vehicles for the four sites (15.16 seconds) is slightly higher than the total average at traffic signals (14.62 seconds). Average headways for the 1st vehicles are longer at traffic signals compared with those at roadwork sites. This increase in perception time may be attributed to drivers at intersections have to take extra care when entering the intersection by ensuring that their path is clear from approaching traffic from other arms compared with the shuttle-lane which has only one opposing direction.

Queue Dispersion

In designing temporary traffic signals for any shuttle-lane, it is essential to know the extent queues are likely to occur and the required time to discharge them. A queue can be defined as the condition when vehicles are either stationary at the selected point or are moving past that point at less than some pre-determined critical speed. The critical speed is considered to be in the range of 5 to 12 mph (or 2.2 to 5.3 m/sec) for most situations as reported and applied by Ham (1967).

<u>Relationship Between Average Journey Time and Site Lengths.</u> Queues were observed before the start of the green period in each cycle and the number of queued vehicles was counted in each direction. A linear regression analysis was undertaken to represent the average travel time through the site length. This would help assessing the required time for queue dispersion. To achieve this, four points have been calculated from the four sites representing different site lengths as shown in Figure 3. Each point represents the average time required by the 1st vehicles from both directions to clear the site length as shown in equation "(1)".

$$J_T = 0.1405\ L + 1.2112 \quad\text{.....................................} \quad (1)$$

Where:

J_T = average travel time for the 1st vehicles to clear the shuttle length (seconds)

L = site length in metres

This equation can be applied to calculate the average speeds within the shuttle-lane for the four sites. These speeds were ranging between 20 and 25 km/h depending on site length (higher speeds for longer site lengths).

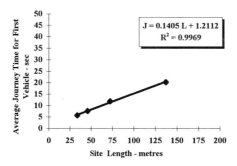

Figure 3. Regression line of average journey times through four sites

<u>Queue Dispersion for Secondary Streams.</u> Data on queue dispersion were only available from the secondary streams for Sites 2 and 4 because of site restrictions and the position of cameras used for data collection. After obtaining the time headways in each cycle for the secondary streams at Sites 2 and 4, it was possible to calculate the travel time required to reach the datum lines by the queued vehicles. These results are as shown in Figure 4 together with the regression lines representing the relationships as given in equations "(2) and (3)". The highest observed queues for Sites 2 and 4 were 11 and 15 vehicles, respectively.

$$J_2 = 2.0478\ Q_2 + 4.5049 \quad\text{.....................................} \quad (2)$$
$$J_4 = 2.0029\ Q_4 + 5.8557 \quad\text{.....................................} \quad (3)$$

Where,

J_2, J_4 = time in seconds required for queue dispersion at Sites 2 and 4 respectively

Q_2, Q_4 = queues at Sites 2 and 4 respectively

Based on equations "(2) and (3)", the time required to disperse queues can be calculated for Sites 2 and 4 as highlighted in Columns 4 and 5, Table 6. For illustration purposes, queues of sizes 1, 5, 10 and 15 vehicles were selected as shown in Column 3.

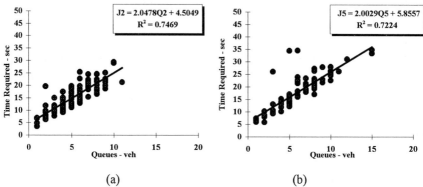

(a)   (b)

Figure 4. Time required to disperse queues for the secondary stream, (a) Site 2 and (b) Site 4

Equations "(2) and (3)" were also used to calculate the time required to disperse queues for the different site lengths. In order to obtain this, use was made of equation "(1)" which gave the relationship between the average journey time for the leading vehicle in a queue (1st vehicle) for the different site lengths. This was calculated and shown in Column 6.

Table 6. Time required to disperse queues for different site lengths

Col. 1	Col. 2	Col. 3	Col. 4	Col. 5	Col. 6	Col. 7
Site	Site length	Queue	Time required to disperse queues			
	(metres)	(No. of vehicles)	Equation 2, Site 2	Equation 3, Site 4	Equation 1, All Sites	Col. 4 minus Col. 5
1	72.00	1	11.92	11.51	11.33	0.41
		5	20.11	19.53	—	0.58
		10	30.35	29.54		0.81
		15	NA	39.56		-
2	33.80	1	**6.55**	6.14	5.96	0.41
		5	**14.74**	14.16	—	0.58
		10	**24.98**	24.17		0.81
		15	**NA**	34.19		-
3	137.20	1	21.08	20.67	20.49	0.41
		5	29.27	28.69	—	0.58
		10	39.51	38.70		0.81
		15	NA	48.72		-
4	46.00	1	8.26	**7.85**	7.67	0.41
		5	16.45	**15.87**	—	0.58
		10	26.69	**25.88**		0.81
		15	NA	**35.89**		-

Note: NA = Not available

Both equations "(2) and (3)" can be generalised and used to calculate the time required to disperse the queues for the different sites by calculating the differences in values shown in Column 6.

These values could be either positive or negative depending upon the site lengths. For example, equation "(2)" for Site 2 can be used for Site 1 by shifting it up by a value equals to [11.33 minus 5.96] which represents the difference in average travel time between the two sites. The results are as shown in Column 4. Similarly equation "(3)" for Site 4 can be used for other sites and the results are given in Column 5.

The differences in the calculated times to disperse queues for the different site lengths given in Columns 4 and 5 were relatively small for the selected queues as shown in Column 7. Therefore, equation "(3)" together with equation "(1)" can be generalised for site lengths between 33 and 137 m and queues between 1 and 15 vehicles to calculate the time required to disperse queues. This information may be useful in choosing a suitable green period in signals setting for different site lengths with different expected queues.

Driver Behaviour and Other Site Observations

Driver's Compliance. Observations from the four sites have shown that some drivers do not comply with traffic signals when the signals change to red. This violation occurred on a few occasions when drivers were seen to be travelling at high speeds. Violations also occurred on one of the sites because a heavy plant used by the contractor was obstructing the passage of the open lane, resulting in drivers failing to comply with the signals. This working activity influenced drivers' behaviour and was causing chaos and congestion in the vicinity of the shuttle-lane.

On another occasion where congested conditions occurred, manual controls were used for a short duration in order to clear the shuttle-lane from any remaining vehicles. This type of control was found to be effective in improving the operation of the shuttle-lane within a relatively short period of time under severe congested conditions.

The proportion of the violations described earlier varied between 12 and 30% of the total number of vehicles which passed in the primary stream during the survey period. Most of these violations were influenced by the congested conditions which affected drivers' behaviour especially during the morning peak hours. These violations could endanger the safety of road users. Further research into accident rates and traffic control devices is therefore necessary to examine the safety and efficiency of the roadworks section.

Signing at Roadworks. Observations have shown that the signs were either insufficient or not completely visible to drivers. Some signs were not secured properly (for example some were moved or knocked down by vandals). There were no marked stop lines in either directions in the position of the temporary signals. This is important from the view point of safety, visibility and manoeuvrability as recommended in the Traffic Signs Manual - Chapter 8 (1991).

<u>Other Observations</u>. It was also observed that parked vehicles near to traffic signals could affect the position of vehicles queuing in the primary stream. This might also create blockage to the oncoming traffic from the secondary stream. On one of the sites, it was found that the "all-red" period was not sufficient to clear the shuttle-lane from the remaining vehicles, while on another, violations were caused by the long cycle time and long "all-red" periods. These could have adverse effects on both safety and capacity of the shuttle-lane.

4. CONCLUSIONS AND RESULTS

1. Results revealed that equivalent saturation flows for secondary streams were relatively higher than that of primary streams.

2. On average, the 1st vehicle queuing in the primary stream in the shuttle-lane required shorter time to clear the stop line compared with that of traffic signals at intersections. This may be attributed to drivers at intersections being more aware of traffic approaching from other arms.

3. The procedure described in Table 6 may be used in choosing a suitable green period in signals setting for different site lengths with different expected queues.

4. Under congested conditions, manual controls were found to be effective in preventing the shuttle-lane from being blocked for longer periods.

5. The proportion of drivers' non-compliance with traffic signals increased under congested conditions and when unusual working activities were taking place within the roadworks site. There was some evidence of non-compliance with the standards for signing at roadworks.

REFERENCES

Briggs, T. (1977). Time headways on crossing the stop-line after queuing at traffic lights. *Traffic Engineering and Control*, **18**, 264-265.

Department of Transport (1991). *Traffic Signs Manual-Chapter 8*, Traffic safety measures and signs for road works and temporary situations. HMSO, London.

Freeman Fox and Associates (1973). *Traffic Behaviour at Roadworks*. Final Report to Transport Road Research Laboratory, Department of the Environment, London.

Ham, R. (1967). Vehicle queue detection using inductive loop detectors. *Traffic Engineering and Control*, **8**, 669-671.

OECD Road Transport Research (1989). Traffic management and safety at highway work zones. *OECD Scientific Expert Group*, Paris.

Samoail, N.A. (1997). *A study into shuttle-lane operation in urban roadworks*. MSc Dissertation, University of Salford.

Summersgill, I. (1981). The control of shuttle working at roadworks. *Transport and Road Research Laboratory*, Research Report 1024, Crowthorne.

THE BENEFITS AND DISBENEFITS TO TRAFFIC AT PRE-SIGNALLED INTERSECTIONS

Jianping Wu and Nick Hounsell,Department of Civil and Environmental Engineering, University of Southampton, Southampton, UK

ABSTRACT

The need to provide efficient public transport services in urban areas has led to the implementation of bus priority measures in many congested cities. Much interest has recently centred on priority at signal controlled junctions, including the concept of pre-signals, where traffic signals are installed at or near the end of a with-flow bus lane to provide buses with priority access to the downstream junction. Although a number of pre-signals have now been installed in the UK, particularly in London, there has been very little published research into the analysis of benefits and disbenefits to both buses and non-priority vehicles at pre-signalised intersections. This paper addresses these points through the development of analytical procedures which allow pre-implementation evaluation of specific categories of pre-signals.

1. INTRODUCTION

The implementation of pre-signals in UK cities is becoming significant. Taking London only, 14 pre-signals are now in place since the first scheme was introduced in Shepherd's Bush in 1993. A further 20-25 pre-signals are planned for the coming years to contribute to London's bus priority network. Many pre-signals have had site-specific aspects and operate with additional traffic detection for real time control. However, there remains a strong need for a methodology for the design and evaluation of pre-signals at the pre-implementation stage, to ensure that optimum deployment is achieved. The research described in this paper focuses on this point.

A pre-signal aims to give buses priority access into a bus advance area of the main junction stop line so as to avoid the traffic queue and reduce bus delay at the signal controlled junction. Three different categories of pre-signal design have been suggested and/or implemented in the UK (Tee et al, 1994; Roberts and Buchanan, 1993; Oakes et al, 1994) each with its own operating characteristics (These are denoted categories A, B and C in this paper). In category A (Figure 1) the pre-signal controls only the non-priority traffic with buses uncontrolled (or subject to 'give way control'). In contrast, in pre-signal category B, buses are also controlled by a pre-signal at the end of the bus lane. When the pre-signal in the bus lane turns red, non-priority vehicles receive right of way, and vice versa. A typical layout of a pre-signal category B is shown in Figure 2. The number of lanes for general traffic in the bus advance area is usually greater than the number of lanes available upstream of the pre-signal. (An equal number of lanes is shown in Figures 1 and 2 for illustration).

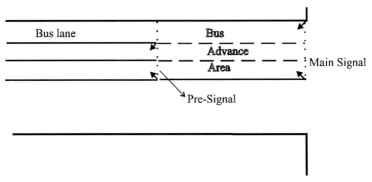

Figure 1. Typical Layout of Pre-signal Category A

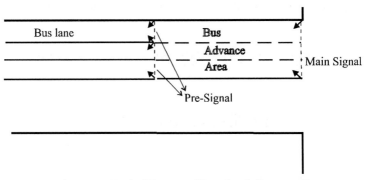

Figure 2. Typical Layout of Pre-signal Category B

A pre-signal category C also has a similar layout to category B, except that detectors are installed in the bus lane. When a bus approaches the pre-signal, a red signal is shown to the non-priority traffic stream to enable the bus to access to the main signal without impedance. The bus lane pre-signal then turns to red, allowing non-priority traffic to fill the bus advance area behind the bus after it has selected the appropriate lane at the main signal. Delays to non-priority vehicles are more complicated to analyse for Category C, and, a microscopic simulation model may be required to study such a VA or semi-VA (only the bus's arrival is detected) controlled junction. In this paper, only the pre-signal categories A and B with fixed time controls are considered.

2. ASSUMPTIONS

When designing pre-signals, we must make sure to release traffic from the pre-signal sufficiently early to make use of green time at the main junction. The following two assumptions were made for this purpose.

2.1 Assumption One

The non-priority vehicles arriving at the pre-signal stop line during the red time of the main signal, r_m, will be fully discharged into the main signal stop line to fill up the bus advance area before the main signal green time starts. This yields

$$r_m \upsilon_d = (r_m - r_p) S_1 N_p \qquad (2.1)$$

where r_p, and r_m are the red time at the pre-signal and main signal respectively (see Figure 3). υ_d is the rate of demand of non-priority vehicles in the approach, (vehicles/second). S_1 is the lane based saturation flow (vehicles/second/lane) and N_p the number of lanes at the pre-signal for non-priority vehicles.

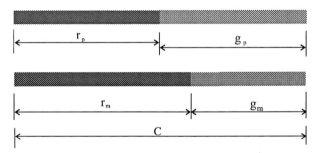

Figure 3. The Relationships of r_p, g_p, r_m, g_m and C

This assumption implies co-ordination of the pre-signal and the main signal for non-priority traffic and ensures no loss of junction capacity due to the pre-signal.

2.2 Assumption Two

Junction capacity equals to demand. In other words, there will be no residual queues at the main signal stop line at the end of green time, g_m. This is given by the following equation.

$$g_m S_1 N_m = C(\upsilon_d + \upsilon_{dB}) \qquad (2.2)$$

where g_m is the green time at the main signal and C the cycle time. N_m is the number of lanes at the main signal stop line and υ_{dB} the rate of bus flow (buses/second). υ_d and S_l are as defined earlier.

3. SIGNALS TIMING

Equations are developed to calculate signal times for pre-signal categories A and B, given the known signal cycle, C, of the main junction.

With equation (2.2), we have

$$g_m = C\frac{(\upsilon_d + \upsilon_{dB})}{S_1 N_m} \qquad (3.1)$$

From Figure 3, we have

$$\begin{aligned} r_m &= C - g_m \\ &= C[1 - \frac{(\upsilon_d + \upsilon_{dB})}{S_1 N_m}] \end{aligned} \qquad (3.2)$$

From equation (2.1), we have

$$r_p = (1 - \frac{\upsilon_d}{S_1 N_p}) r_m \qquad (3.3)$$

From Figure 3, we have

$$\begin{aligned} g_p &= C - r_p \\ &= g_m + \frac{\upsilon_d r_m}{S_1 N_p} \end{aligned} \qquad (3.4)$$

Further, in pre-signal category B, the bus lane is also controlled by a pre-signal which has the following relationship with the non-priority traffic pre-signal.

$$g_{pB} = r_p \qquad (3.5)$$

and

$$r_{pB} = g_p \qquad (3.6)$$

where g_{pB} and r_{pB} are the green and red time for the bus lane pre-signal in pre-signal category B.

4. BENEFITS AND DISBENEFITS TO TRAFFIC AT PRE-SIGNALLED JUNCTIONS

The benefit (delay saving) to buses and non-priority vehicles is defined as the difference in delay with and without pre-signals.

4.1 Traffic Delay Without Pre-signal

Assuming buses and non-priority vehicles arrive randomly, without a pre-signal, they have the same value of average delay d(normal), which is given by

$$d(normal) = \frac{r_m}{2} \tag{4.1}$$

Equation (4.1) is the same as the first term of Webster's expression (Webster & Ellson, 1965) when $\rho = 1$.

4.2 Bus Delay and Delay Saving With Pre-signal Category A

The average bus delay in pre-signal category A, d(CA), derived from the assumptions, has the form

$$d(CA) = \frac{1}{2C(\upsilon_d + \upsilon_{dB})(C\upsilon_d + g_p\upsilon_{dB})}\{\upsilon_d\upsilon_{dB}(2r_mC^2 - r_pr_mC) +$$
$$\upsilon_d^2(r_mC^2 - r_p^2g_m) + \upsilon_{dB}^2(r_mg_pC)\} \tag{4.2}$$

With the delay saving definition, the bus delay saving with pre-signal category A, $\Delta d(CA)$, is

$$\Delta d(CA) = \frac{g_mr_p^2\upsilon_d^2}{2C(\upsilon_d + \upsilon_{dB})(C\upsilon_d + g_p\upsilon_{dB})} \tag{4.3}$$

Equations (4.3) shows that the bus delay saving increases as the parameters g_m and r_p increase.

4.3 Bus Delay and Delay Saving With Pre-signal Category B

Considering the above assumptions, the average bus delay with pre-signal category B, d(CB), has the form

$$d(CB) = \frac{1}{2}(2r_m + g_p - r_p) \tag{4.4}$$

With the delay saving definition, the bus delay saving with pre-signal category B, $\Delta d(CB)$, is

$$\Delta d(CB) = -\frac{1}{2}(r_m + g_p - r_p) \tag{4.5}$$

It can be seen that equation (4.5) would have a negative value. This indicates that buses do not benefit from a delay saving with pre-signal category B.

4.4 Delay and Delay Saving of Non-Priority Vehicles with Pre-Signal Category A

By a similar method, we can get the average delay for non-priority vehicles with pre-signal category A, $d_n(CA)$, which has the form

$$d_n(CA) = \frac{1}{2C(\upsilon_d + \upsilon_{dB})(C\upsilon_d + g_p\upsilon_{dB})}\{r_m C^2 \upsilon_d^2 +$$
$$[r_p^2 C + (2C + r_p)r_m g_p]\upsilon_d\upsilon_{dB} + r_m g_p C\upsilon_{dB}^2\} \tag{4.6}$$

By the delay saving definition, the delay saving of non-priority vehicles with pre-signal category A is

$$\Delta d_n(CA) = -\frac{r_p^2 g_m \upsilon_d \upsilon_{dB}}{2C(\upsilon_d + \upsilon_{dB})(C\upsilon_d + g_p\upsilon_{dB})} \tag{4.7}$$

As we can see, equation (4.7) gives a negative value. This means that the average delay for non-priority vehicle has a little increase with pre-signal category A.

4.5 Delay and Delay Saving of Non-Priority Vehicles with Pre-Signal Category B

The average non-priority vehicle delay with pre-signal category B, $d_n(CB)$, is given by

$$d_n(CB) = \frac{r_m v_d + C v_{dB}}{2(v_d + v_{dB})} \qquad (4.8)$$

The delay savings for the non-priority vehicles with pre-signal category B, $\Delta d_n(CB)$, according to the definition, is

$$\Delta d_n(CB) = -\frac{g_m v_{dB}}{2(v_d + v_{dB})} \qquad (4.9)$$

As we can see, equation (4.9) also gives a negative value. This means an increase in average delay for non-priority vehicles with pre-signal category B.

4.6 The Changes of Total Traffic Delay

The above discussions have shown positive delay saving for buses and negative delay saving for non-priority vehicles with pre-signal category A; and negative delay savings for both buses and non-priority vehicles with pre-signal category B. The following discussion explains the reasons.

With pre-signal category A, buses arriving during time r_p are given priority to access the bus advance area, and take the front positions of the queue at the main signal. As a result, more non-priority vehicles are located to the rear part of the traffic queue compared to the without pre-signal case. Therefore, there is a reduction in average delay to buses, and, meanwhile, an increase to non-priority vehicles. However, for the total traffic (including buses and non-priority vehicles) delay is the same with and without pre-signal category A.

In the case of pre-signal category B, all buses queued in the bus lane pre-signal are discharged into the bus advance area during time g_{pB} ($=r_p$) when the non-priority vehicles are held by the pre-signal red light. These buses will take the front positions of the queue at the main signal and leave all non-priority vehicles to be located to the rear part of the traffic queue. Although no non-priority vehicles will be held to the next cycle at the main signal stop line, the change of the position in the traffic queue adds extra delay to them. Unlike pre-signal category A, the total traffic (including buses and non-priority vehicles) delay with pre-signal category B also increases because that the buses which arrive at the bus lane pre-signal during time r_{pB} ($=g_p$) are not be able to discharge into the bus advance area and leave the main signal stop line until green time g_{pB} ($=r_p$) in the following signal cycle. The extra total vehicle delay (including buses and non-priority vehicles) with pre-signal category B per cycle, $\Delta TD(CB)$, can be calculated by the following equation.

$$\Delta TD(CB) = g_p C \upsilon_{dB} - \frac{g_m C \upsilon_{dB}^{\,2}}{2(\upsilon_d + \upsilon_{dB})} \qquad (4.10)$$

5. APPLICATION EXAMPLES

Three examples are presented here to demonstrate how the delays of buses and non-priority vehicles will be affected with the pre-signals category A and B. Table 1 lists the details of the three example approaches.

Table 1. The Details of the Three Example Approaches

Example Approaches	N_p	N_m	C (seconds)	S_l	υ_d	υ_{dB}
Example 1	1	2	100	2000/3600	1000/3600	120/3600
Example 2	2	2	100	2000/3600	1000/3600	120/3600
Example 3	2	3	100	2000/3600	2000/3600	120/3600

5.1 Delay and Delay Savings

Vehicle delay and delay savings with pre-signal categories A and B are calculated for all the three example approaches and listed in Tables 2 to 5.

Table 2. Bus Delay and Delay Saving with Pre-signal Category A

Examples	Without Pre-Signal (sec/bus)	With Pre-Signal (sec/bus)	Bus Delay Saving (sec/bus)	Delay Saving in Percentage (Delay Saving)/ (Delay Without Pre-signal)
One	36.00	34.50	1.50	4.2%
Two	36.00	32.55	3.45	9.6%
Three	32.50	30.80	1.70	5.2%

Table 3. Bus Delay and Delay Saving with Pre-signal Category B

Examples	Without Pre-Signal (sec/bus)	With Pre-Signal (sec/bus)	Bus Delay Saving (sec/bus)	Delay Saving in Percentage (Delay Saving)/ (Delay Without Pre-signal)
One	36.00	86.00	-50.00	-139%
Two	36.00	68.00	-32.00	-89%
Three	32.50	82.50	-50.00	-154%

Table 4. Non-priority Vehicle Delay and Delay Saving with Pre-signal Category A

Examples	Without Pre-Signal (sec/veh)	With Pre-Signal (sec/ veh)	Delay Saving (sec/ veh)	Delay Saving in Percentage (Delay Saving)/ (Delay Without Pre-signal)
One	36.00	36.18	-0.18	-0.50%
Two	36.00	36.41	-0.41	-1.14%
Three	32.50	32.60	-0.10	-0.30%

Table 5. Non-priority Vehicle Delay and Delay Saving for Pre-signal Category B

Examples	Without Pre-Signal (sec/veh)	With Pre-Signal (sec/ veh)	Delay Saving (sec/ veh)	Delay Saving in Percentage (Delay Saving)/ (Delay Without Pre-signal)
One	36.00	37.50	-1.50	-4.16%
Two	36.00	37.50	-1.50	-4.16%
Three	32.50	33.50	-1.00	-3.08%

It is clear from these examples that the pre-signal category B is not beneficial using fixed time signalling. Active bus detection and signal response are likely to be needed to provide the required benefits.

6. CONCLUSIONS

This paper has illustrated how an analytical approach can be used for the pre-implementation evaluation of pre-signals. The approach presented in this paper covers issues of capacity, signal settings, delay and delay saving estimations for buses and non-priority traffic. Examples have shown how delay savings to buses can be achieved with the category A pre-signal (where buses are unsignalised at the pre-signal) without significant disbenefit to non-priority traffic. Delay savings to buses are highest where there is a long red period at the non-priority traffic pre-signal, which is possible when the proportion of green time at the main signal is low. However, category B pre-signals (where buses are also signalised at the pre-signal) are shown to generally cause disbenefit to buses. This excludes the situation where special bus detectors are installed.

REFERENCES

Oakes J., ThellMann A. M. and Kelly I.T. (1994) Innovative bus priority Measures. *Proceedings of Seminar J, Traffic Management and Road Safety, 22nd PTRC European transport summer annual meeting.* University of WARWICK, ENGLAND. **381** 301-12.

Roberts M. and Buchanan C. (1993) Bus priority - the solution and west London demonstration project. *Proceedings of Seminar C, Highways and Planning, 21st PTRC European transport summer annual meeting.* **365** 49-57.

Tee A., Cuthbertson T. and Carson G. (1994) Public transport initiatives in Surrey. *Traffic Engineering and Control.* **35** 70-73.

Webster F. V. and Ellson P. B. (1965) Traffic signals for high-speed roads. *Road Research Technical Paper No. 74.* Road Research Laboratory, (H.M. Stationery Office), London.

MOMENT APPROXIMATION TO A MARKOV MODEL OF BINARY ROUTE CHOICE

David Watling, Institute for Transport Studies, Leeds University, UK.

ABSTRACT

The paper considers a discrete-time, Markov, stochastic process model of drivers' day-to-day evolving route choice, the evolving 'state' of such a system being governed by the traffic interactions between vehicles, and the adaptive behaviour of drivers in response to previous travel experiences. An approximating deterministic process is proposed, by approximating both the probability distribution of previous experiences—the "memory filter" —and the conditional distribution of future choices. This approximating process includes both flow means and variances as state variables. Existence and uniqueness of fixed points of this process are examined, and an example used to contrast these with conventional stochastic equilibrium models. The elaboration of this approach to networks of an arbitrary size is discussed.

1. INTRODUCTION

Stochastic network equilibrium approaches to modelling driver route choice in congested traffic networks are concerned with predicting network link flows and travel costs, corresponding to a fixed point solution to a problem in which (Sheffi, 1985):

— actual link travel costs are dependent on the link flows; and

— drivers' route choices are made according to a random utility model, their perceptual differences in cost represented by a known probability distribution.

The term *Stochastic User Equilibrium (SUE)* is typically used to describe such a fixed point.

A radically different approach to modelling this interaction between flow-dependent travel costs and travel choices was proposed by Cascetta (1989), and studied further by Davis & Nihan (1993), Cantarella & Cascetta (1995) and Watling (1996). This *Stochastic Process (SP)* approach differs in two major ways from SUE. Firstly, the SP model is specified as a dynamical process of the day-to-day adjustments in route choice made by drivers, in response to travel experiences encountered on previous days. The concept of long-term "equilibrium" is therefore related to an explicit adjustment process. Secondly, in the SP approach flows are modelled as stochastic quantities, in contrast to SUE in which they are regarded throughout as deterministic quantities. Equilibrium in the SP sense, if achieved, relates to an equilibrium probability distribution of network flows, relating to the probabilities of the alternative discrete flow states.

Certainly, since it is well-known that the flows on roads may vary considerably from day-to-day, it is not difficult to make a case that flows are more appropriately represented as stochastic variables. However, it is recognised that the SP approach makes a radical departure from conventional network equilibrium wisdom. Moreover, if we are interested only in the equilibrium behaviour of the SP, there seems no clear way of directly estimating it, without simulating the actual dynamical behaviour of the process (as suggested by Cascetta, 1989).

The work in the present paper was therefore motivated by two broad objectives: firstly, to gain an improved understanding of the relationship between the SUE and SP approaches, and secondly to examine the extent to which the equilibrium behaviour of the SP model may be estimated directly. Due to the limitations on space, and in order to make the analysis more accessible, the paper will restrict attention to the simplest case of a binary choice. (In section 6 the extension to general networks is briefly discussed—further details are available on request from the author).

2. STOCHASTIC PROCESS MODEL: SPECIFICATION AND NOTATION

Consider a network serving a single origin and destination with a fixed integer demand of q over some given time period of the day (e.g. peak-hour) of duration τ hours. Here, q is a known, deterministic, integer quantity that does not vary between days. Suppose that the origin and destination are connected by two parallel links/routes, link 1 and link 2. We wish to examine how the flows on these links will vary over time, i.e. days. We therefore let the random variables $F^{(n)}$ ($n=1,2,\ldots$) denote the flow on link 1 on day n. Clearly the flow on link 2 will then be $q - F^{(n)}$, and so knowledge of $F^{(n)}$ is sufficient to characterise the state of the network on any given day. We suppose that these flows arise from the choices of individual drivers, and so may only take integer values. The demand-feasible flows on link 1 (denoted $f \in D$) are therefore $0,1,2,\ldots,q$.

Related to a flow f in a period of duration $\tau > 0$ hours, we define $f\tau^{-1}$ vehicles/hour to be the *flow rate* (similarly, $q\tau^{-1}$ is the origin-destination demand rate). We shall suppose that the *actual cost* (e.g. travel time) of travelling along link j at a given flow rate $f\tau^{-1}$ on link 1 is given by $c_j\left(f\tau^{-1}\right)$ ($j=1,2$), the $c_j(\cdot)$ being time-independent, known, deterministic functions. Denote the *mean perceived cost* of travel on link j at the end of day n by $U_j^{(n)}$ ($j = 1,2;\ n = 1,2,\ldots$). These are random variables, whose stochasticity is—we shall assume—related entirely to the stochasticity in the flows, from the *learning model*:

$$U_j^{(n-1)} = \frac{1}{m} \sum_{i=1}^{m} c_j \left(F^{(n-i)} \tau^{-1} \right) \quad (j = 1,2; \ n = 1,2,\dots) \tag{2.1}$$

where m is some given, finite, positive integer. That is to say, $U_j^{(n-1)}$ is an average cost over all drivers, based on the actual costs from previous days. Finally, suppose that conditional on the past, the q drivers select a route at the beginning of each day n independently of one another. The probability of a randomly selected driver choosing route 1 on day n, *given* mean perceived costs at the end of day $n-1$ of $(u_1^{(n-1)}, u_2^{(n-1)})$, is then assumed to be given by $p(u_1^{(n-1)}, u_2^{(n-1)})$, where $p(\cdot,\cdot)$ is a time-independent, known, deterministic function.

For a given initial condition $F^{(0)} = f^{(0)} \in D$, the evolution of this process can be written:

$$F^{(n)} \Big| \left\{ F^{(n-i)} : i = 1,2,\dots,m \right\} \sim \text{Binomial}\big(q, p(U_1^{(n-1)}, U_2^{(n-1)}) \big) \tag{2.2}$$

where the $U_j^{(n-1)}$ $(j = 1,2)$ are related to the $\left\{ F^{(n-i)} : i = 1,2,\dots,m \right\}$ by (2.1). Cascetta (1989) has shown that under mild conditions, the above process converges to a unique stationary probability distribution regardless of the initial conditions. These are basically conditions for the process to be irreducible and Markov, for which it is sufficient that m be finite, and $p(\cdot,\cdot)$ give values strictly in the open interval $(0,1)$.

3.   APPROXIMATING DETERMINISTIC PROCESS AND EQUILIBRIUM CONDITIONS

3.1 Approximation of memory filter

Now for any $f \in D$, by standard laws of conditional probabilities we have:

$$\Pr(F^{(n)} = f) = \sum_{f_1 \in D} \sum_{f_2 \in D} \dots \sum_{f_m \in D} \Pr(F^{(n)} = f | F^{(n-1)} = f_1, F^{(n-2)} = f_2, \dots, F^{(n-m)} = f_m) \times$$
$$\Pr(F^{(n-1)} = f_1, F^{(n-2)} = f_2, \dots, F^{(n-m)} = f_m). \tag{3.1}$$

Since the dependence on the past is, from (2.1), only in the form of dependence on the $U_j^{(n-1)}$ $(j = 1,2)$, then if we define the set of implied demand-feasible mean perceived costs as

$$\Omega \equiv \left\{ (u_1, u_2) : u_j = \frac{1}{m} \sum_{i=1}^{m} c_j (f_i \, \tau^{-1}) \ \ (j = 1,2) \ \text{for} \, f_i \in D \ (i = 1,2,\dots,m) \right\}$$

we can write (3.1) as

$$\Pr(F^{(n)} = f) = \sum_{(u_1, u_2) \in \Omega} \Pr(F^{(n)} = f | (U_1^{(n-1)}, U_2^{(n-1)}) = (u_1, u_2)) \, \Pr((U_1^{(n-1)}, U_2^{(n-1)}) = (u_1, u_2)). \tag{3.2}$$

Now, it is proposed that an approximation to the joint distribution of $(U_1^{(n-1)}, U_2^{(n-1)})$ is, as $m \to \infty$,

$$\Pr((U_1^{(n-1)}, U_2^{(n-1)}) = (u_1, u_2)) \approx \begin{cases} 1 & \text{if } (u_1, u_2) = (\mathrm{E}[U_1^{(n-1)}], \mathrm{E}[U_2^{(n-1)}]) \\ 0 & \text{otherwise} \end{cases} . \tag{3.3}$$

Although intuition suggests that, from (2.1), as $m \to \infty$ then $\mathrm{var}(U_j^{(n-1)}) \to 0$ $(j = 1,2)$, which would support the approximation (3.3), a formal proof has yet to be obtained, although some suggestive points should be noted. In stationarity, $F^{(n-1)}, F^{(n-2)}, \dots, F^{(n-m)}$ will have a common marginal probability distribution, and hence so will $c_j\left(F^{(n-1)}\tau^{-1}\right), c_j\left(F^{(n-2)}\tau^{-1}\right), \dots, c_j\left(F^{(n-m)}\tau^{-1}\right)$ (j=1,2). If these were independent, then applying a Central Limit Theorem would establish the required result, but they are clearly correlated.

However,

$$\mathrm{var}(U_j^{(n-1)}) = \frac{1}{m^2} \sum_{i=1}^{m} \mathrm{var}\left(c_j\left(F^{(n-i)}\tau^{-1}\right)\right) + \frac{2}{m^2} \sum_{i=1}^{m} \sum_{\substack{k=1 \\ (k<i)}}^{m} \mathrm{cov}\left(c_j(F^{(n-i)}\tau^{-1}), c_j(F^{(n-k)}\tau^{-1})\right).$$

The first term is the variance that would arise if the costs were independent over time, and the second term is the sum of cost covariances i-k time periods apart (i-k=1,2,…,m-1). Now for a typical choice probability function for $p(\cdot, \cdot)$ such as a random utility model, in which the probability of choosing an alternative is a decreasing function of the cost of that alternative, and cost functions in which $c_j(\cdot)$ is an increasing function of the flow rate on alternative j, we could guarantee that costs *one* time period apart will be negatively correlated. This is because an increased cost of alternative j on day i, will tend to reduce the number of users choosing j on day i+1, which in turn will tend to reduce the cost of alternative j on day i+1. Although this does not extend to other, more distant costs, which may be positively correlated, there would appear to be some possibilities here for future work examining conditions under which cost covariances may decay in magnitude over time, and/or of using the variance obtained in the independent case as some sort of bound.

The proposed approximation (3.3) then gives rise to the approximating process of:

$$\Pr(F^{(n)} = f) \approx \Pr(F^{(n)} = f | (U_1^{(n-1)}, U_2^{(n-1)}) = (\mathrm{E}[U_1^{(n-1)}], \mathrm{E}[U_2^{(n-1)}])). \tag{3.4}$$

3.2 Approximation to the flow probability distribution

The approximation proposed in section 3.1 simplifies the process by saying that we need not condition on the whole cost probability distribution of the previous day (which describes the memory of all previous days) in order to determine approximately the flow probability distribution for the current day. The approximating process (3.4)/(2.1) does, however, still require computation of the evolution of the individual state probabilities, i.e. the whole flow probability distribution. Here we introduce a further approximation, which allows us to consider only the evolution of the first two moments of the flow probability distribution.

Now, by (2.1), and using a second order Taylor series approximation to $c_j(\cdot)$ in the neighbourhood of $E[F^{(n-i)}\,\tau^{-1}]$, we obtain:

$$E[U_j^{(n-1)}] = \frac{1}{m}\sum_{i=1}^{m} E[c_j(F^{(n-i)}\tau^{-1})] \approx \frac{1}{m}\sum_{i=1}^{m}\left(c_j(E[F^{(n-i)}\tau^{-1}]) + \tfrac{1}{2}\mathrm{var}(F^{(n-i)}\tau^{-1}).c_j''(E[F^{(n-i)}\tau^{-1}])\right)$$

$$= \frac{1}{m}\sum_{i=1}^{m}\left(c_j(\mu^{(n-i)}) + \tfrac{1}{2}\phi^{(n-i)}c_j''(\mu^{(n-i)})\right) \tag{3.5}$$

where the mean flow rate, and variance in the flow rate, are given by

$$\mu^{(n)} = E[F^{(n)}\tau^{-1}] = \tau^{-1}E[F^{(n)}] \quad\text{and}\quad \phi^{(n)} = \mathrm{var}(F^{(n)}\tau^{-1}) = \tau^{-2}\,\mathrm{var}(F^{(n)}).$$

Therefore, with such an approximation, the right hand side of (3.4) can be written in terms of the evolution of the flow rate means and variances alone. In other words, in order to know the evolution of this approximating process, there is no need to compute the whole flow probability distribution, only its first two moments. Now, from (2.2) and the approximation (3.4), combined with standard properties of the Binomial, we have on the demand-side:

$$\mu^{(n)} = \tau^{-1}E[F^{(n)}] \approx \tau^{-1}q\,p(E[U_1^{(n-1)}], E[U_2^{(n-1)}])$$
$$\phi^{(n)} = \tau^{-2}\mathrm{var}(F^{(n)}) \approx \tau^{-2}q\,p(E[U_1^{(n-1)}], E[U_2^{(n-1)}])\left(1 - p(E[U_1^{(n-1)}], E[U_2^{(n-1)}])\right) \tag{3.6}$$

and so, letting $\bar{q} = q\tau^{-1}$ denote the origin-destination demand rate and combining (3.5)/(3.6), we end up with the approximating deterministic process given by the linked expressions:

$$\begin{cases} \mu^{(n)} = \bar{q}\,p\left(\frac{1}{m}\sum_{i=1}^{m}h_1(\mu^{(n-i)},\phi^{(n-i)}), \frac{1}{m}\sum_{i=1}^{m}h_2(\mu^{(n-i)},\phi^{(n-i)})\right) \\ \phi^{(n)} = \bar{q}\tau^{-1}\,p\left(\frac{1}{m}\sum_{i=1}^{m}h_1(\mu^{(n-i)},\phi^{(n-i)}), \frac{1}{m}\sum_{i=1}^{m}h_2(\mu^{(n-i)},\phi^{(n-i)})\right)\left(1 - p\left(\frac{1}{m}\sum_{i=1}^{m}h_1(\mu^{(n-i)},\phi^{(n-i)}), \frac{1}{m}\sum_{i=1}^{m}h_2(\mu^{(n-i)},\phi^{(n-i)})\right)\right) \end{cases}$$

$$\tag{3.7}$$

where

$$h_1(\mu,\phi) = c_1(\mu) + \frac{\phi}{2}.c_1''(\mu) \quad\text{and}\quad h_2(\mu,\phi) = c_2(\bar{q}-\mu) + \frac{\phi}{2}.c_2''(\bar{q}-\mu) \tag{3.8}$$

where we have used (in the expression for h_2) the fact that the mean flow rates on the two routes must sum to $\bar{q}$, and that by the binomial assumption the (absolute) flow variances on the two routes must be the same, hence the flow rate variances must also be the same (ϕ).

3.3 Equilibrium conditions

Stationary points of the process (3.7) are obtained by setting

$$\mu^{(n-i)} = \mu^* \quad \text{and} \quad \phi^{(n-i)} = \phi^* \quad (i = 1,2,\ldots,m)$$

and in this case the summations in (3.7) drop out to yield:

$$\begin{cases} \mu^* = \bar{q}\, p\big(h_1(\mu^*,\phi^*), h_2(\mu^*,\phi^*)\big) \\[2mm] \phi^* = \bar{q}\tau^{-1}\, p\big(h_1(\mu^*,\phi^*), h_2(\mu^*,\phi^*)\big)\big(1 - p\big(h_1(\mu^*,\phi^*), h_2(\mu^*,\phi^*)\big)\big) \end{cases}$$

$$(3.9)$$

A flow allocation (μ^*,ϕ^*) satisfying the fixed point conditions (3.9) will be termed a *Second Order Stochastic User Equilibrium*, or *Second Order SUE* for short. It will also be denoted SUE(2), the 2 relating to the 2nd order approximation made in forming (3.5). It is trivial to show that had a first order approximation been made here instead, we would have obtained a conventional SUE for μ^*, i.e. a SUE(1) is a conventional SUE.

4. EXAMPLE

Consider a problem with an origin-destination demand rate of $\bar{q} = 200$ vehicles/hour over a period of duration $\tau > 0$ hours, and relationships between cost and flow rate of the form:

$$c_1\big(f\tau^{-1}\big) = 10\left(\frac{f\tau^{-1}}{100}\right)^6 \qquad c_2\big(\bar{q} - f\tau^{-1}\big) = 2.$$

$$(4.1)$$

(For information, this problem has a unique Wardrop equilibrium, at a flow rate on route 1 of approximately 76.5.) Let us suppose that the choice probability has the logit form:

$$p\big(u_1, u_2\big) = (1 + \exp(\theta(u_1 - u_2)))^{-1} \qquad (\theta > 0)$$

$$(4.2)$$

with dispersion parameter $\theta = 0.3$. This problem has a unique SUE at $f \approx 82.6$.

Now, this problem does not have a unique SUE(2) solution, but rather one that varies with the value of the time period duration τ ; for a given τ, there is a unique SUE(2) solution. Such solutions have been determined numerically, by a fine grid search technique, for various given values of τ, and the resulting mean flow rates are illustrated in Figure 1. The horizontal dashed line in the figure is the SUE solution, which is invariant to τ. As we might have expected from (3.9), as $\tau \to \infty$ then the SUE(2) mean flow rate approaches the SUE solution. This may be regarded as an asymptotic ("law of large numbers") result, in a similar spirit to that of Davis and Nihan (1993), since for fixed $\bar{q}$ by letting $\tau \to \infty$, we are effectively letting the number of drivers tend to infinity.

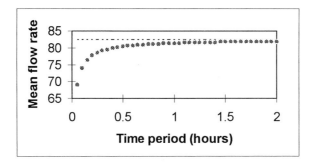

Figure 1: SUE(2) mean flow rate as a function of time period duration τ

5. EXISTENCE AND UNIQUENESS CONDITIONS

It is pertinent to ask under what conditions will there exist solutions to the SUE(2) model proposed in (3.9), and under what conditions will there be a unique solution. The existence question is quite straightforward to answer, using tools similar to those used by Smith (1979) for deterministic user equilibrium and Cantarella & Cascetta (1995) for conventional SUE. Basically, if all the functions involved—i.e. the choice probability function $p(u_1, u_2)$ and the *modified cost-flow functions* given by (3.8)—are continuous, then Brouwer's fixed point theorem (Baiocchi & Capelo, 1984) will ensure the existence of at least one solution to the fixed point problem (3.9), since the mapping implied by the right hand side of (3.9) is to a closed, bounded, convex set. In order for the modified cost-flow functions to be continuous, it is clearly sufficient that the original cost-flow functions are twice continuously differentiable throughout their range.

The uniqueness question needs a little more thought. In the introduction of the SUE(2) problem (expressions (3.7)-(3.9)), it was notationally convenient to express the problem purely in terms of the flow mean and variance on route 1. However, let us now assume the functions h_1 and h_2 given by (3.8) are instead replaced by functions of (μ_1, ϕ_1) and (μ_2, ϕ_2), respectively the flow rate mean and variance on routes 1 and 2, i.e.

$$\tilde{h}_j(\mu_j, \phi_j) = c_j(\mu_j) + \frac{\phi_j}{2} c_j''(\mu_j) \quad (j = 1,2). \tag{5.1}$$

Letting

$$\Omega(\mu, \phi) = p(\tilde{h}_1(\mu_1, \phi_1), \tilde{h}_2(\mu_2, \phi_2)) \text{ where } \mu = (\mu_1, \mu_2) \text{ and } \phi = (\phi_1, \phi_2) \tag{5.2}$$

then the SUE(2) conditions on (μ, ϕ) are:

$$\mu_1^* = \bar{q} \, \Omega(\mu^*, \phi^*) \qquad \phi_1^* = \bar{q}\tau^{-1} \, \Omega(\mu^*, \phi^*)(1 - \Omega(\mu^*, \phi^*)) \tag{5.3}$$

with

$$\mu_2^* = \bar{q} - \mu_1^* \quad \text{and} \quad \phi_2^* = \phi_1^* . \tag{5.4}$$

Writing the condition on ϕ_1^* in (5.3) in terms of μ_1^* (similarly, for ϕ_2^* and μ_2^* from (5.4)), yields:

$$\phi_j^* = \frac{\mu_j^*}{\tau} - \frac{(\mu_j^*)^2}{\bar{q}\tau} \quad (j = 1,2). \tag{5.5}$$

Now, (5.5) holds only at SUE(2) solutions, i.e. at any *solution* to the fixed point problem (3.10). However, if this condition were imposed at all points, i.e.:

$$\phi_j = \frac{\mu_j}{\tau} - \frac{\mu_j^2}{\bar{q}\tau} \quad (j = 1,2) \tag{5.6}$$

then under (5.6), we would then be able to regard the modified cost-flow functions (5.1) as functions of the μ_j only:

$$\frac{\partial \tilde{h}_j}{\partial \mu_j} = c_j'(\mu_j) + \frac{1}{2}\frac{d\phi_j}{d\mu_j}c_j''(\mu_j) + \frac{\phi_j}{2}c_j''(\mu_j) \quad (j = 1,2) \tag{5.7}$$

which, in view of (5.6), may be expressed after some rearrangement as:

$$\frac{\partial \tilde{h}_j}{\partial \mu_j} = \left(c_j'(\mu_j) - \frac{\mu_j}{\bar{q}\tau}c_j''(\mu_j) \right) + \frac{1}{2}\tau^{-1}c_j''(\mu_j) + \frac{\phi_j}{2}c_j''(\mu_j) \quad (j = 1,2). \tag{5.8}$$

Now let us consider separately two possibilities for the cost-flow functions, firstly that they are linear and secondly that they are non-linear. If they are linear and increasing, then all derivatives but the first are zero, and (5.8) reduces to one term which is positive. In the second case, let us assume they are non-linear on link j (say) and satisfy the following conditions:

$$c_j''(x) > 0 \quad \text{for all} \ x > 0 \tag{5.9}$$

$$c_j''(x) \geq 0 \quad \text{for all} \ x \geq 0 \tag{5.10}$$

$$\frac{x\, c_j''(x)}{c_j'(x)} \leq \bar{q}\tau \quad \text{for all} \ x \geq 0 \tag{5.11}$$

(For example, in the case of power-law functions $c_j(x) = \alpha_a + \gamma_a x^{n_a}$, then conditions (5.9)-(5.11) hold if $\alpha_a \geq 0$, $\gamma_a > 0$, and $n_a = 1$ or $2 \leq n_a \leq 1 + \bar{q}\tau$.) Under conditions (5.9)-(5.11), it can be seen that the first (bracketed) term and third term in (5.8) are non-negative, and the second term positive, and so the derivative overall is positive.

Hence, under (5.6), in both the linear and non-linear cases, the modified cost-flow functions are strictly increasing in the mean flow rates, and so the vector function

$$\left(\tilde{h}_1(\mu_1,\phi_1),\tilde{h}_2(\mu_2,\phi_2)\right)=\left(\tilde{h}_1\left(\mu_1,\frac{\mu_1}{\tau}-\frac{{\mu_1}^2}{q\tau}\right),\tilde{h}_2\left(\mu_2,\frac{\mu_2}{\tau}-\frac{{\mu_2}^2}{q\tau}\right)\right)=(\hat{h}_1(\mu_1),\hat{h}_2(\mu_2))\ (\text{say})$$

$$(5.12)$$

is strictly monotonically increasing in (μ_1,μ_2). This holds for any (μ,ϕ) related by (5.6), and so certainly holds at a SUE(2) solution (μ^*,ϕ^*), which is guaranteed to satisfy (5.5). A slight complication is that in view of (5.9), this result only applies to the case where all μ_j^* are non zero ((5.9) notably excludes the case $x=0$). However, this complication may be overcome by requiring the path choice probability function p to produce a result strictly in the open interval (0,1), thus excluding the possibility of an assignment with some $\mu_j^* = 0$ being a SUE(2) solution. Such a condition is satisfied by conventional logit and probit choice models, for example.

With this monotonicity condition, we can apply similar arguments to those used by Cantarella and Cascetta (1995) to establish SUE uniqueness. We shall require the choice probability function $\zeta(\mathbf{u}) = (p(u_1,u_2),1-p(u_1,u_2))$ to be monotonically non-increasing in $\mathbf{u} = (u_1,u_2)$, as satisfied by random utility models that are 'regular' (in the sense that the probability distribution of perceptual errors is independent of (u_1,u_2), as in logit choice or probit choice with a constant covariance matrix). Letting $\hat{\mathbf{h}}(\mu)=(\hat{h}_1(\mu_1),\hat{h}_2(\mu_2))$ be given by the right hand side of (5.12), then let us suppose that two SUE(2) solutions do, on the contrary, exist:

$$\mathbf{h}' = \hat{\mathbf{h}}(\mu') \quad \text{and} \quad \mathbf{h}'' = \hat{\mathbf{h}}(\mu'') \quad \text{where} \quad \mu' = \bar{q}\zeta(\mathbf{h}') \quad \text{and} \quad \mu'' = \bar{q}\zeta(\mathbf{h}'').$$

Then

$$(\mu'-\mu'')^{\mathrm{T}}(\mathbf{h}'-\mathbf{h}'') = \bar{q}\ (\zeta(\mathbf{h}')-\zeta(\mathbf{h}''))^{\mathrm{T}}(\mathbf{h}'-\mathbf{h}'') \le 0 \tag{5.13}$$

since ζ is monotonically non-increasing, by hypothesis. But we also have:

$$(\mu'-\mu'')^{\mathrm{T}}(\mathbf{h}'-\mathbf{h}'') = (\mu'-\mu'')^{\mathrm{T}}(\hat{\mathbf{h}}(\mu')-\hat{\mathbf{h}}(\mu'')) > 0 \tag{5.14}$$

since we have established above that $\hat{\mathbf{h}}(\mu)$ is monotonically increasing (and note that here, we only use the monotonicity condition at a SUE(2) mean flow solution) . Since (5.13) and (5.14) together give a contradiction, we can conclude that only one SUE(2) solution may exist.

6. CONCLUSION AND FURTHER RESEARCH

The SUE(2) model derived, in approximating the equilibrium behaviour of a SP approach, goes some way to achieving the objectives stated in the Introduction. Firstly, it gives the potential to derive efficient algorithms for directly estimating equilibrium behaviour, without regard to the underlying dynamical process. Secondly, by providing an intermediate modelling paradigm, it provides an insight into the relationship between the SP approach and conventional stochastic network equilibrium (an SUE(1)).

The extension of this work to general networks is a natural step to consider. The author has strong evidence that this is achievable, though with rather greater complexity: for example, (2.2) becomes a multinomial route flow distribution, (3.6) then involves route flow covariance terms, and the step to (3.7) (transformation from the route flow to link flow domain) is no longer trivial. Existence and uniqueness of SUE(2) may be established in an analogous way, but with the $\hat{h}$ expressions in (5.12) now being considered functions of the mean link flows disaggregated by origin-destination movement. One efficient, heuristic solution algorithm is derived by repeated application of the "method of successive averages" (Sheffi, 1985) to the modified costs (5.1), conditional on the link flow variances.

In parallel with this extension to more general networks, more elaborate approximations to the SP model may be considered—for example, relating equilibrium conditions to the assumed learning model (a dependence noted by Cascetta, 1989, in the context of the SP approach).

A third area of further research is to extend the modelling capabilities, such as the incorporation of stochastic variation in travel demand, by including a no-travel option with a user-specified choice probability, and the extension to multiple driver classes with different learning and predictive capabilities (as may occur if some drivers are provided with travel information).

ACKNOWLEDGEMENTS

The support of an EPSRC Advanced Fellowship is gratefully acknowledged. I would like to thank Martin Hazelton and Mike Maher for their insightful comments on an earlier draft.

REFERENCES

Baiocchi C. & Capelo A. (1984). *Variational and Quasivariational Inequalities*. John Wiley & Sons, Chichester.

Cantarella G.E. & Cascetta E. (1995). Dynamic Processes and Equilibrium in Transportation Networks: Towards a Unifying Theory. *Transpn Sci* **29**(4), 305-329.

Cascetta E. (1989). A stochastic process approach to the analysis of temporal dynamics in transportation networks. *Transpn Res B* **23B**(1), 1-17.

Davis G.A. & Nihan N.L. (1993). Large population approximations of a general stochastic traffic assignment model. *Oper Res* 41 (1), 169-178.

Sheffi Y. (1985). *Urban transportation networks*. Prentice-Hall, New Jersey.

Smith M.J. (1979). Existence, uniqueness and stability of traffic equilibrium. *Transpn Res* **13B**, 245-50.

Watling D.P. (1996). Asymmetric problems and stochastic process models of traffic assignment. *Transpn Res* **30B**(5), 339-357.

STOCHASTIC NETWORKS AND TRAFFIC ASSIGNMENT

Suzanne P. Evans, Department of Statistics, Birkbeck College, University of London

ABSTRACT

This paper explores the use of some stochastic models for traffic assignment in the case of homogeneous traffic and simple networks. For non-dynamic routing we obtain asymptotic results in the form of paths representing time dependent evolution of traffic over routes. A functional limit theorem gives integral equations for the limiting fluid path which converges to an assignment satisfying Wardrop's first principle as time goes to infinity. For linear cost functions we are able to use the theory of large deviations to examine the way in which rare network overload events occur. In the case of dynamic assignment, we discuss the use of heavy traffic limits and Brownian models to examine the efficiency of network capacity usage when drivers choose routes according to conditions obtaining on entrance to the network. In particular we discuss the phenomenon of resource pooling.

1. INTRODUCTION

Stochastic modelling of the way in which traffic is assigned to a network can be approached in many different ways, depending, among other things, on assumptions about a user's access to information and assumptions about user behaviour. Any choice of model involves much simplification of a complex system and the choice will be valid only if the appropriate features of the real system behaviour are captured. The models discussed in this paper are shown to be closely related to deterministic models for the assignment of traffic according to Wardrop's first principle.

Only very simple networks carrying homogeneous traffic are discussed; extension of the results to more complex networks and to heterogeneous traffic is possible but not considered here. The aim is to explore the use of stochastic processes that evolve over time to model two different types of problem arising in the context of traffic assignment. These problems are characterized by whether users' route choices are made over a much slower time scale than the time scale of short term network dynamics and can therefore be modelled as dependent on a quasi-stationary equilibrium for the current routing pattern, or whether they are made dynamically and depend on the network conditions obtaining at the start of the journey. In both types of problem, the stochastic models quickly become complex and difficult to solve exactly. Different limiting processes may be used to obtain results of different kinds.

Sections 2 and 3 of this paper refer to the situation in which the system is assumed to operate on two time scales. A stochastic model is used only for the slow time scale process of changing route patterns and is referred to as the quasi-static traffic assignment model. We consider two different limiting processes for this model. Section 2 investigates the limit of the original process after scaling by a factor N^{-1}, where the parameter N represents the size of the system in some specified way and is allowed to increase without limit. The resulting limiting process is called the fluid limit. In Section 3 we examine large deviations departures from the fluid limit.

The results of Section 2 show that the limiting process is a deterministic trajectory over time satisfying a system of integral equations. The trajectory is unique in a well defined sense, although if we consider heterogeneous traffic the corresponding integral equations may have many solutions. The limit discussed in Section 3 represents a first, only just stochastic, departure from deterministic behaviour. This is used to examine the probability and frequency of extreme events resulting in network overload. In this context it should be noted that our model only allows consideration of overload due to combinations of events that occur on the slow time scale.

Section 4 is concerned with dynamic traffic assignment. We consider a queueing network model and a heavy traffic limit in which both demand and capacity become large with demand very close to capacity. If the process is scaled by a factor $N^{-1/2}$, where again N represents the size of the system in an appropriate sense, then taking the limit as $N \to \infty$, leads to Brownian models as developed by Harrison (1988) and subsequent authors.

Note that although the results we have obtained are stated formally as theorems, the proofs have been omitted due to lack of space. For the full proof in each case the reader is referred to Evans (1998).

2. QUASI-STATIC TRAFFIC ASSIGNMENT

Consider the very simple network consisting of a single origin-destination pair linked by a set of R non-overlapping routes. We wish to allow for a changing population of identical users who enter the system, generate regular journeys on a chosen route for a certain period of time and eventually leave the system, either completely or to return on a lower cost route. We shall assume that users enter the system according to a Poisson process of rate λ and choose a route r. After an exponentially distributed "holding time" with mean $\mu = 1$ the user departs. Without loss of generality we assume $\mu = 1$. At any time t consider $\mathbf{n}(t) = (n_1(t), n_2(2), \ldots, n_R(t))^T$, where n_r is the number of users currently using route r. Any $\mathbf{n}$

represents an assignment of traffic onto routes $1,...,R$ and $d(t) = \sum_{r=1}^{R} n_r(t)$ represents the current total demand for travel between the single origin and destination. We assume that the process $\mathbf{n}$ changes sufficiently slowly for the underlying fast time scale traffic model to have reached a quasi-stationary distribution that depends on the current $\mathbf{n}$. We represent the averaged fast time scale process as a flow model so that each of the $d(t)$ users at time t is assumed to generate the same flow of traffic per unit time, representing regular journeys on their chosen route. This allows us to associate a route cost function with each route r, expressing the average cost of travel on route r as a function of the number of current users generating flow on r. We suppose that users arriving into the system at time t are aware of the current average journey cost on each route r and choose the route with the smallest average journey cost. If more than one route has the same smallest average cost then the user is assumed to choose at random between them.

The above simple model bears some relation to the Markov modulated fluid flow models used in telecommunications models. In principle, the equilibrium distribution of this Markov process can be computed exactly. However for more complex models, allowing general network topologies and/or heterogeneous users, this is not possible and we take a different approach.

2.1. The Scaled Process and Weak Convergence

Suppose now that the arrival rate of users is $N\lambda$ and consider the weak convergence of the scaled traffic load process $\mathbf{X}^N = N^{-1}\mathbf{n}$ as N tends to infinity. For each route r we let c_r be the appropriate scaled cost function relating average journey cost on r to X_r^N. The c_r are each assumed to be continuously differentiable, strictly increasing, convex functions mapping the real line into itself. Moreover we assume that there exist $\delta > 0$ and $K > 0$ such that:

$$0 < \delta \leq \frac{dc_r}{dx}(x) \leq K \qquad 2.1$$

Note that the upper bound can be relaxed. For any $\mathbf{x}$ let $m(\mathbf{x}) = \min_r \{c_r(x_r)\}$ and set

$$a_r(\mathbf{x}) = \frac{I\{c_r(x_r) = m(\mathbf{x})\}}{\sum_{r'} I\{c_{r'}(x_{r'}) = m(\mathbf{x})\}} \lambda \qquad 2.2$$

Then

$$\mathbf{X}^N \to \mathbf{X}^N + N^{-1}\mathbf{e}_r \text{ at rate } Na_r(\mathbf{X}^N)$$
$$\mathbf{X}^N \to \mathbf{X}^N - N^{-1}\mathbf{e}_r \text{ at rate } NX_R^N$$

where $\mathbf{e}_r$ is the $R \times 1$ vector with one in the r'th position and zeroes elsewhere. Let

$$A_r^N(t) = \int_0^t a_r(X^N(s))ds \qquad 2.3$$

Assume that $E\left[\sum_r X_r^N(0)\right] \leq K_0$ for all N and for some K_0. In the following theorem $\Rightarrow$ denotes weak convergence or convergence in distribution. Roughly speaking we are thinking of a sequence of distributions over a sample space consisting of trajectories in R^R which are right continuous with left limits. These include all possible time dependent evolutions of the jump Markov processes $\mathbf{X}^N$. They also include all continuous trajectories and the unique limiting distribution assigns a probability of one to a single, continuous trajectory, the fluid limit.

Theorem 1. *If $\mathbf{X}^N(0) \Rightarrow \mathbf{x}(0)$, then the scaled Markov process $\mathbf{X}^N \Rightarrow \mathbf{x}$, where for some process $\mathbf{A}$, $(\mathbf{x}, \mathbf{A})$ is a deterministic process satisfying the equations*

$$x_r(t) = x_r(0) + A_r(t) - \int_0^t x_r(s)ds \qquad 2.4$$

$$A_r(0) = 0, \ A_r(t) \ \text{nondecreasing}, \ \sum_r A_r(t) = \lambda t \qquad 2.5$$

$$\int_0^t I\{x_r(s) > m(\mathbf{x}(s))\}dA_r(s) = 0 \qquad 2.6$$

for r=1,...,R. The limit $\mathbf{x}(t)$ is unique and continuous with probability one.

Proof. The proof of this theorem uses martingale methods. It is obtained by modifying the proof of Lemma 5.3 and Corollary 6.1 in Alanyali and Hajek (1998a) and full details may be found in Evans (1988). It is worth noting here however, that the proof shows that every subsequence of $(\mathbf{X}^N, \mathbf{A}^N)$ has a weakly convergent subsequence and that any limit $(\mathbf{x}, \mathbf{A})$ of such a subsequence satisfies equations (2.4-2.6). The final part of the proof shows that $\mathbf{x}$ is unique, thus removing the need to pass to a subsequence for the convergence of $\mathbf{X}^N$. It is *not* true, in general, that $\mathbf{A}$ is unique.$\Box$

Remark A. Equation (2.5) implies that A_r has a density a_r such that $\sum_r a_r = \lambda$ for almost all $t \geq 0$. Thus $\mathbf{x}$ and $\mathbf{A}$ are almost everywhere differentiable and whenever the derivatives exist, $\dot{A}_r(t) = a_r(t)$, $\dot{x}_r(t) = a_r(t) - x_r(t)$ and $I\{x_r(t) > m(\mathbf{x}(t))\}a_r(t) = 0$. Thinking of $a_r(t)$ as the limiting arrival *rate* to route r at time t, these equations are intuitively obvious as defining the fluid limit $\mathbf{x}(t)$.

Remark B. From equations (2.4-2.5) we have, $\sum_r x_r(t) = \sum_r x_r(0) + \lambda t - \int_0^t \sum_r x_r(s)ds$, whence $\sum_r x_r(t) = e^{-t}\sum_r x_r(0) + (1 - e^{-t})\lambda$. In particular, $\sum_r x_r(t) \to \lambda$ exponentially fast as $t \to \infty$, and uniformly for all $\mathbf{x}(0)$ in bounded subsets.

2.2. The Limiting Process and Wardrop's First Principle

Consider the minimization problem: Minimize $\sum_r \int_0^{q_r} c_r(x)dx$ subject to $\sum_r q_r = \lambda$ and $q_r \geq 0$ for all r. Clearly the objective function is strictly convex and the feasible region is closed and bounded so that the minimum is attained at some unique point $\mathbf{q}^*$. The unique point satisfies $q_r^* = 0$ whenever $c_r(q_r^*) > m(\mathbf{q}^*)$. Thus there exists c^* such that if $q_r^* > 0$ then $c_r(q_r^*) = c^*$ and so $\mathbf{q}^*$ corresponds to an equilibrium assignment satisfying Wardrop's first principle. We now examine the limiting trajectory $\mathbf{x}(t)$ and see what happens as $t \to \infty$.

Theorem 2. *If $(\mathbf{x}, \mathbf{A})$ is a solution to the fluid equations (2.4-2.6), then $\lim_{t \to \infty} \|\mathbf{x}(t) - \mathbf{q}^*\|_{\sup} = 0$ uniformly for all initial $\mathbf{x}(0)$ in bounded subsets.*

Proof. The proof of this theorem is a modification of similar proofs in Alanyali and Hajek (1998a). Full details are given in Evans (1998). $\square$

Thus, starting from any initial traffic load, the limiting process $\mathbf{x}(t)$ converges over time to the unique assignment $\mathbf{q}^*$ satisfying $\sum_r q_r^* = \lambda$ and Wardrop's first principle. The rate of convergence depends on the cost functions c_r. It can also be shown that the *equilibrium distribution* of $\mathbf{X}^N$ converges to the deterministic distribution concentrated at $\mathbf{q}^*$ as $N \to \infty$.

3. LARGE DEVIATIONS IN QUASI-STATIC TRAFFIC ASSIGNMENT

The weak convergence result of Theorem 1 represents a law of large numbers - a *functional* law of large numbers. However, notwithstanding this convergence to the deterministic process $\mathbf{x}(t)$, there exists the possibility that the process $\mathbf{X}^N(t)$ will make occasional large excursions from $\mathbf{x}(t)$ and visit rare states. This is of interest if, for example, we are interested in quantifying the probability that a route r becomes loaded beyond its design capacity. The theory of large deviations can give us some insights into this. However the Markov process $\mathbf{X}^N(t)$ has discontinuous statistics across boundaries corresponding to equal journey costs on subsets of routes. That is to say the transition rates $a_r(\mathbf{X}^N)$, given by equation (2.2) and representing arrivals onto the routes r, will be discontinuous across such boundaries. Large deviations results are not yet available for other than linear boundaries and so we obtain some results for linear cost functions c_r using the results of Alanyali & Hajek (1998b & 1998c). We consider just two routes and assume that we have linear cost functions c_r such that $c_1(0) < c_2(0), q_1^* > 0$ and $q_2^* > 0$. Suppose that, for each route r, $N\kappa_r$ represents the designated capacity in terms of the number of users regularly generating journeys on r.

Remark C. If we let $\mathbf{x}(0)$ be a vector of route loads such that $c_1\big(x_1(0)\big) = c_2\big(x_2(0)\big)$ then direct verification yields that $(\mathbf{x}, \mathbf{A})$, defined by $x_r(t) = x_r(0)e^{-t} + q_r^*(1 - e^{-t})$ and $A_r(t) = q_r^* t$, is a solution to the fluid equations (2.4-2.6). For each r, $\dot{x}_r(t) = q_r^* - x_r(t)$.

We define the *route r overflow time*, T_r to be the first time the load on route r reaches $N\kappa_r$ starting from just above the system equilibrium point $N\mathbf{q}^*$. The *network overflow time* is the minimum of the overflow times for routes 1 and 2. We obtain estimates of:

$$F_r(\kappa_r) = -\lim_{T \to \infty} \lim_{N \to \infty} N^{-1} \log P\big(T_r \le T | \mathbf{n}(0) = N\mathbf{q}^* +\big)$$

Roughly speaking, this says that for a large but fixed time T

$$P\big(T_r \le T | \mathbf{n}(0) = N\mathbf{q}^* +\big) \approx \exp\big(- NF_r(\kappa_r)\big)$$

Thus the larger $F_r(\kappa_r)$ the longer the time to overflow on route r. Clearly

$$F(\kappa) = -\lim_{T \to \infty} \lim_{N \to \infty} N^{-1} \log P(\text{network overflow time} \le T | \mathbf{n}(0) = N\mathbf{q}+) = \min_r F_r(\kappa_r)$$

We can also describe the typical path, or trajectory over time, taken by the scaled process $\mathbf{X}^N$ to reach overload on route r if the time to overload is unconstrained. The full technical descriptions and proofs of these results are quite lengthy, although the results themselves may be stated reasonably simply (Sections 3.1 and 3.2). For details of related proofs see Alanyali and Hajek (1998c); details of the proofs for the problem considered here are given in Evans (1998).

3.1. Large Deviations for a Single Route

In the case of a single route the results are well-known. We describe them here since our results for the two route problem make use of them. The basis for any large deviations analysis of jump Markov processes is the *local rate function* ℓ, which for the single route problem takes the form:

$$\ell_\lambda(x,y) = \sup_\theta\{\theta y - g(x,\theta)\}, \text{ with} \tag{3.1}$$

$$g(x,\theta) = \lambda(e^\theta - 1) + x(e^{-\theta} - 1), \tag{3.2}$$

where $g(x,\theta)$ is the logarithmic moment generating function of the single route process. Clearly $\ell_\lambda(x,y) \geq 0$. For fixed T the *rate function* $I_\lambda(r,x)$, evaluated at paths $r(t)$ mapping the time interval $[0,T]$ into R_+ with $r(0) = x$, is given by:

$$I_\lambda(r,x) = \begin{cases} \int_0^T \ell_\lambda(r(t),\dot{r}(t))dt & \text{if } r(0) = x \text{ and } r \text{ is absolutely continuous} \\ +\infty & \text{otherwise} \end{cases} \tag{3.3}$$

$I_\lambda(r,x) \geq 0$ and is a measure of the *unlikelihood* of path $r(t)$. We now allow T to be arbitrarily large and consider paths $r:[0,T] \to R_+$. For $\lambda < x < y$ we define $H_\lambda(x,y)$ as

$$\inf\left\{I_\lambda(r,x):T \geq 0, r:[0,T] \to R_+ \text{ absolutely continuous with } r(0) = x, \sup_{0 \leq t \leq T} r(t) \geq y\right\},$$

that is the infimum of $I_\lambda(r,x)$ over all feasible paths r that start at $x > \lambda$ and rise to y or above. It can be shown that:

$$\ell_1(r,\dot{r}) = \dot{r}\log\left((\dot{r} + \sqrt{\dot{r}^2 + 4rl})/2l\right) + r + l - \sqrt{\dot{r}^2 + 4rl} \tag{3.4}$$

$$H_\lambda(x,y) = \int_x^y \log\frac{z}{\lambda}dz \text{ for } \lambda < x < y. \tag{3.5}$$

We interpret $H_\lambda(x,y)$ as a measure of how unlikely it is for the normalized load, starting at level $x > \lambda$, to reach level $y > x$ within a fixed long time interval.

Remark D. From equation (3.4) it follows that for any path r, $\ell_\lambda(r,\dot{r}) = 0$ if and only if $\dot{r}(t) = \lambda - r(t)$. Thus the most likely path in general always moves towards λ with slope $\lambda - r(t)$. In particular, if $r(t) > \lambda$ then $\dot{r}(t) < 0$. If $r(t) > \lambda$ and $\dot{r}(t) > 0$, as would be required for an unusually high load to be reached, then $\ell_\lambda(r,\dot{r})$ takes its minimum possible value if and only if $\dot{r}(t) = r(t) - \lambda$. Thus, provided there is no constraint on the time, the most likely path to overload traverses the optimal path *from* overload *towards* equilibrium, but in reverse time.

3.2. Large Deviations for Two Routes

For the two route case the transition rates of the scaled 2-dimensional load process $\mathbf{X}^N$, and hence the log moment generating function, change across the linear boundary defined by $c_1(X_2^N) = c_2(X_2^N)$. The large deviations analysis in this case is much more complex. Full details and proofs of the following theorems are given in Evans (1998) and a similar analysis is given in Alanyali and Hajek (1998c). Theorem 3 gives the rate function for the 2-dimensional process.

Theorem 3. *For fixed T the rate function $I(\mathbf{r}, \mathbf{x})$ for the 2-dimensional process, evaluated at paths $\mathbf{r}(t)$ mapping $[0, T]$ into R_+^2 with $\mathbf{r}(0) = \mathbf{x}$, is given by:*

$$I(\mathbf{r}, \mathbf{x}) = \begin{cases} \int_0^T \ell(\mathbf{r}(t), \dot{\mathbf{r}}(t))dt & \text{if } \mathbf{r}(0) = \mathbf{x} \text{ and } \mathbf{r} \text{ is absolutely continuous} \\ +\infty & \text{otherwise} \end{cases}$$

and $\ell(\mathbf{r}(t), \dot{\mathbf{r}}(t))$ satisfies

$$\ell(\mathbf{r}(t), \dot{\mathbf{r}}(t)) = \begin{cases} \ell_\lambda(r_1, \dot{r}_1) + \tilde{\ell}(r_2, \dot{r}_2) & \text{if } c_1(r_1) < c_2(r_2) \\ \ell_{q_1^*}(r_1, \dot{r}_1) + \ell_{q_2^*}(r_2, \dot{r}_2) & \text{if } c_1(r_1) = c_2(r_2) \\ \tilde{\ell}(r_1, \dot{r}_1) + \ell_\lambda(r_2, \dot{r}_2) & \text{if } c_2(r_2) < c_1(r_1) \end{cases}$$

where $\tilde{\ell}(r, \dot{r})$, defined for $r > 0$ and $\dot{r} < 0$, is given by $\tilde{\ell}(r, \dot{r}) = r\log(-r/\dot{r}) + r + \dot{r}$. $\square$

Theorem 4 gives the result of the variational optimization problem for overflow on route 1. For $\mathbf{x} = \mathbf{q}^* +$ on the equal journey costs boundary it identifies $F_1(\kappa_1)$ defined as:

$$\inf\left\{ I(\mathbf{r}, \mathbf{x}): T \geq 0, \ \mathbf{r}:[0, T] \to R_+^2 \text{ absolutely continuous with } \mathbf{r}(0) = \mathbf{x}, \ \sup_{0 \leq t \leq T} r_1(t) \geq \kappa_1 \right\}.$$

A comparable result holds for route 2 from which $F(\kappa)$ is obtainable.

Theorem 4. *For each $\kappa_1 > q_1^*$, $F_1(\kappa_1) = H_{q_1^*}(q_1^*, \kappa_1) + H_{q_2^*}(q_2^*, \gamma_2)$, where γ_2 satisfies $c_2(\gamma_2) = c_1(\kappa_1)$.* $\square$

Remark E. From Theorem 4 we see that route 1 reaches overload from $\mathbf{q}^* +$ along the equal journey costs boundary. Remarks C and D tell us that the path to overload on route 1 traverses in reverse time the solution to the fluid equations for $\mathbf{x}(0) = (\kappa_1, \gamma_2)$.

4. DYNAMIC TRAFFIC ASSIGNMENT

To consider dynamic traffic assignment we assume a queueing network model in which each user, on arrival into the network, chooses the shortest expected delay route calculated on the basis of the network state at the moment of arrival. Although the network state may change as the user moves through the network, the *snapshot relation* (see, for example, Reiman (1984)) suggests that in heavy traffic delays estimated at the time of entry to the network will be a good approximation to the delay experienced. We shall concentrate on results for heavy traffic obtained via the use of the large body of recent work based on Brownian network models (see, for example, Harrison (1988)). For a useful survey and a number of illuminating examples see Kelly and Laws (1993).

Dynamic routing can have a dramatic effect on the performance of a queueing network. A well studied example illustrates this. Suppose traffic arrives in a Poisson arrival stream of rate λ and can be served by either one of two identical single server queues with exponential service times of mean μ^{-1}. We assume that $\rho = \lambda / 2\mu < 1$. The solution to a quasi-static assignment problem is probabilistically equivalent to each arrival tossing a fair coin and choosing queue 1 or queue 2 accordingly. The two queues then behave independently as M/M/1 queues with arrival rate $\lambda / 2$ and service rate μ - a simple product form solution. Clearly this solution equalises average waiting times at the two queues. The dynamic policy of "join the shorter queue" is optimal in the sense that it minimizes the mean delay experienced by users. It also equalises average waiting times in the heavy traffic limit. In fact, as $\rho \uparrow 1$, the two single servers act as a single pooled resource, that is as an M/M/2 queue. The two dimensional queue length process collapses to a one dimensional process and average waiting times are halved. Thus dynamic routing produces significant dependencies between different parts of the network and product form solutions do not apply. In general optimal network performance is achieved by dynamic routing (and scheduling) policies that result in the pooling of network resources in the heavy traffic limit. However, as we see in the remaining short discussion, non-optimal policies can achieve the same resource pooling, since the key feature of resource pooling is in avoiding idleness in one part of the network while another part is overloaded. In this case neither server can be allowed to become idle while there is still work in the system.

For general service time distributions "join the shortest queue" is *not* optimal in general. However, at the level of detail captured in the heavy traffic limit theorems of Reiman (1983), the optimality extends to the case of a general renewal arrival stream and general service time distributions. We briefly outline this result since it validates the Brownian network approximation for this case by providing the necessary weak convergence results.

Assume that the arrival stream is a renewal process with interarrivals having mean λ^{-1} and variance a. Service times at server k, $k = 1,2$, are independent random variables with mean μ^{-1} and variance s_k. Consider a sequence of such systems, indexed by N, such that $\lambda(N) \rightarrow \lambda$, $\mu(N) \rightarrow \mu$ and the heavy traffic condition $N^{1/2}(\lambda(N) - 2\mu(N)) \rightarrow \theta < \infty$ is satisfied.   Let $Q_k^N(t)$ be the number of customers in the queue k at time t, including any customer in service. Let $\mathbf{Q}^N(t) = \left(Q_1^N(t), Q_2^N(t)\right)$, and define the scaled queue length process as $\mathbf{Z}^N(t) = \mathbf{Q}^N(Nt)/N^{1/2}$. Note the time as well as space scaling. Reiman (1983) shows that:

$$\sup_{0 \leq t \leq 1} \left| Z_1^N(t) - Z_2^N(t) \right| \to 0 \text{ in probability as } N \to \infty \qquad 4.1$$

$$Z_1^N(t) + Z_2^N(t) \Rightarrow Z(t) , \qquad 4.2$$

where $Z(t)$ is the reflected Brownian motion that would obtain if all arrivals joined a single queue with two servers: that is a Brownian motion with drift θ and variance $\lambda^3 a + \mu^3(s_1 + s_1)$, reflected at the origin.   Assuming $\theta < 0$ the mean of the stationary distribution of $Z(t)$ is $(1/2)(\lambda^3 a + \mu^3(s_1 + s_2))|\theta|^{-1}$.   If $s_1 = s_2 = s$, this compares with the corresponding mean $(1/2)(\lambda + \lambda^3 a + 4\mu^3 s)|\theta|^{-1}$ for the coin tossing policy.

Equation (4.1) shows that the two queue lengths are kept approximately equal at each time t using the dynamic policy of "join the shortest queue". However Kelly and Laws (1993) show that a threshold policy, while not optimal, can achieve the same resource pooling reflected in equation (4.2) provided the threshold is sufficiently large to ensure that a server does not become idle while there is substantial work in the other queue. In the case of more general networks their work suggests that shortest expected delay routing, while not optimal, has the desired resource pooling property in the heavy traffic limit.

Consider an M/M/R system with service rates $(\mu_1,...,\mu_R)$ and compare it with a system of R independent M/M/1 queues with the corresponding service rates μ_r, $r = 1,...,R$. If the arrivals to the set of independent queues are split among the queues using optimal probabilities rather than dynamically then the mean number in the M/M/R system is smaller by a factor of

$$\left(\sum_r \mu_r^{1/2}\right)^2 / \sum_r \mu_r$$

(See Foschini (1977)). In the heavy traffic limit this improvement can be obtained by any policy keeping all servers busy when there is substantial work in the system. This includes the policy of choosing routes according to a shortest expected delay criterion. Note that the shortest expected delay policy is not optimal in general here.

5. CONCLUSION

This paper has explored stochastic models for traffic assignment for very simple situations. Many extensions are possible and could prove useful.

Both the fluid limit and the large deviations limit for quasi-static traffic assignment describe the time dependent evolution of traffic over routes.   There are a number of reasons for being interested in traffic evolution over time. We may be interested in the time scale of response to changes in the network, for example; this can be modelled using a fluid limit. In the case of heterogeneous traffic types, the fluid equations may have multiple solutions corresponding to multiple Wardrop equilibria. We may then be interested in movement between system regimes corresponding to different equilibria; this can be modelled using large deviations limits. In addition large deviations limits are useful for comparing different assignment models for which equilibrium system performance is indistinguishable on the fluid model scale.

There is now a large body of work on the use of Brownian network models for evaluating system performance and for identifying optimal dynamic control policies with respect to routing and sequencing. There is considerable scope for extending and applying this theory to the analysis and control of transport networks.

REFERENCES

Alanyali, M. and B. Hajek (1998a). Analysis of simple algorithms for load balancing. *Mathematics of Operations Research*, to appear.

Alanyali, M. and B. Hajek (1998b). On large deviations of markov processes with discontinuous statistics. *Annals of Applied Probability*, to appear.

Alanyali, M. and B. Hajek (1998c). On large deviations in load sharing networks. *Annals of Applied Probability*, to appear.

Evans, S.P. (1998). The analysis of limit processes for stochastic traffic assignment models in simple networks. In preparation.

Foschini, G.J. (1977). On heavy traffic diffusion analysis and dynamic routing in packet switched networks. In: *Computer Performance* (K.M. Chandy and M. Reiser eds.), pp499-513. North Holland, Amsterdam.

Harrison, J.M. (1988). Brownian models of queueing networks with heterogeneous customer populations. In: *Stochastic Differential Systems, Stochastic Control Theory and Applications* (W. Fleming and P.-L. Lions eds.), pp.147-186. Springer-Verlag, New York.

Kelly, F.P. and C.N. Laws (1993). Dynamic Routing in Open Queueing Networks: Brownian Models, Cut Constraints and Resource Pooling. *Queueing Systems: Theory and Applications*, **13**, 47-86.

Reiman, M.I. (1983). Some diffusion approximations with state space collapse. In: *Modelling and Performance Evaluation Methodology* (F. Bacelli and G. Fayolle eds.), Lecture Notes in Control and Information Sciences, number 60. INRIA, Springer-Verlag.

TOWN CENTRE TOPOLOGY AND TRAFFIC FLOW

P.A. Wackrill, Transport Management Research Centre, Middlesex University, London, U.K.

ABSTRACT

Town centre redevelopment provides an opportunity to redesign the topology of the road network. Environmental and safety considerations will usually entail some streets being reserved for pedestrians only and others for access to car parks and service bays close to the pedestrian precinct. The effect of the new topology on traffic flow can and should be an input to the planning process. One way to appraise alternative schemes is to find the extent to which traffic could be routed to reduce the amount of conflict there will be between the different streams of traffic at the junctions; conflict at junctions is a simple measure indicating the potential for both accidents and congestion with its attendant pollution.

The problem of finding an assignment of flow such that conflict is minimised is a constrained optimisation problem. The main constraints are provision of routes for a demand specified by an origin-destination (OD) matrix. The objective function which measures the amount of conflict between the different streams is simple to formulate; it depends only on the topology of the network and the OD matrix of demand. But it is a quadratic function of a type which is not amenable to existing programming methods. However, the author has developed an iterative method for obtaining good, if not necessarily optimal, solutions.

The purpose of the paper is to demonstrate the appraisal of hypothetical networks involving four external zones. The OD matrix has been chosen to model some through as well as mostly inbound traffic flow. Various alternative schemes are appraised ranging from those which merely allow one access point to a car park within a ring of streets surrounding the pedestrian precinct, to those which retain one link across the ring and up to two car parks. This is a pilot study for a larger one involving the systematic analysis of hypothetical town centre topologies. So far it has shown the advantage of two access points over one, and the greater advantage of two separate car parks each with a single access point which allows each car park to be in a separate cell of the road network. The relative positions of the access points and the link across the ring also affect the amount of conflict.

1.   INTRODUCTION

We can all bring to mind a town centre we know which has been redeveloped. We may even have been aware of public consultation involving an exhibition of alternative schemes. We probably appraised the schemes by considering the effect they would have on our usual drive to the town and the parking place. Once the scheme is in operation it soon becomes obvious that traffic queues build up on the approaches to the car parks. The entrances become bottlenecks because traffic is being concentrated at a few points in the network.

Consider a development consisting of a ring of streets surrounding a pedestrian precinct. The ideal situation, from the traffic point of view, would be for traffic arriving at the ring to have access to a car park straight ahead. Queueing theory tells us that a single queue to many servers is better than a queue for each server. This theory informs the current practice in banks and post offices. In car park terms the servers are the parking spaces and the routes through the car park allow each car to access any vacant space. This suggests that a single car park accessible to all vehicles would be best. (See Figure 1.) But this would effectively cut the pedestrian area into four sectors, and create a conflict between the needs of pedestrians moving between sectors and vehicles entering the car park. Such conflicts can be catered for with pedestrian crossings or bridges between upper floors of a shopping mall. However, the problem of conflict can be addressed while designing the network tolopogy.

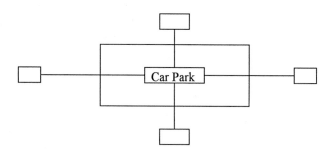

Figure 1. Single car park accessed from four points

A balance has to be struck between what is ideal for vehicle traffic and what is desirable for pedestrians. The paper assesses one range of topologies based on a single car park, and another with two pedestrian sectors and two car parks

The paper consists of three more sections and a conclusion. The first section explains how conflict at junctions is an appropriate measure for the inconvenience to traffic. The second section explains the way the mathematical model of the network and the demand for routes through it are used to find an assignment of flows which reduces conflict as much as possible. The third section covers the analysis of various hypothetical town centre topologies. The paper concludes with a summary of what the analysis indicates in terms of design principles.

2. MEASURE OF INCONVENIENCE TO TRAFFIC

Drivers are assumed to want to minimise their journey times. In the congested urban situation the delay at junctions has a larger effect on journey time than the distance travelled between them. So inconvenience can reasonably be measured in terms of delay at junctions. While there are formulae for this delay in terms of signal settings, geometry, and turning flows, a simple proxy for delay is the amount of conflict between the different streams of traffic. We have to consider how appropriate this proxy is.

When one is attempting to assess inconvenience for a range of topologies, one does not have an existing network with its management measures and observed flows. However, whether one is managing traffic with signals or priority junctions, the delay to any movement will depend on the amount of traffic expected to be making conflicting movements; during a long wait at traffic lights one can usually observe the heavy traffic in the conflicting streams. At a priority junction conflict measures the delay caused by the relatively large amount of traffic on a major road delaying traffic on the minor road. However, conflict would also suggest that delay was caused by the small amount of traffic on the minor road; priority does not allow this to happen so the total amount of conflict will slightly overestimate delay.

3. FLOW ASSIGNMENT

The problem of finding an assignment of flow such that conflict is minimised is a constrained optimisation problem. The variables, representing flow, can be defined in terms of two parameters. One is either the origin of the flow or its destination, and the second indicates a particular turning direction at a particular junction. The reasons for choosing between the first two alternatives is not elaborated here; the latter has been chosen. For the latter, the constraints are that the flow for a particular destination out of each origin should match the corresponding OD matrix element, and that the total flow to each destination should be conserved through all intermediate junctions. The objective function is the sum of the products of the flows on all pairs of conflicting movements at all the junctions; greater detail is appropriate here. For a T-junction three of the pairs represent conflicts where streams of traffic cross, and the other three pairs represent merging conflicts. It would be possible to weight crossings differently to mergings but in the congested situation giving way while waiting to merge causes as much delay as waiting for a gap in crossing traffic. At a cross roads three streams are merging at each exit so there are twelve pairs of merging streams altogether. Each straight across stream has to cross the paths of two other straight across streams; these give rise to four pairs of conflicts. Each right-turning stream crosses two straight ahead streams and two right-turning streams; these give rise to twelve pairs of conflicts. The U.K. Highway Code specifies that pairs of opposing right-turners should follow interlocking paths. However, at big junctions the road will often be marked to allow one or both pairs of opposing right-turners to pass near-side to nearside thus avoiding interlocking paths. This fine detail is relevant at a late stage in the design process; for the research reported in this paper each pair of opposing right-turners is counted as one conflict.

The objective function therefore consists of terms which are products of pairs of different variables, each being flow to a destination on a pair of different turning directions. So it is a quadratic function of a type which is not amenable to existing programming methods.

However, the author has developed an iterative method for obtaining good, if not necessarily optimal, solutions. The procedure involves solving an optimisation subproblem for each destination in turn. In each subproblem the variables are the flows into that destination in each turning direction at each junction. This makes the objective a linear function because the conflicting flows experienced will be computed from the current values of the flows for all the other destinations and will appear as constant coefficients. The initial values for these coefficients can all be chosen to be 1. The Out-of-Kilter algorithm (for finding minimum cost flows) has been adapted for these subproblems. The method is described in detail in Wackrill (1992). It is incorporated in a software suite: the CROWN design tool.

4.   HYPOTHETICAL TOWN CENTRE TOPOLOGIES

The purpose of the paper is to demonstrate the use of the CROWN design tool to appraise hypothetical networks involving four external zones. We need a matrix to model the expected flow between the various origins and destinations of traffic. It needs to represent some through as well as mostly inbound traffic flow. The OD matrix shown below is a simple model. First of all we assume that flow between each pair of external zones, the through traffic, is of the same order of magnitude for each pair; this is taken as the base unit of flow. It includes traffic which passes through the town to edge of town retail or business parks. Based on this we assume that four units of flow are entering from each external zone and needing to park in the town while one unit is flowing between each pair of external zones. In addition we assume that a unit of flow would be leaving the car park(s) for each external zone.

To zones From zone	1	2	3	4	Car park(s)
1	0	1	1	1	4
2	1	0	1	1	4
3	1	1	0	1	4
4	1	1	1	0	4
Car park(s)	1	1	1	1	0

The appraisal is done by finding routes through the network with the objective of minimising the total amount of conflict between streams of traffic at the junctions, measuring the corresponding amount of conflict, in squared units, and ranking the different topologies according to this measurement.

A previous study (Wackrill and Wright 1994) reported that conflict could be reduced more when the car park was inside the ring of streets surrounding the pedestrian area than when it was outside the ring. From the pedestrian point of view, a car park inside the ring is a good idea because it is convenient to have footpaths from the car park which do not have to cross the ring road in order to reach the pedestrian area. The topologies considered in this paper are therefore restricted to ones with car parks within the ring road surrounding the pedestrian area.

First we consider a series of topologies with one car park having between one and four access points. The more access points there are the more the pedestrian area is cut up, but the less conflict there is for the traffic. It is worthwhile going into some detail by considering twelve topologies in this series.

There are eight topologically distinct points on the ring from which the car park could be accessed. (See Figure 2.) The twelve topologies appraised consist of two with four access points, four with three access points, four with two access points, and two with one access point. The results are shown in Table 1.

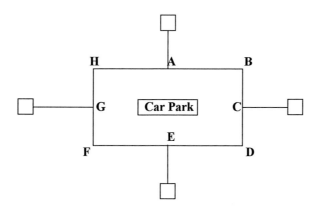

Figure 2. Single car park with 8 possible access points

Table 1 shows a general trade off between fragmentation of the pedestrian area and vehicular conflict. Notice that both A and E, and B and F are on opposite sides of the ring and give a better result than A and C, or B and D which are closer together. This concludes the series of topologies with one car park, so we proceed to topologies with two car parks.

Table 1. Conflict results for topologies with one car park

Car park accessed from:	Amount of conflict	Number of pedestrian sectors
A C E G	80	4
B D F H	80	4
A C E	100	3
B D F	100	3
A D F	100	3
A C F	100	3
A E	120	2
B F	120	2
A C	125	2
B D	125	2
A	202	1
B	205	1

Table 2. Conflict results for topologies with two car parks

Car park accessed from	Road across ring	Amount of conflict
A and C	BE	115
A and C	BF	109
A and C	BG	108
A and E	CG	110
A and E	BF	110
A and E	DH	106
H and D	AE	110
H and D	CG	106
H and D	BF	102

The topologies with two car parks are separated into three where the car parks are separately accessed from A and C, three for access from A and E, and three for access from H and D. Earlier work not detailed here showed that having a road across the ring which put both car parks in the same sector did not reduce conflict. In the set of nine topologies with two car parks reported here we use the advantage of having separate car parks by having a road across the ring so as to pass between the car parks thus reducing the conflict while only separating the pedestrian precinct into two sectors.

One would expect topologies with two car parks to need an OD matrix with an extra column and an extra rowfor the second car park. And in the implimentation of the redevelopment scheme it will be reasonable to direct incoming traffic to a particular car park. This raises the question of which traffic should be directed to which car park. One can start with letting the flow optimisation procedure find routes which minimise conflict and having the flow from each origin go to either of the car parks; one merely labels both car parks with the same zone number and uses the OD matrix with only one car park. In about half of the nine networks assessed, an analysis of the optimum flows showed that, for the pair of flows between an external zone and the car park, the inbound and outbound flows used the same car park. So, having used signposts to direct each inbound flow to a particular car park, traffic flowing out of that car park would return to the same external zone. In the other half of the networks, a car park which received traffic from one pair of external zones discharged it to a different pair, for example traffic from zones 1 and 2 entered the car park and traffic left the same car park but bound for zones 1 and 4. Although there may be journeys which start in one external zone, park in the town, and leave for another external zone, it is unreasonable to let one's assessment be based on an assumption that such journeys specified by the flow optimisation procedure would be the norm. As the major flows were inbound, this difficulty was overcome by letting the flow optimisation procedure choose the car park for each inbound flow, re-specifying the network with the car parks labelled with different zone numbers, and then having a matrix which also differentiated the car parks and specified flows which returned traffic to the zone from which it originated. As expected, this increased conflict slightly; the conflict numbers given by the diagrams of the networks in Figures 3 to 5 and in Table 2 correspond to the situation where traffic in each car park returns to its origin.

All the toplogies with two car parks assessed here have two pedestrian sectors and they all have fewer conflicts than those with one car park and two pedestrian sectors. The first conclusion we can draw from Table 2 is that it is better to have the car parks accessed from points opposite each other on the ring rather than close to each other (Figures 4 and 5 rather than Figure 3). Secondly, we see (in Figures 4 and 5) the advantage of shifting the road across the ring so as to stagger the cross roads into two T-junctions and thus remove the interlocking movement between one pair of opposing right-turners. The best topology is the bottom one in Figure 5; it has no cross roads.

CONCLUSION

The appraisal of hypothetical topologies involving four external zones has been demonstrated using a series of twelve with one car park , and a series of nine with two car parks.

This pilot study has demonstrated the trade off between fragmentation of the pedestrian area and vehicular conflict. For topologies with two pedestrian areas it has shown the advantage of having two car parks, one in each of two pedestrian areas and each accessed from one point, over a single car park with two access points. This kind of conclusion might be used as the first criterion for choosing between alternative schemes for town centre redevelopment. The assessment of which alternative roads to retain for vehicle traffic in order to provide a link across the ring surrounding the pedestrian area can be made by reference to the conflict results for the different topologies assessed here.

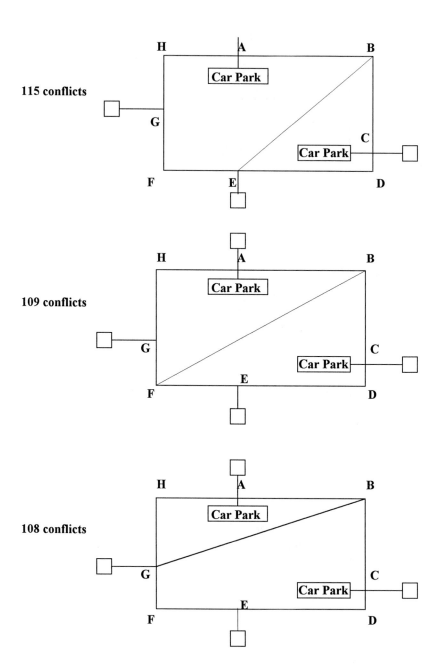

Figure 3. Two car parks with access points at A and C

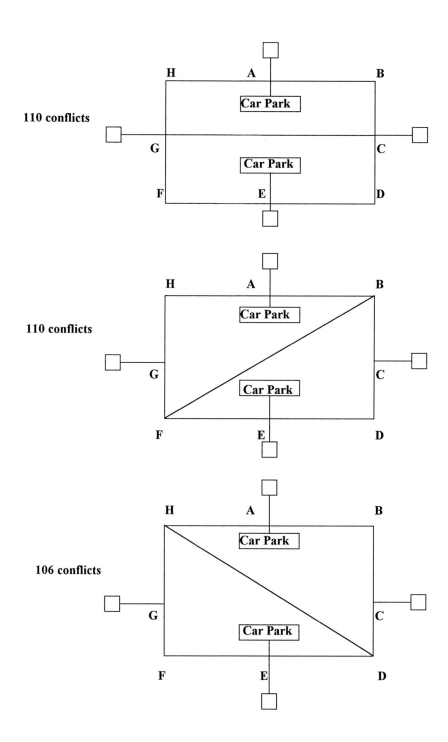

Figure 4. Two car parks with access points at A and E

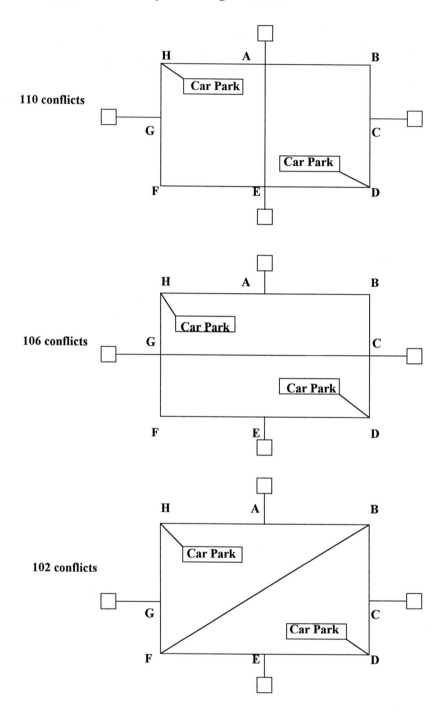

Figure 5. Two car parks with access points at H and D

Having selected a good topology from the research described here, one can only reap the benefit by using routes complying with the flow assignment giving the lowest amount of conflict. These can be identified from the output from the CROWN design tool. They are needed so that drivers can be directed to use the best routes to their destinations whether they be another external zone or a car park.

REFERENCES

Wackrill, P. (1992) A design tool to identify those elements of a road network necessary to support an efficient traffic routeing pattern. *Institute of Mathematics and its Applications Conference Series* New Series No. 38 Mathematics in Transport Planning and Control 1992. 231-239

Wackrill, P.A. and Wright, C.C. (1994) Network Design to Reduce Conflict. *Proceedings of the Second International Symposium on Highway Capacity.* Australian Road Research Board Ltd. and Transportation Research Board Committee A3A10 Vol 2 697-706

A REACTIVE DYNAMIC ASSIGNMENT SCHEME

M.M. Khoshyaran[2] and J.P. Lebacque[1]
[1] *CERMICS-ENPC (Ecole Nationale des Ponts et Chaussées),* FRANCE
[2] *TASC Consultant,* FRANCE

ABSTRACT

In this paper we consider two traffic control strategies relying on user response to information and/or flow restriction. Ultimately, the control strategies are designed to function in real time, hence provide command values based on actual conditions and requiring little computational effort. The proposed control strategies are based on the idea that the network load, as measured by instantaneous travel times for instance, should be shared as equally as possible between paths. In order to achieve such an aim, the commands are designed to make the system state converge towards a state in which instantaneous travel times of paths relative to any given OD tend to be equal.

1. INTRODUCTION

The focus of this paper is to study the controllability of traffic flow through implementation of user information systems. Controling a system means optimizing it. It is therefore not surprising that optimality of assignment (whether user- or system-) eventually combined with coercive means (traffic lights, ramp control) should have been intensively investigated. Without any attempt at exhaustivity, let us mention the works of Merchant and Nemhauser 1978, Friez *et al* 1993, Huang and Lam 1995, Yang *et al* 1994, Ran and Boyce 1996, Heydecker and Addison 1996. Some authors have approached the problem of control in a different spirit: Kuwahara and Akamatsu 1997 for instance assign traffic on the shortest path (according to the instantaneous travel time) whereas Elloumi *et al* 1994 determine assignment coefficients optimizing a collective criterion, which raises the question of the implementabi-lity of the optimal command.

The idea of the present paper is different. Rather than calculating or trying to implement an assignment optimal in some sense, we investigate whether it is possible to achieve some measure of control of a network by using directly the user response to travel time information or simple traffic management measures. The user response model we consider is the Logit model, mainly by virtue of its simplicity; other models would be acceptable too. The aim of the control strategy we investigate is to promote an evolution of the network state towards *a desirable state* , which we choose to be a state in which travel times (instantaneous or eventually experienced) on competing paths are be equal. There is no optimality concept involved: the idea here is simply that the load on the network, as measured by the travel times, should be divided in a fair way. In the sequel, we shall exhibit two (actually related) strategies achieving such an aim.

2. PATH CHOICE, TRAVEL TIMES

User Paths Choice

Considering the present work is exploratory in its nature, we shall be considering very simple situations, typically one origin, one destination, and I paths joining them:

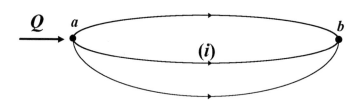

Actually, we shall not consider intersections on those paths, meaning that the paths will essentially be modeled as simple links. The user path choice generation will be modeled in a straightforward manner, by the Logit model. We assume that at any given time t, users entering point a choose path (i) with a probability p_i (t) given by:

$$p_i(t) = \frac{\exp\left(-\theta \Im_i(t)\right)}{\sum_j \exp\left(-\theta \Im_j(t)\right)} \qquad (1)$$

with $\Im_i$ (t) the *information* relative to the travel time of path (i) given to the user. $\Im_i$ (t) in our control strategies will generally differ from the actual travel time $T_i(t)$ of path (i) as estimated at time t by the underlying traffic model. We define the command variables ω_i (t) as:

$$\Im_i(t) = T_i(t) + \omega_i(t) \qquad (2)$$

It should be noted that $T_i(t)$ itself might well differ from the travel time experienced by users: this fact constitutes an additional difficulty.

Travel Time Definition

Let us consider the trajectory of a vehicle entering a path $a \rightarrow b$ at time t, exiting it at time u. u-t will be equal to $PTT(t)$: the *predicted travel time* as it should be estimated at time t ; it is also equal to $ETT(u)$, the *experienced travel time* of the vehicle which can be estimated at the exit time u of the vehicle. The $ITT(t)$ (*instantaneous travel time* at time t) is defined as:

$$ITT(t) = \int_a^b \frac{d\xi}{V(\xi,t)} \qquad (3)$$

with $V(\xi,t)$ the velocity field at time t .This definition is consistent with that given for instance in Ran and Boyce 1996: *the instantaneous travel time is the travel time that would be experienced by vehicles when prevailing traffic conditions remain unchanged.* The $ITT(t)$ can be considered as an index of the state of the path at time t .

For control purposes, both the $ITT(t)$ and the $PTT(t)$ may be relevant. It must be noted that the above definition of the $ITT(t)$ is only valid if the time scale of the traffic flow model is sufficiently large for traffic interruptions (for example those due to traffic lights) to be neglected and the speed always to be considered non zero. Otherwise more complicated models of ITTs should be considered (see Lebacque 1996).

Time derivatives of some travel times. Examples.

It is our contention that the time derivative of path travel times can be expressed as a function, among others, of the path inflow, and that this fact implies the possibility to control at least simple systems such as the ones considered in this paper.

<u>General case: time derivative of the PTT</u>

For a path such as the one depicted hereafter, with inflow qI and outflow qE ,

$$qI \qquad\qquad\qquad\qquad\qquad qE$$
$$a \qquad\qquad\qquad\qquad\qquad b$$

it is well-known (see for instance Astarita 1996) that, under the FIFO hypothesis, the following identity holds:

$$\frac{d}{dt} PTT(t) = \frac{qI(t)}{qE(t + PTT(t))} - 1 \qquad (4)$$

The main drawback of this formula is of course that it requires a *prediction* of the future (at time t) outflow of the path, qE (t+$PTT(t)$) . This formula nevertheless may be applied in some simple cases. In any case it highlights the dependence on the inflow qI (t) of the time derivative of PTT (t) .

Point queue model

The model will be recalled briefly in the following simple setting (paths (*i*) without intersections):

In the model, an exit maximum capacity μ_i constrains the outflow. With α_i designating the freeflow travel time and f_i (*t*) designating the number of vehicles in the queue (assumed to have no physical extension) it follows:

$$qA_i(t+\alpha_i) \;=\; qI_i(t)$$

$$
\begin{aligned}
qE_i(t+\alpha_i) &= \mu_i && \text{if } f_i(t+\alpha_i)\geq 0 \\
&&& \text{or } f_i(t+\alpha_i)=0 \text{ and } qA_i(t+\alpha_i)\geq\mu_i \\
qE_i(t+\alpha_i) &= 0 && \text{if } f_i(t+\alpha_i)=0 \text{ and } qA_i(t+\alpha_i)\leq\mu_i
\end{aligned}
$$

Since $\qquad f_i(t+\alpha_i)=0 \Leftrightarrow T_i(t)=\alpha_i \qquad$ the above is clearly equivalent to the following model, expressed in terms of the predictive travel-time T_i *(t)* :

$$
\begin{aligned}
\frac{dT_i}{dt}(t) &= \frac{qI_i(t)}{\mu_i}-1 && \text{if } T_i(t)>\alpha_i \\
&&& \text{or } T_i(t)=\alpha_i \text{ and } qI_i(t)\geq\mu_i && (5)\\
\frac{dT_i}{dt}(t) &= 0 && \text{if } T_i(t)=\alpha_i \text{ and } qI_i(t)\leq\mu_i
\end{aligned}
$$

which yields the time derivative of the travel-time.

The LWR model

Let us recall briefly some of the features of the well-known Lighthill-Whitham-Richards model (Lighthill and Whitham 1955, Richards 1956). This model, although more close to reality than the preceding one, retains enough simplicity to allow for analytical calculations. The fundamental variables K, Q, V (density, flow, velocity) are related by the fundamental trivial relationships:

- $\dfrac{\partial K}{\partial t}+\dfrac{\partial Q}{\partial x}=0$ (conservation of vehicles)

- $Q=K\,V$ (definition of the velocity)

and by the behavioral equation: $\qquad V = V_e\,(K,x)$ which implies $Q = Q_e\,(K,x) = K\,V_e\,(K,x)$ with V_e and Q_e the equilibrium speed- resp. flow-density relationship. The resulting fundamental conservation equation is:

$$\frac{\partial K}{\partial t} + \frac{\partial}{\partial x}Q_e\,(K,x) = 0 \qquad (6)$$

whose entropy solutions can be calculated analytically according to the well-known characteristics method.

An important point: let us consider the trajectory of a single vehicle along a path (a,b), and let us consider the corresponding space-time diagram with the characteristics. We assume for illustration's sake a future supply restriction.

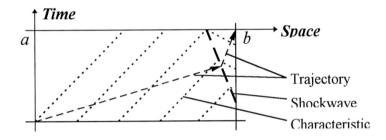

As the diagram shows, the future trajectory of a vehicle at a given time depends only on present traffic conditions on the path and on future downstream supply (at the path exit and eventually at intersections). Indeed, in the LWR model the characteristics speed is always less than the vehicle speed. As a consequence, calculating the time derivative of the predictive travel time requires a *prediction* of future downstream supply, and for control purposes we shall prefer the instantaneous travel time. Let us calculate the time derivative of the ITT, making the same simplifying assumptions as previously: homogeneous path, no intersections, non zero speed. It follows:

$$\frac{d}{dt}ITT(t) \overset{def}{=} \int_a^b \frac{\partial}{\partial t}\left(\frac{d\xi}{V_e\big(K(\xi,t)\big)}\right) \qquad (7)$$

Using $\dfrac{\partial K}{\partial t} = -\dfrac{\partial Q}{\partial x} = -\dfrac{\partial Q_e}{\partial K}\dfrac{\partial K}{\partial x}$, it follows after some trivial algebra that:

$$\frac{d}{dt}ITT(t) = \left[\log V_e(\kappa) + \int d\kappa\, \kappa\left(\frac{V_e'}{V_e}\right)^2\right]_{K(a,t)}^{K(b,t)} $$
$$= \Phi\big(K(b-,t)\big) - \Phi\big(K(a+,t)\big) \qquad (8)$$

with $\quad \Phi(\kappa) \overset{def}{=} \log V_e(\kappa) + \int d\kappa\, \kappa\left(\frac{V_e'}{V_e}(\kappa)\right)^2$

(the symbols + and - refer as usual to upper lower limits). This formula can be easily expressed using the path inflow qI ; this point will be clarified later on in this paper. The function Φ has the following aspect:

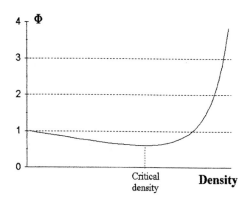

with $\Phi = \infty$ for jam density. For instance if $Q_e\ (\kappa) = 4\ Q_{max}\ (1 - \dfrac{\kappa}{K_{max}}\ )\ \dfrac{\kappa}{K_{max}}$ (Greenshields model) it follows:

$$\Phi(\kappa) = \frac{1}{1 - \kappa / K_{max}} + \log\!\left(1 - \kappa / K_{max}\right)$$

3. THE FIRST CONTROL STRATEGY

Principle

In order to control a system, basic ingredients must be defined and/or calculated: the system state, the command and the criterion. The state of the system (comprising one OD and I paths as already mentioned) is defined as the set of travel times T_i , plus whatever variables which might be necessary to estimate them. The control variables will be the $\mathfrak{I}_i$ or alternatively the ω_i (recall that: $\mathfrak{I}_i = T_i + \omega_i$), with $\mathfrak{I}_i\ (t)$ the travel time announced to users at time t for path (i). Instead of adopting an optimization criterion in the usual sense, we shall try to promote the evolution of the system towards a desirable state, characterized conventionally by the equality of travel-times T_i . It is clear that other options are possible; the idea here is to make the traffic flow homogeneous on the set of competing paths. In the case of two competing paths (the case of I competing paths is not much more complicated), we shall try to implement the following differential equation (the *command equation*):

$$\frac{d}{dt}(T_1 - T_2) = -\alpha(T_1 - T_2) \qquad (9)$$

which should force the convergence of the system towards $T_1 = T_2$ with a convergence time-scale $1/\alpha$, to be chosen. As we shall see, it is not always possible to find commands realizing the above condition. A more easily satisfied condition would be:

$$\frac{d}{dt}(T_1 - T_2) \times (T_1 - T_2) < 0 \qquad (10)$$

Let us examine how this program can be implemented for the models recalled in the preceding section.

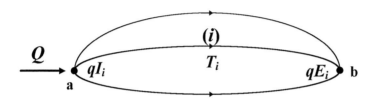

The basic equations are:

$$\left[\begin{array}{l} \dfrac{d}{dt}(T_1 - T_2) = -\alpha(T_1 - T_2) \\[2ex] qI_i(t) = Q(t) \dfrac{\exp\left(-\theta(T_i + \omega_i)\right)}{\sum_j \exp\left(-\theta(T_j + \omega_j)\right)} \quad \text{for} \quad i = 1,2 \end{array} \right.$$

in which $\frac{d}{dt}(T_1 - T_2)$ should be expressed as a function of qI_1 and qI_2 and exogeneous quantities.

The Point Queue Model

Let us recall that according to (5), the time-derivative of the travel-time of path (i) is given by:

- $\dfrac{dT_i}{dt} = \dfrac{qI_i(t)}{\mu_i} - 1$ if $T_i(t) > \alpha_i$ or: $T_i(t) = \alpha_i$ and $qI_i(t) \geq \mu_i$

- $\dfrac{dT_i}{dt} = 0$ if $T_i(t) = \alpha_i$ and $qI_i(t) \leq \mu_i$

with α_1, , α_2 the free flow travel-times of paths (i) and μ_1 , μ_2 the maximum outflow at point b, and let us determine the solution of the command equation (9). We can assume that: $\alpha_1 < \alpha_2$. If

$$T_2 = \alpha_2 \geq T_1 \geq \alpha_1$$

then there is no point in correcting the state of the system, since T_2 cannot be reduced (one might still try to reduce T_1 while keeping T_2 constant, a possibility which will not be analyzed here).

Let us now consider the case: $\qquad\qquad T_2 > \alpha_2 > T_1 = \alpha_1$

then, $\qquad\qquad \dfrac{dT_1}{dt} = \begin{bmatrix} \dfrac{qI_1}{\mu_1} - 1 & \text{if} \quad qI_1 \geq \mu_1 \\ 0 & \text{if} \quad qI_1 \leq \mu_1 \end{bmatrix}$

and it follows: $\qquad \dfrac{dT_1}{dt} - \dfrac{dT_2}{dt} = \begin{bmatrix} \dfrac{qI_1}{\mu_1} - \dfrac{qI_2}{\mu_2} & \text{if} \quad qI_1 \geq \mu_1 \\ 1 - \dfrac{qI_2}{\mu_2} & \text{if} \quad qI_1 \leq \mu_1 \end{bmatrix}$

Let us recall that: $\qquad\qquad qI_1(t) = \dfrac{Q(t)}{1+u} \quad , \quad qI_2(t) = \dfrac{uQ(t)}{1+u}$

with $\quad u = \exp[\,\theta(T_1 - T_2 + \omega)\,]$ The difference $T_1 - T_2$ is negative. In the case $qI_1 \leq \mu_1$ it follows that

$$\frac{1}{1+u} = 1 - \frac{\mu_2}{Q} - \alpha \frac{\mu_2}{Q}(T_1 - T_2) \leq \frac{\mu_1}{Q}$$

requiring that: $\qquad Q \leq \mu_1 + \mu_2 + \alpha\mu_2(T_1 - T_2)$. In the case $qI_1 \geq \mu_1$ it follows that:

$$\frac{1}{1+u} = \frac{\mu_1}{\mu_1 + \mu_2}\left(1 - \alpha\frac{\mu_2}{Q}(T_1 - T_2)\right) \geq \frac{\mu_1}{Q} \tag{11}$$

requiring that: $\qquad Q \geq \mu_1 + \mu_2 + \alpha\mu_2(T_1 - T_2)$.

Actually, <u>the last possible case</u>:

$$T_1 > \alpha_1 , T_2 > \alpha_2$$

yields the same solution, as in the preceding case $qI_1 \geq \mu_1$ (equation (11)) since:

$$\frac{dT_1}{dt} - \frac{dT_2}{dt} = \frac{qI_1}{\mu_1} - \frac{qI_2}{\mu_2} = -\alpha(T_1 - T_2) .$$

At the limit $\quad \alpha \to 0$, the proportionality of qI_i to μ_i follows for $i = 1,2$, a natural result.

The LWR Model

A similar analysis can be carried out with the LWR model. We combine the command equation (9) with the expression (8) of the time-derivative of the ITTs in order to obtain:

$$-\Phi_1(KI_1) + \Phi_1(KE_1) + \Phi_2(KI_2) - \Phi_2(KE_2) = -\alpha(T_1 - T_2)$$

(with KI_i and KE_i the inflow and outflow densities) which yields the following system:

$$\begin{bmatrix} \Phi_1(KI_1) - \Phi_2(KI_2) &=& \Phi_1(KE_1) - \Phi_2(KE_2) + \alpha(T_1 - T_2) \\ qI_1(t) &=& \dfrac{Q(t)}{1+u} \\ qI_2(t) &=& \dfrac{uQ(t)}{1+u} \\ u &=& \exp\!\left[\theta(T_1 - T_2 + \omega_1 - \omega_2)\right] \end{bmatrix}$$

It is necessary to express the inflow densities KI_i as functions of the inflows themselves. This is easy, considering that the inflow densities should be subcritical. The exception to this would be if there is some queue spillover, i.e. the traffic supply at the entry of the system is less than $Q(t)$. It follows that:

$$KI_i = \Delta_e^{-1}(qI_i;i)$$

with Δ_e^{-1} the inverse demand function of path (i), given by:

$$\kappa = \Delta_e^{-1}(q;i) \quad \Leftrightarrow \quad \kappa \leq K_{crit,i} \quad \text{and} \quad q = Q_e(\kappa;i)$$
$$\Leftrightarrow \quad q = \Delta_e(\kappa;i)$$

Finally it follows:

$$\Phi_1\left[\Delta_e^{-1}\left(\frac{Q}{1+u};1\right)\right] - \Phi_2\left[\Delta_e^{-1}\left(\frac{Qu}{1+u};2\right)\right] = \alpha(T_1 - T_2) + \Phi_1(KE_1) - \Phi_2(KE_2)$$

The left-hand side of this equation is an increasing function of u, increasing from $\Phi_1\left[\Delta_e^{-1}(Q;1)\right] - 1$ to $1 - \Phi_2\left[\Delta_e^{-1}(Q;2)\right]$. Here again we note that there is, when possible, a unique solution depending continuously on the data, and that the inflow $Q(t)$ has to be sufficiently large for a control value to exist.

Remarks:

1. As the preceding examples show, the command, when it exists, is unique, and depends continuously on the data (travel times, outflows, total demands...). Nevertheless, it is not always possible to determine an adequate command, even by reducing the parameter α. This notably happens when $Q(t)$ is too small for the assignment at time t to influence positively the evolution of travel times. In this case the only possible course in keeping with the spirit of the algorithm is to minimize $\frac{d}{dt}(T_1 - T_2) \times (T_2 - T_1)$.

2. We are also faced with a problem of practical implementation of the command. Two possibilities can be considered. If the ω_i are to be considered as biases in the information relayed to the drivers, these might rapidly react adversely to it, by say, distrusting the information they receive. The other possibility consists in actually imposing delays equal to ω_i on each path, at for instance the entry point of the path (ramp control). Is it possible to implement such a control strategy? The successive values of $\omega_i(t)$ must then be consistent with the actual dynamics of a queue.

For instance with the LWR model, for describing the traffic on a path (i), the following system would result (with another traffic flow model the equation in T_i should be modified):

$$\left[\begin{array}{rcl} \dfrac{d\omega_i}{dt} &=& \dfrac{qI_i}{s_i} - 1 \\[2ex] \dfrac{dT_i}{dt} &=& \Phi_i\left(\Delta_e^{-1}(qI_i;i)\right) - \Phi_i\left(\Delta_e^{-1}(qE_i;i)\right) \\[2ex] qI_i &=& Q(t)\dfrac{\exp\left(\vartheta(T_i + \omega_i)\right)}{\sum_j \exp\left(\vartheta(T_j + \omega_j)\right)} \end{array} \right.$$

whose stability should then be studied. The equation for ω_i is then the regular equation for queue delay (as in the queue point model) with s_i the saturation flow of the queue. If we considered s_i as time-dependent, we should express ω_i directly as a function of the queue length, and the resulting system would be very close to what we shall describe in the next section.

4. THE SECOND CONTROL STRATEGY

The control strategy presented in this section does not rely on a modification or a bias of the travel time information; the control variables are flow restrictions, hence easy to implement in practice. A drawback of this concept: additional delays may be imposed on users in order to achieve a suitable network state. Let us consider a path (a,c), divided into two subpaths (a,b) and (b,c) as depicted hereafter:

$$\xrightarrow{\quad Q(t) \quad} \overset{\displaystyle a}{\bullet} \overset{\displaystyle R}{\underset{\displaystyle b}{\bullet}} \overset{\displaystyle S}{\underset{\displaystyle c}{\bullet}}$$

with instantaneous travel times R and S on subpaths (a,b) and (b,c) respectively. Total travel time of path (a,c) is $T=R+S$ (instantaneous travel times are additive). The traffic is assumed to be described by the LWR model. It follows from (8) that:

$$\frac{dT}{dt} = \frac{dR}{dt} + \frac{dS}{dt} = \Phi\big(K(c,t)\big) - \Phi\big(K(a,t)\big) + \Delta\Phi(b)$$

with:

$$\Delta\Phi(b) \overset{(def)}{=} \Phi\big(K(b-,t)\big) - \Phi\big(K(b+,t)\big)$$

the symbols $+$ and $-$ representing as usual left- and right-side limits respectively. If a control is exerted at point b, taking the form of a capacity restriction limiting the throughput at that point to $q(b)$, we deduce that:

$$
\begin{aligned}
K(b-,t) &= \Sigma_e^{-1}\big(q(b)\big) \\
K(b+,t) &= \Delta_e^{-1}\big(q(b)\big)
\end{aligned}
$$

where the inverse demand function Δ_e^{-1} has already been defined and Σ_e^{-1} is the inverse supply function defined as:

$$\Sigma_e^{-1}(q) = \kappa \Leftrightarrow q = Q_e(\kappa) \quad \text{and} \quad \kappa \geq K_{crit}$$

It follows that $\Delta\Phi(b)$ is a decreasing function of $q(b)$ varying from $+\infty$ to 0 as $q(b)$ varies from 0 to Q_{max}. Of course, for these relationships to hold, the traffic demand say, δ upstream of point b, has to be greater than the command $q(b)$, otherwise $\Delta\Phi(b) = 0$. The upstream demand δ can be defined as the flow value carried by characteristic lines passing through point b at time t.

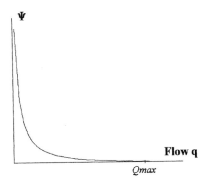

Let $\Psi(q(b))$ be defined by:

$$\Psi(q(b)) = \Delta\Phi(b) \Leftrightarrow \begin{bmatrix} K(b+) & = & \Delta_e^{-1}(q(b)) \\ K(b-) & = & \Sigma_e^{-1}(q(b)) \\ \Delta\Phi(b) & = & \Phi(K(b-)) - \Phi(K(b+)) \end{bmatrix}$$

It is normal that $\Delta\Phi(b)$ should always be positive. Let us now consider the following assignment situation:

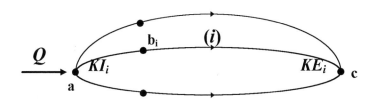

The time derivative of the instantaneous travel-time of path (i) is given by:

$$\frac{dT_i}{dt} = \Phi_i(KE_i) - \Phi_i(KI_i) + \Delta\Phi_i(b_i) = \Phi_i(KE_i) - \Phi_i\left(\Delta_e^{-1}(qI_i;i)\right) + \Psi_i(q(b_i))$$

with:

$$qI_i = Q(t)\frac{\exp(\theta T_i)}{\sum_j \exp(\theta T_j)}$$

a given quantity, and $q(b_i)$ the commands. For example, with two competing paths, we would impose as usual the command equation (9): $\frac{d}{dt}(T_1 - T_2) = -\alpha(T_1 - T_2)$ and deduce:

$$-\alpha(T_1 - T_2) - \Phi_1(KE_1) + \Phi_2(KE_2) + \Phi_1\left(\Delta_e^{-1}(qI_1;1)\right) - \Phi_2\left(\Delta_e^{-1}(qI_2;2)\right) = \Psi_1(q(b_1)) - \Psi_2(q(b_2))$$

This equation yields command values for $q(b_1)$ and $q(b_2)$. Indeed, the difference Ψ_1 (q_1)-Ψ_2 (q_2) takes all possible real values on the subset $\{\delta_1\} \times [0,\delta_2] \cup [0,\delta_1] \times \{\delta_2\}$ of the plane (q_1 , q_2), with δ_i being the upstream demands at points b_i , a result which stems from the continuity of the ψ_i and the fact that $\psi_i(0) = +\infty$.

A similar study could be carried out in the case of the point queue model, with an additionnal difficulty: the travel-times to be used are the predictive travel times. Both strategies could also be studied with the travel time function model (Friesz *et al* 1993, Ran and Boyce 1996).

5. CONCLUSION

This study was essentially one of feasibility. We showed that it is possible to use the drivers path choice in response to travel-time information in order to achieve some control of the system. Certain elements, such as the drivers behavioral model or the objective of the control strategy, could easily be replaced by alternatives, without changing the basic ideas outlined in this paper. What remains to be done? Nearly everything: extension to actual networks (this implies intersection modeling and multiple ODs), numerical and simulation experiments. This is the object of ongoing research.

REFERENCES

Astarita, V. (1996). A continuous-time link model for dynamic network loading based on travel time function. In: Transportation and traffic theory (Ed: J.B. Lesort). 79-102. Pergamon.

Daganzo, C. (1995). Properties of link travel time functions under dynamic loads. *Transportation Research* **29 B**: 95-98.

Elloumi, N., H. Hadj-Salem, and M. Papageorgiou (1994). METACOR, a macroscopic modelling tool for urban corridors. *TRISTAN II Int. Conf.* Capri.

Friesz, T.L., J. Luque, T.E. Smith, R.L. Tobin, and B.W. Wie (1993). A variational inequality formulation of the dynamic network user equilibrium model. *Operations Research* **41**: 179-191.

Heydecker, B.G., and J.D. Addison (1996). An exact expression of dynamic equilibrium. In: Transportation and traffic theory (Ed: J.B. Lesort). 359-384. Pergamon.

Lam, W.H.K., and H.J. Huang (1995). Dynamic user optimal traffic assignment model for many to one travel demand. *Transportation Research* **29 B**: 243-259.

Kuwahara, M., and T. Akamatsu (1997). Decomposition of the reactive dynamic assignments with queues for a many-to-many origin-destination pattern. *Transportation Research* **31 B**: 1-10.

Lebacque, J.P. (1996). Instantaneous travel times computation for macroscopic traffic flow models. *CERMICS report 59-96.*

Lighthill, M.H., and G.B. Whitham (1955). On kinematic waves II: A theory of traffic flow on long crowded roads. *Proc. Royal Soc. (Lond.).* **A 229**: 317-345.

Merchant, D.K., and G.L. Nemhauser (1978). 1. A model and an algorithm for the dynamic traffic assignment problems. 2. Optimality conditions for a dynamic traffic assignment problems. *Transportation Science* **12(3)** : 183-207.

Ran, B., and D.E. Boyce (1996). Modeling dynamic transportation network models. 2nd revised edition. Springer Verlag.

Richards, P.I. (1956). Shock-waves on the highway. *Operations Research* **4**: 42-51.

Yang, H., S. Yagar, Y. Iida, and Y. Asakura (1994). An algorithm for the inflow control problem on urban freeway networks with user-optiml flows. *Transportation Research* **28 B**: 123-258.

SYSTEM OPTIMAL ASSIGNMENT WITH DEPARTURE TIME CHOICE

J.D. Addison and B.G. Heydecker, University of London Centre for Transport Studies, University College London

ABSTRACT

This paper investigates the temporal inflow profile that minimises the total cost of travel for a single route. The problem is formulated to consider the case in which the total demand to be serviced is fixed. The approach used here is a direct calculation of the first order variation of total system cost with respect to variations in the inflow profile. Two traffic models are considered; the bottleneck with deterministic queue and the kinematic wave model. For the bottleneck model a known solution is recovered. The wave model proves more difficult and after eliminating the possibility of a smooth inflow profile the restricted case of constant inflow is solved. As the space of possible profiles is finite dimensional in this case, the standard techniques of calculus apply. We establish a pair of equations that are satisfied simultaneously by the optimal inflow and time of first departure.

1. INTRODUCTION

In this paper we investigate the temporal inflow profile to a route that gives a minimal total system cost. The cost experienced by all traffic has two components: the cost of travel along the route when account is taken of the congestion caused by this traffic and a cost associated with deviation from a desired time of arrival. We suppose that the total amount of traffic to be assigned, E, is exogenous in order to avoid the trivial solution of no travel.

Let $q(s)$ be an inflow such that the first traffic enters at time s_0. The accumulated entry flow, $A(s)$, is given by

$$A(s) = \int_{s_0}^{s} q(u) \, du \qquad (1.1)$$

Except on intervals where $q(s) = 0$, $A(s)$ is a monotonic increasing function of s and is therefore inevitable. Let $s(A)$ denote this inverse: we have that $s(0) = s_0$ and $s(A)$ is the time at which the "A^{th}" vehicle enters the route. Similarly let $\tau(A)$ denote the time at which the accumulated efflux from the route is A. Note that τ depends on the choice of q. From (1.1) it follows that

$$\frac{dA(s)}{ds} = q(s) \text{ and so } \frac{ds(A)}{dA} = \frac{1}{q[s(A)]}.$$ (1.2)

Whenever the entry time function $s(A)$ is given, the inflow profile q is also available. The arrival time τ then depends on $s(A)$. When we wish to emphasise this dependence we use τ_s to denote the arrival time.

The individual cost incurred is the travel time $\tau(A) - s(A)$ plus the arrival time-specific cost $f(\tau(A))$ so the total system cost, $\Theta(s)$ is (Jauffred and Bernstein, 1996 also use this form)

$$\Theta(s) = \int_0^E \tau(A) - s(A) + f(\tau(A)) dA$$ (1.3)

With $s(0) = s_0$ and $s(E) = T$ the relation $dA(s)/ds = q(s)$ allows a simple change of variable to give the perhaps more familiar expression in terms of flow-weighted time-specific costs:

$$\Theta(s) = \int_{s_0}^T [\tau(u) - u + f(\tau(u))] q(u) du.$$

2. SOME BACKGROUND

The technique used in this paper is to investigate the variation in the total cost $\Theta(s)$ for small variations $h(A)$ in the inflow profile specified by $s(A)$ for each of the bottleneck and the wave traffic models. We assume that $s(A)$ is a piecewise smooth function on $[0, E_0]$. The variations that we consider are to be sufficiently smooth that they are small in the norm $\|h\| = \max_x \{|h(x)| + |h'(x)| + |h''(x)|\}$. For certain inflow profiles denoted by $s(A)$, we find the first order variation in $\Theta(s)$. From this, using suitable choices for the first order variation $h(A)$, we can make deductions concerning the form of the system optimal inflow profile.

3. USING THE BOTTLENECK MODEL

In this section, we suppose that the route has a long free-flow section and a capacity that is determined by a single bottleneck. The free-flow travel time is denoted by ϕ and the service rate of the bottleneck by b. Given $s(A)$, we suppose that there is single period during which a queue forms. Denote by A_0 the value for which the queue begins to form and A_1 the value for which the queue disappears. Assuming that s is smooth we have $s'(A_0) = 1/b$ and $s''(A_0) < 0$. The value A_1 at which the queue disappears satisfies

$$A_1 = A_0 + b[s(A_1) - s(A_0)].$$ (3.1)

The arrival time τ is given by

$$\tau_s(A) = \begin{cases} s(A) + \phi & (A \le A_0, A \ge A_1) \\ s(A_0) + \phi + \dfrac{(A - A_0)}{b} & (A_0 \le A \le A_1) \end{cases} \tag{3.2}$$

Now consider a small variation h of s. In what follows we will add a superior tilde $\ \tilde{}\ $ to values associated with $s + h$, so for example $\tilde{A}_0$ and $\tilde{A}_1$ denote the new values for which the queue appears and disappears. The new value $\tilde{A}_0$ is given by $s'(\tilde{A}_0) + h'(\tilde{A}_0) = 1/b$ which when expanded to first order gives

$$\tilde{A}_0 - A_0 = -\frac{h'(A_0)}{s''(A_0)} + O\left(\|A_0 - A_0\|^2\right) + O\left(\|h\|^2\right). \tag{3.3}$$

Similarly, an expansion of (3.1) leads to

$$\tilde{A}_1 - A_1 = \frac{h(A_1) - h(A_0)}{s'(A_0) - s'(A_1)} + O\left(\|A_1 - A_1\|^2\right) + O\left(\|h\|^2\right). \tag{3.4}$$

The two equations (3.3) and (3.4) show that the variations in $\tilde{A}_0 - A_0$ and $\tilde{A}_1 - A_1$ are first order in h. To find the variation in $\tilde{\tau} = \tau_{s+h}$

$$\tilde{\tau}(A) = \begin{cases} s(A) + h(A) + \phi & (A \le \tilde{A}_0, A \ge \tilde{A}_1) \\ s(\tilde{A}_0) + h(\tilde{A}_0) + \dfrac{A - \tilde{A}_0}{b} + \phi & (\tilde{A}_0 \le A \le \tilde{A}_1) \end{cases} \tag{3.5}$$

There are four possibilities to consider according as $A_0 > \tilde{A}_0$ and $A_1 > \tilde{A}_1$. These divide the interval $[0, E]$ into five subintervals: 1. $\left[0, \min(A_0, \tilde{A}_0)\right]$; 2. $\left[\min(A_0, \tilde{A}_0), \max(A_0, \tilde{A}_0)\right]$; 3. $\left[\max(A_0, \tilde{A}_0), \min(A_1, \tilde{A}_1)\right]$; 4. $\left[\min(A_1, \tilde{A}_1), \max(A_1, \tilde{A}_1)\right]$; 5. $\left[\max(A_1, \tilde{A}_1), E\right]$. Equations (3.3) and (3.4) show that the length of intervals 2 and 4 are of order $\|h\|$. The cases that remain can separated according to the existence or non-existence of a queue. In intervals 1 and 5 there is no queue either before or after perturbation, so $\tilde{\tau}(A) = \tau_s(A) + h(A) + O\left(\|h\|^2\right)$. In interval 3 there is a queue in both cases and the variation is found to be $\tau_{s+h}(A) = \tau_s(A) + h(A) + O\left(\|h\|^2\right)$. The variation in $\Theta(s)$ is found from

$$\Theta(s + h) = \int_0^E \tau_{s+h}(A) - s(A) - h(A) + f\left(\tau_{s+h}(A)\right) dA \tag{3.6}$$

Because the durations of the intervals between A_0 and $\tilde{A}_0$, and A_1 and $\tilde{A}_1$ are first order and the perturbation of the integrand is also of first order they can be ignored. Thus

$$\Theta(s + h) = \int_0^{A_0} \tau_s(A) + h(A) - s(A) - h(A) + f(\tau_s(A)) + f'(\tau_s(A))h(A)\, dA \ +$$

$$+ \int_{A_1}^E \tau_s(A) + h(A) - s(A) - h(A) + f(\tau_s(A)) + f'(\tau_s(A))h(A)\, dA +$$

$$+ \int_{A_0}^{A_1} \tau_s(A) + h(A_0) - s(A) - h(A) + f(\tau_s(A)) + f'(\tau_s(A))h(A_0)\, dA + O\left(\|h\|^2\right)$$

The first order variation is

$$\Delta\Theta(s) = \int_0^{A_0} f'(\tau_s(A))h(A)\,dA + \int_{A_1}^{E} f'(\tau_s(A))h(A)\,dA + \int_{A_0}^{A_1} h(A_0) - h(A) + f'(\tau_s(A))h(A_0)\,dA$$

At a minimum, the variation must be positive for all choices of h: this is not possible unless $A_0 = 0$ and $A_1 = E$. We deduce that the outflow will always be equal to the capacity, b, so there will always be a queue though possibly one of zero length. The variation now becomes

$$\Delta\Theta(s) = \int_0^{E} [h(0) - h(A)]\,dA + h(0)\int_0^{E} f'(\tau_s(A))\,dA.$$

The queue is busy on the whole of the interval $[0, E]$ so $s(A) \le s(0) + A/b$. If a non-zero queue arose, we would have $s(A) < s(0) + A/b$ on some interval and could then choose h to make $\Delta\Theta(s)$ negative while still retaining a busy queue, so we must have $s(A) = s(0) + A/b$. For this choice of s all variations that maintain the queue have $h(0) - h(A) \ge 0$.

At the minimum, the integral in the second term must be zero otherwise a constant variation could be chosen to make $\Delta\Theta(s)$ negative. We have $s(A) = s(0) + A/b$ so that $\tau_s(A) = s(0) + \phi + A/b$. A change of variable in this expression shows that

$$\int_0^{E} f'(s(0) + \phi + A/b)\,dA = b[f(\tau_s(E)) - f(\tau_s(0))].$$

Thus at the optimum $f(\tau_s(0)) = f(\tau_s(E))$ showing that the first and last traveller experience identical arrival costs. For any uniminimal arrival time cost function f, this determines $s(0)$.

4 THE WAVE MODEL

4.1 Introduction

We next consider the Lighthill and Whitham (1955) kinematic wave model. Denote by υ the speed of traffic, which is a function of the density k, and by ω the wave speed which is also a function of the density. These speeds are related by $\omega(k) = \upsilon'(k)k + \upsilon(k)$. The density k is a function of the flow q, the two being related by $q(k) = \upsilon(k)k$.

The instantaneous flow into the route is given by $q(s) = 1/\dfrac{ds}{dA} = \dfrac{dA}{ds}$. According to this model,

υ and ω are functions of q which is itself a function of A. Thus $\upsilon'(k)$ will mean the derivative with υ' with respect to k and similarly with other functions.

Provided that $s(A)$ is smooth, for given $s(A)$ we can construct a parametric representation of the accumulated efflux G. For the wave model we know (Newell, 1993) that for a each A and writing s for $s(A)$, $G(s + l / \omega(k)) = A - \upsilon'(k) k^2 \, l / \omega(k)$. Thus G is a parametric curve with parameter A which will give a single value provided that no shock-waves occur.

4.2 The Wave Model With Smooth Inflow

First we establish that a system-optimal inflow profile that is otherwise smooth must start abruptly. In this analysis we suppose that the inflow profile is smooth everywhere, in particular at time $s(0)$, and that no shock waves occur. We show that for any such inflow profile, one can be constructed that gives lower total cost.

The arrival time $\tau(A)$ satisfies $G(\tau(A)) = A$. To find $\tau(A)$, first find A_i such that, writing s_i for $s(A_i)$, we have

$$A_i - \frac{\upsilon'(q(s_i)) k^2 (q(s_i)) l}{\omega(q(s_i))} = A. \tag{4.1}$$

Then $\qquad \tau(A) = s_i + l / \omega(q(s_i))$. $\tag{4.2}$

We now consider the effect on the total cost of a variation $h(A)$:

$$\Theta(s + h) = \int_0^E \left[\tau_{s+h}(A) - s(A) - h(A) + f(\tau_{s+h}(A)) \right] dA.$$

We express τ_{s+h} in terms of τ_s and h. For fixed A let A_i be as in (4.1) and let $\tilde{A}_i$ be the corresponding value for $s+h$ with $\tilde{s}_i = (s + h)(\tilde{A}_i)$. Then $\tilde{A}_i - \dfrac{\upsilon'(\tilde{q}(\tilde{s}_i)) k^2 (\tilde{q}(\tilde{s}_i)) l}{\omega(\tilde{q}(\tilde{s}_i))} = A$ and

$\tau_{s+h}(A) = s(\tilde{A}_i) + h(\tilde{A}_i) + \dfrac{l}{\omega(\tilde{q}(\tilde{s}_i))}$. The inflow for $s+h$ is $\tilde{q}(A) = q(A) - q(A)^2 \, h'(A) + O(\|h\|^2)$

so that $\qquad \tilde{q}(A) - q(A) = -q(A)^2 \, h'(A) + O(\|h\|^2) \tag{4.3}$

To find the variation in density k we use a Taylor expansion on the right hand side of $\tilde{q} = \upsilon(\tilde{k}) \tilde{k}$ to obtain $\tilde{q} = q + \omega(k)(\tilde{k} - k) + O(\|\tilde{k} - k\|^2)$. Combining this with (4.3) we get

$$\tilde{k} - k = -\frac{q^2(A) h'(A)}{\omega(k(A))} + O(\|\tilde{k} - k\|^2). \tag{4.4}$$

Thus $\tilde{k} - k = O(\|h\|)$. It is then straightforward to find

$$\upsilon'(\tilde{k}) = \upsilon'(k) + \upsilon''(k)(\tilde{k} - k) + O(\|h\|^2) \tag{4.5}$$

$$\omega'(\tilde{k}) = \omega'(k) + \omega''(k)(\tilde{k} - k) + O(\|h\|^2) \tag{4.6}$$

$$1 / \omega'(\tilde{k}) = 1 / \omega'(k) - \frac{\omega''(k)}{\omega(k)^2}(\tilde{k} - k) + O(\|h\|^2) \tag{4.7}$$

To find the variation in A_i recall from (4.1) that A_i is determined, for a suitable choice of function g, by $A = A_i + g\left(s'\left(A_i\right)\right)$ and so $\tilde{A}_i$ is given by $A = \tilde{A}_i + g\left(s'\left(\tilde{A}_i\right) + h\left(\tilde{A}_i\right)\right)$. To simplify the appearance of later expressions we write s_i, s_i' and s_i'' for the values of s, s' and s'' evaluated at A_i. Expanding in terms of $\tilde{A}_i - A_i$ gives

$$A = A_i + \tilde{A}_i - A_i + g(s_i') + g'(s_i')h(A_i) + g'(s_i')s''(A_i)(\tilde{A}_i - A_i) + O\left(\left\|\tilde{A}_i - A_i\right\|^2\right).$$

This then gives
$$\tilde{A}_i - A_i = -\frac{g'(s_i')h(A_i)}{1 + g'(s_i')s_i''} + O\left(\left\|h\right\|^2\right) \tag{4.8}$$

Under our supposition that no shock waves occur, the variation $\tilde{A}_i - A_i$ will be of order $\left\|h\right\|$.

Similarly for suitable H, $\tau(A) = s_i + H\left(s_i'\right)$ and it is a straightforward calculation to show that

$$\tilde{\tau}(A) - \tau(A) = h(A_i) + H'(s_i')h(A_i) + \left(s_i' + H''(s_i')s_i''\right)\left(\tilde{A}_i - A_i\right) + O\left(\left\|h\right\|^2\right). \tag{4.9}$$

To find $g'(s')$ and $H'(s')$ recall that $q = 1/s'$ so $\dfrac{dq}{ds'} = -1/s'^2$. Now $g(s') = \overline{g}(q) = -\dfrac{\upsilon'(k)k^2 l}{\omega(k)}$

and $H(s') = \overline{H}(q) = \dfrac{1}{\omega(k)}$, where k is a function of q. Then $g'(s') = \overline{g}'(q)/(s')^2$ and similarly

for H. This gives $g'(s) = \dfrac{\omega'(k)l}{(s'\omega(k))^3}$ and $H'(s') = -\dfrac{\omega'(k)l}{(s')^2\omega(k)^3}$ as $\upsilon(k)k = q = (s')^{-1}$. (4.10)

Substituting these into (4.8) gives $\tilde{A}_i - A_i = \dfrac{\omega'(k_i)lh(A_i)}{\left(\omega(k_i)s_i'\right)^3 - \omega'(k_i)ls_i''}$. (4.11)

The coefficient of $\tilde{A}_i - A_i$ in (4.9) becomes $s_i' + H'(s_i')s_i'' = \dfrac{\left(s_i'\omega(k_i)\right)^3 - \omega'(k_i)ls_i''}{s_i'^2\omega(k_i)^3}$ so the term

involving $\tilde{A}_i - A_i$ simplifies to $\left(s_i' + H'(s_i')s_i''\right)\left(\tilde{A}_i - A_i\right) = \dfrac{\omega'(k_i)lh(A_i)}{s_i'^2\omega(k_i)^3} = -H'(s_i')h(A_i)$

which cancels with the preceding term giving

$$\tilde{\tau} - \tau = h(A_i) + O\left(\left\|h\right\|^2\right). \tag{4.12}$$

The variation in $\Theta(s)$ is then $\Delta\Theta(h) = \int_0^E h(A_i) - h(A) + h(A_i)f'(\tau(A)) dA$. (4.13)

We can now choose a variation so that, if A_1 is the value of A_i for E

$$h(A) \begin{cases} = 0 & (0 \le A \le A_1) \\ > 0 & (A_1 < A \le E) \end{cases}$$

With such a choice of variation, $\Delta\Theta(h) < 0$ so that we have established by construction an inflow profile of lower cost than $\Theta(s)$. Consequently, an inflow with $q(0) = 0$ cannot be optimal, so according to the wave model, any optimal inflow will start with a step increase.

4.3 Constant inflows

In this section we investigate a restricted case of the total cost minimisation problem: we assume that the influx is at a constant rate throughout the period in during which it occurs. We require that the total throughput be equal to an exogenous value E, so there is a trade-off between inflow rate and duration, and a balance of start and end times. In this case the total system cost $\Theta(s)$ can be expressed as a function of two variables: the time s_0 at which flow starts and the rate of influx, q. The optimum for this restricted problem then satisfies

$$\frac{\partial \Theta}{\partial s_0} = 0 \quad \text{and} \quad \frac{\partial \Theta}{\partial q} = 0 \tag{4.14}$$

Because changing s_0 simply relocates the flow in time and has no effect on the travel time $\tau - s$ we have $\dfrac{d\tau}{ds_0} = 1$ and so

$$\frac{\partial \Theta}{\partial s_0} = \int_0^E f\left(\tau(A)\right) dA. \tag{4.15}$$

For the second equation,

$$\frac{\partial \Theta}{\partial q} = \int_0^E \left[\frac{\partial \tau}{\partial q} - \frac{\partial s}{\partial q} + f\left(\tau(A)\right) \frac{\partial \tau}{\partial q} \right] dA. \tag{4.16}$$

For a steady flow q with associated density k, there are two regions of distinct character in the time-distance diagram: an initial fan shaped region of varying density is followed by a region of constant density. The first wave of the constant density, k, leaves the route at time $\tau_1 = s_0 + l/\omega(k)$. The first traffic to experience homogeneous conditions enters at time

$$s_1(q) = s_0 + \frac{l}{\omega(k)} - \frac{l}{\upsilon(k)}. \tag{4.17}$$

Let $E_1(q)$ denote the corresponding value of the accumulated inflow so that $s(E_1(q)) = s_1(q)$. From now on we specialise to Greenshields' speed-density relationship

$$\upsilon(k) = V - \alpha k \tag{4.18}$$

In this case, the arrival time is given by

$$\tau(A) = \begin{cases} \dfrac{l}{V} + \dfrac{2\upsilon'(k)A}{V^2} + \dfrac{2\sqrt{\alpha A^2 - Vl\alpha A}}{V^2} + s_0 & \left(0 \le A \le E_1(q)\right) \\[4mm] \dfrac{A}{q} + \dfrac{l}{\upsilon(k)} + s_0 & \left(E_1(q) \le A\right) \end{cases} \tag{4.19}$$

so that $\dfrac{\partial \tau}{\partial q} = \begin{cases} 0 & \left(0 \le A \le E_1(q)\right) \\[3mm] -\dfrac{A}{q^2} - \dfrac{l\upsilon'(k)}{\upsilon(k)^2 \omega(k)} & \left(E_1 \le A\right) \end{cases}$ since $\dfrac{dk}{dq} = 1/\omega(k)$ \hfill (4.20)

The departure time function is $s(A) = s_0 + \dfrac{A}{q}$ and so $\dfrac{\partial s}{\partial q} = -\dfrac{A}{q^2}$. Using this and (4.20) in (4.16), and assuming that some traffic experiences homogeneous conditions, ie $E_1 \le E$, gives

$$\frac{\partial \Theta}{\partial q} = \int_0^{E_1(q)} -\frac{A}{q^2}\, dA + \int_{E_1(q)}^{E} -\frac{l\upsilon'(k)}{\upsilon(k)^2 \omega(k)} + \left(-\frac{A}{q^2} - \frac{l\upsilon'(k)}{\upsilon(k)^2 \omega(k)}\right) f'(\tau(A))\, dA$$

$$= \frac{A^2}{2q^2}\bigg|_0^{E_1} - \frac{A\upsilon'(k)l}{\upsilon(k)^2 \omega(k)}\bigg|_{E_1}^{E} - \int_{E_1}^{E}\left(\frac{A}{q^2} + \frac{l\upsilon'(k)}{\upsilon(k)^2 \omega(k)}\right) f'(\tau(A))\, dA \qquad (4.21)$$

Note that if $E_1 \geq E$ then $\dfrac{\partial \Theta}{\partial q} = \int_0^{E_1(q)} \dfrac{A}{q^2}\, dA > 0$ so that the value of $\Theta(s)$ can be reduced by

decreasing q. This is because the arrival cost is fixed but the delay will decrease as a result of reduced congestion on the route.

For A in the range $[E_1, E]$ we see from (4.19) that $\tau(A)$ is a linear function. For the change of variable $\tau = \tau(A)$ with $\tau_f = \tau(E)$ and $\tau_1 = \tau(E_1)$ the integral in (4.21) becomes

$$\int_{\tau_1}^{\tau_f} f'(\tau)\left(\frac{\tau}{q} - \frac{l}{\upsilon(k)q} - \frac{s_0}{q} + \frac{l\upsilon'(k)}{\upsilon(k)^2 \omega(k)}\right) q\, d\tau = \int_{\tau_1}^{\tau_f} f'(\tau)\tau\, d\tau - \left(s_0 + \frac{l}{\omega(k)}\right)\int_{\tau_1}^{\tau_f} f'(\tau)\, d\tau.$$

Integrating by parts, noting that $\tau_1 = s_0 + l/\omega(k)$ and substituting into (4.21), the equation $\partial \Theta / \partial q$ $= 0$ becomes

$$\frac{E_1^2}{2q^2} + \frac{E_1 \upsilon'(k)l}{\upsilon(k)^2 \omega(k)} - \frac{E\upsilon'(k)l}{\upsilon(k)^2 \omega(k)} - f(\tau_f)(\tau_f - \tau_1) + \int_{\tau_1}^{\tau_f} f(\tau)\, d\tau = 0 \qquad (4.22)$$

Turning our attention to $\partial \Theta / \partial s_0$ given by (4.15) we have that

$$\int_0^{E} f'(\tau(A))\, dA = \int_0^{E_1} f'(\tau(A))\, dA + \int_{E_1}^{E} f'(\tau(A))\, dA \qquad (4.23)$$

In the range of the second term $\tau(A)$ is, as we saw earlier, a linear function of A so that the same change of variable allows integration and we get

$$\int_0^{E} f'(\tau(A))\, dA = \int_0^{E_1} f'(\tau(A))\, dA + q\big[f(\tau_f) - f(\tau_1)\big]. \qquad (4.24)$$

To proceed further, we specialise to a particular arrival-cost function. Following Vickrey (1969), we use a 2-part piecewise linear function with $m_0 < 0$, $m_1 > 0$, and ideal time of arrival

0:    $$f(t) = \begin{cases} m_0 t & (t \leq 0) \\ m_1 t & (t \geq 0) \end{cases} \qquad (4.25)$$

Note that we must have $s_0 + \phi < 0$ otherwise (4.15) would be strictly positive so that the time of first arrival is before the ideal one. The cases with $\tau_1(q)$ negative and positive are now considered separately. In the former case, $f'(\tau(A)) = m_0$ on $[0, E_1(q))$ and we get from (4.25) and (4.14) the equation for $\partial \Theta / \partial s_0 = 0$ which, together with the special form (4.25) gives

$$m_0 E_1(q) + q\, m_1 \tau_f - q\, m_0 \tau_1 = 0. \qquad (4.26)$$

Because $E_1(q)$ is the first traffic to experience homogeneous conditions, we find that for Greenshields' speed-density function $E_1(q) = \left((V - \omega(k))^2\right)\big/\left(4\alpha\,\omega(k)\right)$. Using this, $\tau_1(q) = s_0 + l/\omega(k)$ and $\tau_f = s_0 + E/q + l/\upsilon(k)$ we obtain

$$s_0 = \frac{1}{m_0 - m_1}\left(\frac{m_0 E + m_1 E_1}{q} + \frac{m_1 l}{\upsilon(k)} - \frac{m_0 l}{\omega(k)}\right) \qquad (4.27)$$

Now consider the case $\tau_1(q) \geq 0$. Let E_2 be the index of the traffic that arrives at the destination at the desired arrival time 0. The integral in (4.25) evaluates to $m_0 E_2 + m_1(E_1 - E_2)$ and (4.25) becomes

$$(m_0 - m_1) E_2 + m_1 E_1 + q\, m_1 (\tau_f - \tau_1) = 0 . \tag{4.28}$$

During the initial phase when the density at the exit varies the accumulated efflux is given by $G(\tau) = \left(V(\tau - s_0) - 1\right)^2 / \left(4\alpha(\tau - s_0)\right)$; E_2 is the value when $\tau = 0$ so that

$$E_2 = -\frac{(V s_0 + 1)^2}{4\alpha s_0} . \tag{4.29}$$

Also E_1 is the accumulated flow at the end of this period of variable efflux. The end is at time $\tau_1 = s_0 + l/\omega(k)$ so that

$$E_1 = \frac{l(V - \omega(k))^2}{4\alpha\omega(k)} = \frac{\alpha k^2 l}{\omega(k)} . \tag{4.30}$$

We note that E_2 is independent of q. Using $\tau_f = E/q + l/\upsilon(k)$ and $\tau_1 = s_0 + l/\omega(k)$, substituting the explicit expressions for $\omega(k)$ and $\upsilon(k)$ shows that the expression $m_1 E_1 + q m_1 (\tau_f \cdot \tau_1) + m_1 E$ is independent of k and hence of q. This substitution leads to the quadratic equation

$$(m_1 - m_0) V^2 s_0^2 + \left[2\, V l (m_1 - m_0) - 4\alpha m_1 E\right] s_0 + (m_1 - m_0)\omega(k) l^2 = 0 . \tag{4.31}$$

Similarly (4.23) gives rise to a pair of equations:

$$\frac{E_1^2}{2q^2} - \frac{\upsilon'(k)l}{\upsilon(k)^2 \omega(k)}(E - E_1) - \frac{m_1}{2}(\tau_f - \tau_1)^2 - \frac{m_0 - m_1}{2}\tau_1^2 = 0 \qquad (\tau_1(q) \leq 0) \tag{4.32}$$

$$\frac{E_1^2}{2q^2} - \frac{\upsilon'(k)l}{\upsilon(k)^2 \omega(k)}(E - E_1) - \frac{m_1}{2}(\tau_f - \tau_1)^2 = 0 \qquad (\tau_1(q) \geq 0) . \tag{4.33}$$

Gathering together the various equations that arise from $\dfrac{\partial\Theta}{\partial s_0} = 0$ and $\dfrac{\partial\Theta}{\partial q} = 0$ give the specification of the system optimal inflow rate and start time for the present restricted problem for a single route with specified total throughput. This the optimal assignment is given in the case that $\tau_1(q) \leq 0$ by (4.27) and (4.32) as

$$\frac{E_1^2}{2q^2} - \frac{\upsilon'(k)l}{\upsilon(k)^2 \omega(k)}(E - E_1) - \frac{m_1}{2}(\tau_f - \tau_1)^2 - \frac{m_0 - m_1}{2}\tau_1^2 = 0$$

$$s_0 - \frac{m_1}{m_0 - m_1}\left(\frac{E}{q} + \frac{l}{\upsilon(k)}\right) + \frac{m_0}{m_0 - m_1}\left(\frac{E_2}{q} - \frac{l}{\omega(k)}\right) = 0 \tag{4.34}$$

and case that $\tau_1(q) \geq 0$ by (4.31) and (4.33) as

$$\frac{E_1^2}{2q^2} - \frac{\upsilon'(k)l}{\upsilon(k)^2 \omega(k)}(E - E_1) - \frac{m_1}{2}(\tau_f - \tau_1)^2 = 0$$

$$(m_1 - m_0) V^2 s_0^2 + \left[2\, V l (m_1 - m_0) - 4\alpha m_1 E\right] s_0 + (m_1 - m_0)\omega(k) l^2 = 0 . \tag{4.38}$$

4. CONCLUSIONS

The analysis presented in this paper has established several results that describe the assignment of a specified amount of traffic to a single route that minimises the sum of the total travel and arrival-time costs. When the bottleneck model is used to describe a congested route, the optimal inflow rate is equal to the capacity so that the outflow rate is maximal but queueing does not occur; the first departure is timed so that the arrival-time cost incurred by the first and the last departures are identical. When the more detailed kinematic wave model is used, the optimal inflow rate is shown to have a step increase from 0 at the time of first departure. Explicit equations have been established for the optimal constant inflow rate and associated time of first departure when Vickrey's 2-part piecewise linear arrival cost function is adopted.

ACKNOWLEDGEMENTS

The authors are grateful to Richard Allsop for his encouragement and discussions on the topic of this paper. Thanks are also due to an anonymous referee for his helpful comments. This work was funded by the UK Engineering and Physical Science Research Council.

REFERENCES

Jauffred, F.J., and Bernstein, D. (1996) An alternative Formulation of the Simultaneous Route and Departure Time Choice Equilibrium Problem. *Transportation Research*, **4C**(6), 339-357.

Lighthill, M.J. and Whitham, G.B. (1955) On kinematic waves: II. A theory of traffic flow on long crowded roads. *Proceedings of the Royal Society*, **229A**, 317-45.

Newell, G.F. 1993. A simplified theory of kinematic waves in highway traffic, Part I: General theory. *Transportation Research*, **27B**(4), 281-7.

Vickrey, W.S. (1969) Congestion theory and transport investment. *The American Economic Review*, **59**, 251-61.

INCREMENTAL TRAFFIC ASSIGNMENT: A PERTURBATION APPROACH

D. Kupiszewska and D. Van Vliet, Institute for Transport Studies, University of Leeds, Leeds LS2 9JT, UK

ABSTRACT

This paper develops a new algorithmic approach to equilibrium road traffic assignment which, by directly estimating <u>differences</u>, can more accurately estimate the impact of (relatively) small traffic schemes or changes in the demand pattern. Comparing the outputs of two independent traffic assignments to "with" and "without" scheme networks very often masks the effect of the scheme due to the "noise" in the resulting solutions. By contrast an incremental approach attempts to directly estimate the changes in link flows - and hence costs - resulting from (relatively) small perturbations to the network and/or trip matrix. The algorithms are based firstly on "route flows" as opposed to "link flows", and secondly, they use a variant of the standard Frank-Wolfe algorithm known as "Social Pressure" which gives a greater weight to those O-D path flows whose costs are well above the minimum costs as opposed to those which are already at or near minimum. Tests on a set of five "real" networks demonstrate that the Social Pressure Algorithm is marginally better than Frank-Wolfe for single assignments but is very much faster and more accurate in predicting the impact of small network changes.

1. INTRODUCTION

The economic and operational evaluation of road schemes requires the comparison of several alternative schemes both between themselves and with "do-nothing" or "do-minimum" scenarios. Currently each network tested is assigned "from scratch" so that one assignment cannot make use of any information obtained in another. A common problem with relatively small scale schemes is that the uncertainty ("noise") in the model outputs can exceed the true scheme-induced differences and the actual impacts are effectively masked.

This paper describes alternative network assignment algorithms based on a "perturbation" or "incremental" approach whereby the network <u>differences</u> (in terms of both flows and travel times) are predicted directly, so as to eliminate noise. The algorithms follow the general structure of the Frank-Wolfe algorithm but: (a) are based on path flows as opposed to link flows; and (b) calculates their "descent direction" using a different principle based on "Social Pressure".

Although most assignment algorithms calculate minimum cost routes and load trips using path flows, they frequently only store the resulting link flows. Algorithms entirely based on path flows demand greater computer storage capacity and have not been so popular in large network applications, although we can find in the literature algorithms that use path flows throughout e.g., Schittenhelm (1990) and Larson and Patriksson (1992). We show here that the memory requirements are not excessive by modern standards and that improved convergence rates may be obtained with a path-based approach, in particular in the "perturbation mode".

2. PROBLEM DEFINITIONS

The Wardrop Equilibrium Assignment problem and the notation used in this paper are as follows. Given a trip matrix of origin-destination flows T_{ij} to be assigned to a network of links indexed by 'a' the trips are to be divided into path flows T_{pij} such that at equilibrium:

$$C_{pij} = C^*_{ij} \quad \text{if} \quad T_{pij} > 0 \tag{2.1}$$
$$C_{pij} \geq C^*_{ij} \quad \text{if} \quad T_{pij} = 0$$

where: C_{pij} is the cost of travel on path pij

C^*_{ij} is the minimum cost of travel from origin i to destination j.

Path costs are the sum of constituent link costs C_a and may be written as:

$$C_{pij} = \sum_a C_a \, \delta^a_{pij} \tag{2.2}$$

where $\delta^a_{pij} = 1$ if path pij "uses" link a, 0 otherwise.

The costs of travel on each link are functions of travel flows, generally assumed to be "separable" functions of the form:

$$C_a = C_a(V_a) \tag{2.3}$$

although more complex "non-separable" forms are frequently necessary.

As is well known if costs are separable Wardrop Equilibrium is equivalent to the following minimisation problem:

$$\min Z\,(T_{pij}) = \sum_a \int_0^{V_a} C_a\,(v)dv \tag{2.4}$$

subject to: $T_{pij} \geq 0$ (2.5)

$$\sum_p T_{pij} = T_{ij} \tag{2.6}$$

$$V_a = \sum_{pij} T_{pij}\,\delta^a_{pij} \tag{2.7}$$

As expressed here Z is a function of the path flows T_{pij} although its calculation only involves the more aggregate quantities, the link flows V_a. Correspondingly the algorithms used to solve for Wardrop Equilibrium may be divided into "path-based" and "link based" depending on whether, as we shall see below, they explicitly retain information on both T_{pij} and V_a or whether the path flows are only used to generate link flows and then discarded. Traditionally the latter approach is favoured by computer applications since the memory required to store path flows is considerably in excess of that required to store link flows.

3. ALGORITHMS: FRANK-WOLFE AND SOCIAL PRESSURE

Both the Frank-Wolfe and Social Pressure algorithms have an iterative structure whereby, at each iteration, a certain proportion of flow is transferred from current routes to cheaper alternatives in an attempt to approximate equilibrium. Both can be described as a sequence of the following steps:

1. Initialize iteration number (n = 1)
2. Calculate minimum cost paths for the unloaded network and set current path flows equal to the all-or-nothing flows: $T^{(1)}_{pij} = T_{ij}$
3. Load all current path flows onto the network to obtain link flows $V_a^{(n)}$ and calculate link costs $C_a(V_a^n)$.
4. Calculate the cost of travelling along all used paths (C_{pij}) and determine the minimum cost paths pij*.
5. Test stopping conditions and terminate if a stopping criterion has been achieved; otherwise:
6. Calculate a descent direction (A_{pij})
7. Calculate the optimum step length (λ)
8. Calculate new route flows: $T^{n+1}_{pij} = T^n_{pij} + \lambda A_{pij}$
9. Increment iteration number n and return to step 3.

In Frank-Wolfe, step 8 combines the current set of flows (T^n) with the current all-or-nothing minimum cost assignment to obtain a new set of path flows:

$$T^{(n+1)}_{pij} \begin{cases} = (1-\lambda)T^n_{pij} & pij \neq pij* \\ = (1-\lambda)T^n_{pij} + \lambda T_{ij} & pij = pij* \end{cases}$$

(3.1)

At each iteration the path flows increase on the minimum cost route (pij*) and decrease on all other routes. The rate of flow transfer is proportional to: (1) the current flow level T_{pij} ; and (2) the step length λ, which is common for all paths and O-D pairs. The algorithm is insensitive to aspects such as the cost difference between paths.

Thus consider, for example, an O-D pair ij which has flow on routes 1 and 2 and the costs are virtually equal, therefore in near Wardrop equilibrium. Assume 2 is marginally cheaper than 1.

Frank-Wolfe, by transferring flows from 1 to 2, will tend to decrease the costs on 1 and increase costs on route 2 and therefore actually drive that O-D pair beyond equilibrium unless λ is very small. On the other hand, trips that are transferred between routes with very large cost differences will tend to promote equilibrium and require large values of λ.

A problem with Frank-Wolfe is that it fails to distinguish between these two situations and the optimum value of λ represents a compromise. The Social Pressure algorithm was devised to present a more sensitive answer to this problem. In this method the rate at which trips are transferred to the cheapest route is not a constant factor of the route flows, but a variable rate proportional to a "Social Pressure Factor". The Social Pressure concept is intuitively based on the idea that drivers travelling along relatively expensive routes are more strongly inclined to shift routes than those with travel times closer to the minimum. It can be viewed as a social measure of dissatisfaction for each route. For example we could define a social pressure as:

$$P_{pij} = C_{pij} - C_{ij}^{*} \tag{3.2}$$

The updated flows within iteration n of the Social Pressure algorithm are now calculated according to:

$$T_{pij}^{(n+1)} \begin{cases} = (1 - \lambda P_{pij}) T_{pij}^{n} & pij \neq pij^{*} \\ \\ = T_{pij}^{(n)} + \sum_{q \neq p} \lambda P_{qij} T_{qij}^{n} & pij = pij^{*} \end{cases} \tag{3.3}$$

Where there is more than one possible minimum cost route, one is selected arbitrarily. Flows on alternative minimum cost routes are unchanged, since their social pressure term is zero.

The step length λ at each iteration is chosen in order to minimize the value of the objective function, Z when applied to networks with separable costs, but more general formulations are possible. It should be noted that λ must have units of inverse pressure, and is therefore no longer bounded in the range $(0,1)$ as in Frank-Wolfe, but must lie in the range:

$$0 \leq \lambda \leq 1/\max (P_{pij}) = \lambda_{max}$$

Note that at $\lambda = \lambda_{max}$ the flow on the path(s) with the maximum social pressure will have been reduced to zero. However on all other paths with positive social pressure (and non-zero flow) the flows will be reduced but still non-zero.

The modifications introduced to the descent direction and step length are intended to improve convergence. They increase the contribution of elements with greater cost differences and reduce the effect of flow shifts in routes whose costs are closer to the minimum, inducing a more homogeneous distribution of costs along the different paths.

The ideas described in this section are based on concepts proposed in algorithm D by Smith (1984). One significant difference between this proposition and algorithm D is the fact that through this method we only transfer trips to the minimum cost route. There is no transfer to any other less expensive routes, even if potentially beneficial.

4. PERTURBATION ALGORITHMS

A perturbation algorithm in the context of this paper is defined to be one where the initial solution is obtained from a previous assignment rather than an all-or-nothing assignment as in step 2 of the standard Frank-Wolfe algorithm above. An essential consideration here is that the initial solution be <u>feasible</u> for the problem in hand. In terms of path flows feasibility requires:

1)   that all paths exist
2)   that all path flows are non-negative – condition (2.5)
3)   that the path flows sum to the correct O-D totals – condition (2.6)

Condition 1) may be violated if the network from which the initial solution has been obtained was topologically different (e.g. links have been removed) while 3) will be violated if the trip matrix has been altered. Condition 2) does not generally cause problems.

At the level of link flows those conditions are implicit rather than explicit and herein lies a basic problem with perturbation algorithms when we have link flows but not a corresponding set of path flows. Thus if we change one element in the trip matrix from T_{ij} to S_{ij} it is easy to produce a new set of feasible path flows, for example by factoring each T_{pij} by the ratio S_{ij}/T_{ij} and the new link flows may be obtained by summation (2.7) as per normal. However without path information we have no way of knowing to which links the extra trips should be assigned and therefore no way to obtain feasible link flows.

The problem is less acute if $S_{ij} > T_{ij}$, in which case we could do an incremental all-or-nothing assignment of $(S_{ij} - T_{ij})$, but if $S_{ij} < T_{ij}$ this may violate the non-negativity restriction on path flows. It is also possible to correctly adjust link flows if the new matrix is a uniform factor of the old matrix, $S_{ij} = \alpha T_{ij}$, in which case the link flows may be factored by α as well.

The path-based perturbation algorithms described in this paper use the following rules to update previously assigned path flows for each ij pair:

1. If all paths for which $T_{pij} > 0$ still exist then factor by S_{ij}/T_{ij}
2. If none of the used paths exist then carry out an all-or-nothing assignment based on free flow costs and set $T_{pij*} = S_{ij}$
3. If some are valid and some are not then set the path flows for the invalid paths to zero and divide those trips (plus any new trips) pro rata amongst the valid paths.

5. RESULTS

The Social Pressure Algorithm and perturbation processes have been tested on five networks of varying size and complexities based on UK cities – Cambridge, York, Chester and two networks cordoned off from a full Leeds network representing Headingley and Leeds North-West. Relevant size statistics are given in Table 1; "K Bytes" gives the RAM required by a program running these networks using the Social Pressure path-based algorithm described above.

Table 1. Network Dimensions

	Zones	Nodes	Links	No. of Tij>0	K Bytes
York	176	1246	2329	27574	7140
Cambridge	141	1690	2795	7575	3509
Chester	132	1356	2104	3066	1071
Leeds North West	70	913	1406	3222	836
Headingley	29	73	188	252	20

The first series of tests compares the relative convergence behaviour of the Social Pressure and Frank Wolfe algorithms in the five test networks, comparing both the number of iterations and the amount of cpu time required to reach pre-specified target levels of convergence. Note that all Social Pressure-based algorithms were coded using path flows whilst the Frank Wolfe algorithm used link flows only.

Figure 1 compares the convergence rate of three algorithms for the York network in terms of the "uncertainty in Z" (technically the relative difference between Z and a lower bound) as a function of the number of iterations. It can be seen that the Social Pressure produces lower uncertainties than Frank-Wolfe over the whole range of iterations and in particular converges much faster as it nears convergence. The perturbation algorithm plotted here (and described in more detail below) starts with flows from a previous solution to a slightly different network and therefore has a distinct "headstart" over the other two which started "from scratch".

However the number of iterations is not necessarily the best criterion to use; cpu time may be a more realistic basis of comparison since one iteration of a path-based algorithm (Social Pressure) is likely to take much longer than one iteration of a link-based algorithm (Frank Wolfe). Table 2 lists both the number of iterations and the cpu time to reach 1% and 0.2% uncertainty values. As already demonstrated in Figure 1 Social Pressure requires fewer iterations, particularly for 0.2% convergence. However Frank-Wolfe requires less cpu times in all but one case.

Table 2. Cpu Time/Iterations to achieve 1% and 0.2% Uncertainty

Network	1% Uncertainty		0.2% Uncertainty	
	S.P.	**F.W.**	**S.P.**	**F.W.**
York	74.4 / 32	34.3 / 69	230 / 102	127 / 263
Cambridge	13.3 / 12	8.3 / 17	52.0 / 45	31.8 / 69
Chester	7.4 / 15	4.9 / 17	22.9 / 47	18.8 / 70
Leeds NW	16.4 / 58	8.6 / 78	33.0 / 116	32.0 / 306
Headingley	0.3 / 9	0.3 / 20	0.6 / 23	2.1 / 135

Perturbation tests were created by modifying the properties of a single link in each of the five test networks, generally a reduction in capacity. For example in the Cambridge network we removed one lane from a bypass link, the end effect being to add roughly one extra minute of travel time on that link once the re-assignment had been carried out. The total benefit (or strictly disbenefit) of this lane closure may be calculated as the difference in total generalised cost between the "before" and "after" networks. Table 3 lists the cost changes per network in percentage terms (ΔC).

In the non-perturbation mode the change in total costs can be calculated by assigning both networks from an initial all-or-nothing assignment for the same total number of iterations and subtracting one total cost from the other. Perturbation starts the "after" test from the end of the "before" as explained in Section 4.

Figure 2 plots the total costs in the before and after Leeds North-West networks as a function of number of iterations; the "correct" costs (obtained after a very large number of iterations) are indicated. The level of noise as well as the slow rate of convergence are obvious. Note that at several points the "sign" of the cost changes would be incorrectly calculated since the before costs actually exceed the after costs.

Table 3. Cpu time and number of iterations to achieve differences in the total costs ΔC to within a specified accuracy of the "correct" values

Network	ΔC (%)	Accuracy	Frank-Wolfe	Social Pressure	SP Perturbation
York	-0.325	25%	>400	312.9/136	39.8/20
Cambridge	0.173	10%	82.8/183	166.4/141	25.7/21
Chester	0.0927	25%	>200	>200	78.2/161
Leeds NW	0.125	25%	>200	>200	13.0/52
Headingley	7.68	10%	0.55/30	0.33/14	0.01/1

Figure 3 plots the difference in costs directly as a function of number of iterations with the "correct" value indicated. Both the Frank Wolfe and the "pure" Social Pressure (i.e. where both networks start from free flow) exhibit large amounts of noise but are confined within converging envelopes towards the correct result. However the solid line shows the result of running Social Pressure in a "perturbation" mode where:

a)   Social Pressure is run to convergence on the "before" network, and then;

b)   It is run on the "after" network but starting with the "before" link flows.

In this case the perturbation algorithm starts with a cost difference of approximately 2/3 the correct value which then converges relatively smoothly towards the correct final value.

We may "quantify" the convergence of the difference in costs by determining the minimum number of iterations necessary to consistently obtain the correct value to an accuracy of, say, $\pm10\%$. These results, along with the associated cpu times, are given in Table 3. Thus in the Cambridge tests all cost differences <u>after</u> 141 iterations of the Social Pressure Algorithm starting from scratch were within $\pm10\%$ (as were some before 141 but not all). In three of the five networks it was not possible to guarantee consistent cost differences with either of the non-perturbation methods even with the accuracy relaxed to 25% and performing over 200 iterations. This simply demonstrates the difficulty in separating out genuine changes from the background noise when the changes are very small, e.g. of the order of $\pm 0.1\%$ in terms of total vehicle costs.

The advantages of the Perturbation plus Social Pressure approach are clear. It was able to obtain consistent results even when the standard methods failed. When the standard methods did converge within the required limits the pertubation approach gave substantial reductions in both the number of iterations <u>and</u> cpu time.

The reasons for this improvement are two fold. First, the perturbation method has an advantage by starting with a better initial solution than a free-flow all or nothing. Secondly, the Social Pressure terms (3.2) themselves can be expected to be greatest on those O-D paths which use the altered link and therefore experience relatively large increases in their path costs as opposed to those paths which are remote from the changes and would have a pressure near zero since they are already in equilibrium. Thus the pressure term differentiates automatically between path flows that are at or near equilibrium and those that are not and to make the maximum changes to the paths genuinely affected by the scheme.

6. APPLICATIONS OF PERTURBATION ASSIGNMENTS

Perturbations in traffic assignment applications may be broadly sub-divided into (1) changes to the trip matrix, and (2) changes to the network; where the network changes may be further sub-divided into those that leave the network topology unaltered (but change its properties) and those that change the topology.

Examples of small changes to the trip matrix include linking the assignment with various forms of demand models such as combined assignment and modal split (or, more generally, with an ij-specific elastic demand model, assignment plus distribution, assignment plus departure time choice or assignment plus trip matrix estimation from counts (ME2). In all these cases the assignment model and the "demand" model are run iteratively so that at each assignment iteration the trip matrix to be assigned is not that different from the previous trip matrix, in which case a perturbation approach will considerably speed up all assignments after the first.

A non-iterative example of small trip matrix changes would be a study of new trip generations (e.g. a housing estate or shopping centre) where the new trips are added incrementally to an existing trip matrix.

Small network changes, e.g. add or remove a link, redesign a junction, etc., are of course "bread and butter" jobs for traffic engineers which could benefit from perturbation techniques. A further example of changes to network properties with fixed topology is the combined signal optimisation/network assignment problem which again is generally solved iteratively with small changes to the signal settings leading to small changes in successive assignments.

7. DISCUSSION

These tests have demonstrated that an algorithm which is (a) based on path flows and (b) uses a Social Pressure rule for determining descent directions achieves a more stable and more rapid convergence for small perturbations to the network and/or trip matrix than traditional Frank-Wolfe algorithms. There are a large number of practical situations in which these conditions arise as discussed in Section 6.

There are also a number of other important practical benefits from using path-based assignment algorithms (with or without Social Pressure). They allow greater flexibility in post-assignment analysis (e.g. select link analysis) and the possibility of reduced cpu times. They also simplify the transfer of data to micro-simulation models such as DRACULA whose basic inputs are path flows or to models such as ME2 which estimate trip matrices from traffic counts.

One disadvantage of path-based algorithms is that they require greater internal computer memory (RAM), particularly for networks where "most" T_{ij} cells are non-zero as tends to be the case with synthesised trip matrices although not with observed matrices. Required values for our test networks are given in Table 1. The maximum required, 7 Mbytes, is not small but well within the limits of a "standard" new pc. Given the rapid increase in memory limits, this is not felt to be a serious detriment to running larger networks.

ACKNOWLEDGEMENTS

It is a pleasure to be able to acknowledge many fruitful discussions over the years with – and ideas pinched from! - Mike Smith, Helmut Schittenhelm, Michael Patriksson, Helena Cybis, Judith Wang and Sameiro Carvalho. This research was funded by a grant from EPSRC.

REFERENCES

Larsson, T. and Patriksson, M. (1992). Simplicial Decomposition with Disagregated Representation for the Traffic Assignment Problem. *Transportation Science*, **26**(1), 4-17

Schittenhelm, H. (1990). On the integration of an effective assignment algorithm with path and path flow management in a combined trip distribution and assignment algorithm. *18th PTRC Summer Annual Meeting*, Seminar H.

Smith, M.J. (1984). A Descent Algorithm for Solving a Variety of Monotone Equlibrium Problems. *Proc. Ninth International Symposium on Transportation and Traffic Theory*, Utrecht, 273-297

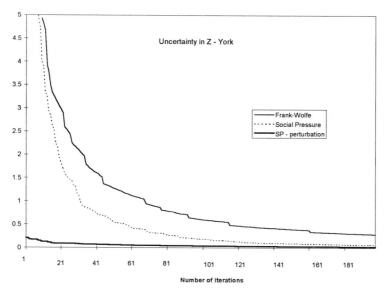

Figure 1. Uncertainty in the objective function. York network.

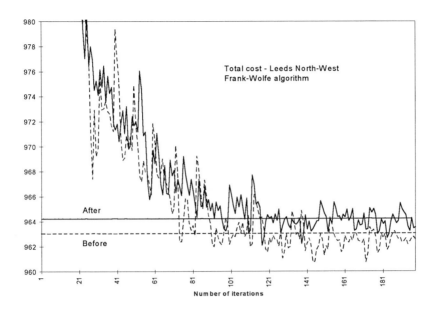

Figure 2. Total cost for the "before" and "after" networks using Frank-Wolfe, and the "correct" values. Leeds North-West.

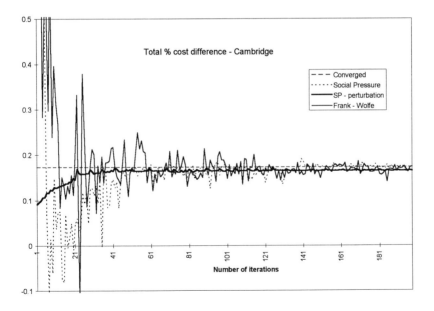

Figure 3. Total percentage cost difference between the "before" and "after" Cambridge networks.

New Methods For Solving The SUE Assignment Problem

P.C. Hughes and M.J. Maher, Napier University, Edinburgh

Abstract

The traffic assignment problem aims to predict driver route choice, and is typically applied in the assessment of road schemes. The authors have previously published an SUE (Stochastic User Equilibrium) assignment algorithm, i.e. one which models variation in driver perception, and cost variation due to congestion. The algorithm works by minimising a function given by Sheffi and Powell (1982); in this paper the three terms of the function are investigated separately, and the possibility explored of constructing more sophisticated versions of the SUE algorithm.

1. Introduction

There have traditionally been two classes of assignment which aim to model the phenomenon of multi-routeing. The first class consists of models which assume "perfect" network knowledge, but incorporate the effect of congestion (**User Equilibrium** or UE models). The second class consists of probabilistic or **stochastic** methods, which model the variations in driver perceptions or preferences. Models incorporating both congestion and driver variation are called stochastic user equilibrium (SUE) models; by definition they should be able to model more diverse situations than either stochastic or UE models on their own. The authors have developed a robust SUE model, described in several papers (e.g. Maher & Hughes (1997a and 1997b)); a stochastic method is combined with cost-flow functions. The stochastic loading method can be either the logit model, or a probit model (SAM) first described in Maher (1992), which uses Clark's approximation to predict routes for a whole network without path enumeration.

The SUE model works by moving an amount (the "step size") from the current flow pattern towards the auxiliary flow pattern, the auxiliary being found by a stochastic loading. The value of the step size has a large effect on the efficiency of the algorithm. The Method of Successive Averages (MSA) uses predetermined step sizes and is known to converge, but does so very slowly. Our model is based on the optimisation of a three-term objective function due to Sheffi and Powell (1982); the current algorithm relies on evaluating the function at least twice for each iteration. As the MSA requires only one evaluation, each iteration of the optimisation method has a cost in cpu time roughly double that of MSA. However, the gain in efficiency outweighs this extra cost. Clearly, a method which could calculate an optimised step length using only one stochastic loading per iteration would capture the best features of both methods. It would estimate the second derivative of the cost function with respect to flows, and use the Newton-Raphson formula to give an optimised step length. This paper investigates the possibility of estimating the second derivative and constructs such a method, comparing it with the available alternatives.

2.   SUE ALGORITHMS

In this section a brief description is given of the SUE algorithms already available. The stochastic model is either the logit method based on the STOCH algorithm of Dial (1971), with the addition of a more thorough treatment of networks containing loops, or the SAM probit method described first in Maher (1992). The latter scans through the network in a similar way to the STOCH algorithm, using Clark's approximation (1961) to calculate the cost distributions from the origin to intermediate nodes in the network and the splits of traffic entering nodes by different links. For both the logit and probit models, we can find the expected perceived minimum travel cost between each O-D pair r-s (S_{rs}, the so-called "satisfaction" function), as part of the loading process. These stochastic algorithms can now be extended to the SUE case by introducing capacity restraint, using separable link cost-flow functions. If x_a is the flow on link a, and $c_a(x_a)$ is the cost of travel along that link depending only on the flow x_a, the flow vector x gives rise to a cost vector $c(x)$. Carrying out a stochastic loading based on these costs will give an auxiliary flow pattern y, which in general will not equal x. Only when costs and flows are consistent will x will equal y; this is the SUE solution. The condition for SUE is thus:

$$x_a = y_a \ \forall a \qquad (1)$$

A mathematical program for the fixed demand SUE problem is given in Sheffi and Powell (1982); it is necessary to minimise the following function of x:

$$z = \sum_a x_a c_a(x_a) - \sum_a \int_0^{x_a} c_a(x)dx - \sum_{rs} q_{rs} S_{rs}(x) \qquad (2)$$

The fact that the satisfaction values S_{rs} are readily available from the stochastic loading process can thus be used to calculate z, and hence to construct an optimisation method, similar to the Frank-Wolfe (1956) method for UE. The partial derivatives of z with respect to x_a are:

$$\frac{\partial z}{\partial x_a} = (x_a - y_a)\frac{dc_a}{dx_a} \qquad (3)$$

and will be zero at the SUE solution. At iteration n we carry out a line search between the current flow vector $x^{(n)}$ and its auxiliary $y^{(n)}$. The aim is find the value of the step length $\lambda^{(n)}$ which minimises

z, or equivalently for the which the gradient is zero:

$$g = \frac{\partial z}{\partial \lambda} = -\sum_a (y_a^{(n)} - x_a^{(n)}) \frac{dc_a}{dx_a} (y_a^{(\lambda)} - x_a^{(\lambda)}) \tag{4}$$

By carrying out stochastic loadings at x ($\lambda = 0$) and y ($\lambda = 1$) we can calculate g_0 and g_1, the values of the gradient along the search direction at these two points. The value λ at which g is zero can then be estimated by linear interpolation: $\lambda = -g_0 / (g_0 + g_1)$. Note here that, if we knew the second derivative of z with respect to λ, we could estimate λ using the Newton-Raphson formula: $\lambda = -g_0 / \left(\frac{\partial^2 z}{\partial \lambda^2} \right)$, potentially using only one stochastic loading. This step length is the value $\lambda^{(n)}$ which is used to update the current solution, before proceeding to the next iteration:

$$x^{(n+1)} = x^{(n)} + \lambda^{(n)}(y^{(n)} - x^{(n)}) \tag{5}$$

We define a convergence statistic R to measure the "distance" between current and auxiliary:

$$R = \sqrt{\frac{1}{N} \sum_a \left(\frac{x_a - y_a}{\frac{1}{2}(x_a + y_a)} \right)^2} \tag{6}$$

where N is the number of links. In order to examine convergence near the solution, another useful measure is defined by taking the log of the ratio of the current value of R with the value R^0 at the first iteration; this measure will tend to minus infinity:

$$J = \log\left(\frac{R}{R^0}\right) \tag{7}$$

The method can be compared with the well-known alternative of the Method of Successive Averages (MSA), where $\lambda^{(n)}$ is set using a predetermined formula (typically $(n + 1)^{-1}$). The MSA only uses one stochastic loading per iteration, and often performs well in early iterations. However as the step lengths continuously decrease, it converges very slowly, and is overtaken by the 2-loading method described above. An ideal method would be able to choose an efficient step length, using the Sheffi and Powell objective function, but only needing one stochastic loading per iteration. The following sections explore this possibility.

3.   THE THREE TERMS

We now consider the three terms of the objective function (3) in a little more detail (equation 8). They each have an interpretation in terms of network costs. The first is the sum of all (deterministic) costs, and is minimised when the System Optimum (SO) solution is reached. The second is the objective function minimised at UE (Beckmann *et al* (1956)). The third is the stochastic term, the sum of all *perceived* minimum travel times in the network. This term can be evaluated only by carrying out a stochastic loading, thus incurring a cost in cpu.:

$$z_1 = \sum_a x_a c_a(x_a)$$

$$z_2 = \sum_a \int_0^{x_a} c_a(x) dx \qquad\qquad (8)$$

$$z_3 = \sum_{rs} q_{rs} S_{rs}(x)$$

The derivatives of the three terms are given below; Sheffi (1985) shows the result for the third term. Summing (9) with the correct sign will give equation (4):

$$\frac{\partial z_1}{\partial \lambda} = \sum_a (y_a^{(n)} - x_a^{(n)})(x_a^{(\lambda)} \frac{dc_a}{dx_a} + c_a(x_a^{(\lambda)}))$$

$$\frac{\partial z_2}{\partial \lambda} = \sum_a (y_a^{(n)} - x_a^{(n)}) c_a(x_a^{(\lambda)}) \qquad\qquad (9)$$

$$\frac{\partial z_3}{\partial \lambda} = \sum_a (y_a^{(n)} - x_a^{(n)}) \frac{dc_a}{dx_a} y_a^{(\lambda)}$$

The derivatives of all three terms, and hence of z are given by a closed formula. The second derivatives of the first two terms are also easy to compute, but $\partial^2 z_3 / \partial \lambda^2$ is much more difficult: the authors are not aware of an analytical formula which does not involve complete path enumeration. Note however that it is possible to estimate $\partial^2 z / \partial \lambda^2$ by evaluating the 1st derivative at two points along the step length, using two stochastic loadings. The estimate improves depending on how close together the points are.

4.   COMPARISONS OF FIRST DERIVATIVES

In this section we give some plots of the derivatives of the three terms over typical step lengths for the well-known Sioux Falls network, which has 76 links and 24 nodes. Figure 1 shows the gradients of the three terms over the step length for iteration 1; Figure 2 shows the same data for iteration 5. The most obvious point to note is that all three gradients are much more linear, and smaller in magnitude, for the later iteration. We expect this: the current and auxiliary flow patterns become closer, so a step length represents a smaller "distance" in later iterations. The first graph also shows that the gradient of the third term is not related to the other two by a simple formula, and not in a way that carries through from iteration to iteration. On the other hand, the gradients of the first two terms do vary in a similar way.

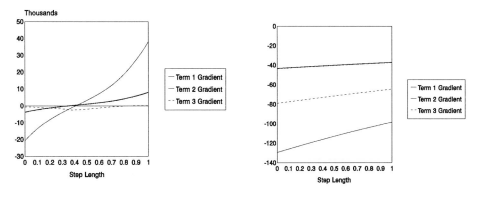

Figure 1: Gradients, Iteration 1

Figure 2: Gradients, Iteration 5

The next series of figures shows some data on second derivatives, again for the Sioux Falls network, as the iterations progress. All the derivatives are now taken at the current flow pattern, pointing towards the auxiliary flow pattern. Figure 3 shows the second derivatives for the SO and UE terms (notated g2(SO) and g2(UE)), together with the estimated second derivative for the whole function (g2(SUE est)), calculated by evaluating the gradient at step lengths of zero and 0.01. Figure 4 shows the ratios of the second derivatives of the SO and UE terms to that of the whole function, while Figure 5 shows the log-ratios. The second derivatives of the SO and UE terms can in fact be easily shown to be proportional, for the BPR cost-flow functions which were used in these tests; this is borne out by the graphs. However, the second derivative of the third term is related in a more complicated way to the others. Figure 5 shows that, as the algorithm nears the solution, the log-ratios decrease linearly; hence there is a relationship between the second derivatives and the iteration number. This gives support for a method which uses information on the second derivatives gained in previous iterations, to estimate the second derivative in the current iteration. Such a method is outlined in the next section.

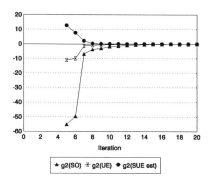

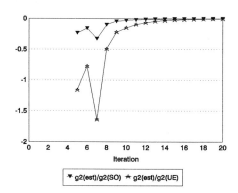

Figure 3: 2nd Derivatives

Figure 4: 2nd Derivative Ratios

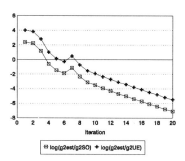

Figure 5: 2nd Derivative Log-ratios

5.   ESTIMATING THE SECOND DERIVATIVE

In this section we give a possible method for calculating the step length which is based on estimating the second derivative, while only using one stochastic loading per iteration. The method is based on the fact that an iteration can give us some information about the gradients in the previous search direction. At any iteration n we carry out a stochastic loading from the costs at $x^{(n)}$ to find the auxiliary $y^{(n)}$; this also gives us the information to calculate the gradient at $x^{(n+1)}$ in the direction towards $y^{(n)}$. We then choose a step length $\lambda^{(n)}$ between $x^{(n)}$ and $y^{(n)}$, to find the new current flows $x^{(n+1)}$. Then at iteration $(n+1)$ we need to carry out a stochastic loading from the costs at $x^{(n+1)}$, to find the auxiliary $y^{(n+1)}$. This new stochastic loading means that we have the information to calculate easily the gradient at $x^{(n+1)}$ in the new search direction, but also, if we choose, in the old search direction between $x^{(n)}$ and $y^{(n)}$. Thus we have two gradient evaluations in the search direction for iteration n, and we can use these to give an estimate of the second derivative. This in turn can be used as an estimate of the second derivative for the next iteration. We now summarise the above procedure:

1. Calculate costs $c(x^{(n)})$ for the current flows $x^{(n)}$. Carry out a stochastic loading, to find the auxiliary flows $y^{(n)}$, and calculate the gradient at $x^{(n)}$ in the direction $x^{(n)} - y^{(n)}$.
2. Calculate the gradient at $x^{(n)}$ in the direction $x^{(n-1)} - y^{(n-1)}$. We already have the gradient at $x^{(n-1)}$ in the direction $x^{(n-1)} - y^{(n-1)}$, so we can estimate the second derivative.
3. Choose the step length $\lambda^{(n)}$ between $x^{(n)}$ and $y^{(n)}$ by using the Newton-Raphson formula, and the estimate of the second derivative.
4. If not converged, update n and go to 1.

The estimate of the second derivative should be better for later iterations; therefore a method is proposed whereby MSA is used for an initial series of iterations, after which the second derivative is estimated from the previous iteration, and the Newton-Raphson step length calculated.

6. ALTERNATIVE SUE ALGORITHMS

We now list a number of alternative SUE algorithms for comparison, and give some test results. Firstly, we have two "benchmarks", one which we certainly hope to improve on, the other being the best we could hope to achieve. The MSA is reliable but slow, and we have shown in previous papers that it can be improved on. On the other hand, if we postulate a method that was able to calculate the second derivative at the cost of only one loading per iteration, and thus find the Newton-Raphson step length, that should be the very best performance we could expect. In the tests which follow, we show results for this idealised method by estimating the second derivative accurately at each iteration (evaluating the gradient at step lengths of zero and 0.01), but basing the comparisons on one loading used per iteration. A third method which is a kind of benchmark in simple-mindedness consists of taking a constant step length of 0.5. Clearly, there is no formal reason why this method should converge, and for the UE problem it certainly would not converge, as the auxiliary flow pattern is always an extreme point, and the step lengths must decrease as the algorithm reaches the solution. However, in the SUE case, where the current and auxiliary flow patterns become equal at the solution, there is no reason for the step length to be small, so it interesting to see whether the choice of step length has a large effect, or whether in practice an arbitrary value could be taken. We also have, of course, the SUE method which was described in previous papers and uses two loadings per iteration to estimate where the gradient where will be zero. Finally, any of the non-MSA methods can be used with a preliminary series of MSA iterations, MSA being efficient early on.

In Figure 6 we show comparisons of these methods for the Sioux Falls network, using the J statistic defined earlier; the faster that J tends to minus infinity, the better the convergence. As expected,

MSA does poorly after the first series of iterations, while "free" 2nd derivative information gives the best performance. The two-loading method does much better when combined with 15 iterations of MSA than on its own; of the other alternatives, there is little to choose - both the estimation method and the constant step length converge for this network.

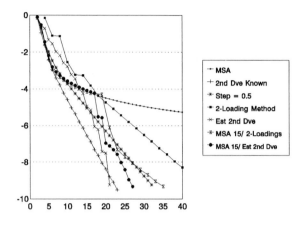

Figure 6: Convergence for Sioux Falls network

7. THE HEADINGLEY NETWORK

The authors are indebted to Dirck Van Vliet and David Watling of the Institute of Transport Studies, Leeds University, for the use of data for the Headingley network; it has 188 one-way links, 73 nodes, and 29 origins. As we will see, it proved to be a more searching test of the algorithms than the Sioux Falls network. In Figures 7-9 the convergence measure J is again shown for the different algorithms. Figures 7 and 8 show the MSA compared with the estimation method (Figure 7) and with the constant step length (Figure 8); both are unstable, even when moderated by using MSA for the first 15 iterations. The estimation method actually reached the convergence criterion when combined with 15 iterations of MSA, and hence stopped, but the graph gives little confidence in the robustness of the method. Figure 9 compares the more stable alternatives: we see that the 2-loading method eventually does better than MSA, but does much better when combined with 15 iterations of MSA, getting quite close to the baseline comparison, where the 2nd derivative is known.

We conclude from these graphs that the two-loading method is still likely to be the best alternative, being more robust than the 2nd derivative method, and efficient, especially when combined with an initial series of MSA iterations. The simple-minded approach of a constant step length was also unstable, vindicating the effort in developing more sophisticated algorithms.

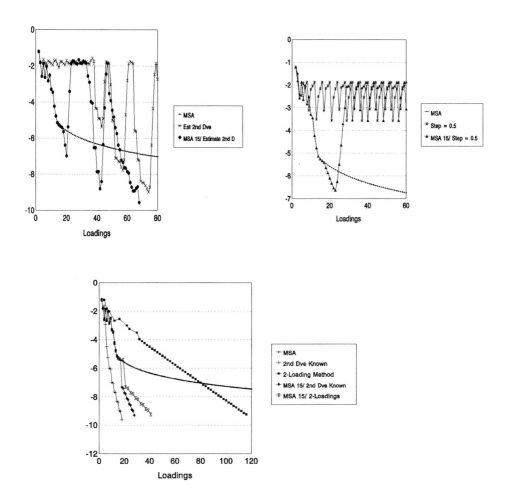

Figures 7-9: Convergence for Headingley network

8. SUMMARY AND CONCLUSIONS

This paper has investigated possible alternatives to the SUE algorithm previously published, while coming to the eventual conclusion that this method still represents the best alternative. After an introduction, the objective function was described, and the relative difficulty of evaluating the three terms, and their first and second derivatives, was discussed. The second derivative of the third term is difficult to calculate, and not easily approximated by the other two terms. A possible alternative method was then described where the second derivative of the whole function is estimated by using information from the previous iteration. The following two sections gave test results for two test networks, Sioux Falls and Headingley. While the method of estimating the second derivative appeared promising in the case of Sioux Falls, it gave an unstable performance in the case of Headingley. A simple-minded approach of choosing a constant step length of 0.5 was also shown to be very unstable for the Headingley network. The two-loading method appears to give the best results over the two networks, especially when combined with a series of MSA iterations at the beginning of the process.

9.   REFERENCES

Beckmann M., McGuire C.B. and Winsten C.B. (1956) *Studies in the Economics of Transportation.* Cowles Commission Monograph, Yale University Press, New Haven.

Clark C.E. (1961) The greatest of a finite set of random variables. *Operations Research* **9**, 145-162.

Dial R.B. (1971) A probabilistic multipath traffic assignment model which obviates path enumeration. *Transportation Research* **5**, 83-111.

Frank M. and Wolfe P. (1956) An algorithm for quadratic programming. *Naval Research Logistics Quarterly* **3**(1-2), 95-110.

Maher M.J. (1992) SAM - A stochastic assignment model. In: *Mathematics in Transport Planning and Control* ed. J.D. Griffiths, Oxford University Press, 121-132.

Maher M.J. and Hughes P.C. (1997a) A probit-based stochastic user equilibrium assignment model. *Transportation Research* B, **31**(4), 341-355.

Maher M.J. and Hughes P.C. (1997b) An Algorithm for SUEED - Stochastic User Equilibrium with Elastic Demand. Presented at the 8th IFAC Symposium on Transportation Systems, Crete.

Sheffi Y. and Powell W.B. (1982) An algorithm for the equilibrium assignment problem with random link times. *Networks* **12**(2), 191-207.

Sheffi Y. (1985) *Urban transportation networks: equilibrium analysis with mathematical programming methods*, Prentice-Hall, Englewood Cliffs, New Jersey.

AN ALGORITHM FOR THE SOLUTION OF BI-LEVEL PROGRAMMING PROBLEMS IN TRANSPORT NETWORK ANALYSIS

Xiaoyan Zhang and Mike Maher, Department of Civil and Transportation Engineering, Napier University, Edinburgh, Scotland

ABSTRACT

This paper deals with two problems in transport network planning and control: trip matrix estimation and traffic signal optimisation. These two problems have both been formulated as bi-level programming problems with the User Equilibrium assignment as the second-level programming problem. One currently used method for solving the two problems consists of alternate optimisation of the two sub-problems until mutually consistent solutions are found. However, this alternate procedure does not converge to the solution of the bi-level programming problem. In this paper, a new algorithm will be developed and will be applied to two road networks.

1. INTRODUCTION

In this paper, we deal with two mathematically similar problems in transport network analysis: trip matrix estimation and traffic signal optimisation. These two problems are of great importance in transport planning, scheme appraisal and traffic management. Both problems have been formulated as bi-level programming problems in which the upper-level problem is matrix estimation (ME) or signal optimisation (SO), and the lower-level problem a User Equilibrium (UE) assignment model. The UE assignment has been included so as to model congestion effects in the network.

The bi-level programming approach has been used in several transport system analysis problems (Fisk, 1984), including the two combined problems mentioned here. In game theory, a bi-level programming problem is known as a Stackelberg game or leader-follower game, in which the leader chooses his variables so as to optimise his objective function, taking into account the response of the follower who tries to optimise his objective function according to the leader's decisions. Let $\mathbf{x} \in D_x \subset R^m$ be the vector of leader's decision variables, $\mathbf{y} \in D_y \subset R^n$ be the vector of follower's decision variables, and $\mathbf{M}(\mathbf{x})$ be the follower's response for a given $\mathbf{x}$. Then a bi-level programming problem can be written as

$$\min_{\mathbf{x} \in D_x} Z(\mathbf{x}, \mathbf{M}(\mathbf{x}))$$

or alternatively,

$$\min_{\mathbf{x} \in D_x} Z(\mathbf{x}, \mathbf{y})$$

where $\mathbf{y}$ is obtained by solving

$$\min_{\mathbf{y} \in D_y} z(\mathbf{x}, \mathbf{y})$$

An optimal solution, $[\mathbf{x}^* \ \mathbf{y}^*]$, where $\mathbf{y}^* = \mathbf{M}(\mathbf{x}^*)$, of the bi-level problem must satisfy

$$Z(\mathbf{x}^*, \mathbf{M}(\mathbf{x}^*)) \leq Z(\mathbf{x}, \mathbf{M}(\mathbf{x})), \quad \mathbf{x} \in D_x$$
$$z(\mathbf{x}^*, \mathbf{M}(\mathbf{x}^*)) \leq z(\mathbf{x}^*, \mathbf{y}), \quad \mathbf{y} \in D_y$$

In the bi-level problem, the evaluation of the upper-level objective function requires solving the lower-level problem whose functional form is generally unknown. In addition, the lower-level problem acts like a nonlinear constraint so that the bi-level problem is in general a nonconvex problem. It is clear that any solution algorithm that requires excessive objective function evaluations (such as in line searching) or analytical expression of gradients of constraints (such as in direction searching) may involve considerable computational effort and may become impractical to apply to general network problems. What is more, nonconvexity in the bi-level problem means the potential existence of local minimum solutions and so a global minimum may be difficult to find.

The techniques currently used in commercial software such as the SATURN suite (for the ME/UE problem) and INTRESS (linking the TRANSYT signal optimisation program and the CONTRAM assignment model for solving the SO/UE problem) includes alternate optimisation of the two sub-problems. The process is continued until convergence is achieved (Hall *et al.*, 1980; Fisk, 1984, 1988). However, it has been indicated (Fisk, 1984) that this procedure does not solve the leader-follower game, but rather may converge to the solution of Nash noncooperative games, in which each of the two players tries to minimise his own objective function only without considering the reaction of the other player. Some more formally-based methods have also been proposed for solving the two bi-level problems (Heydecker and Khoo, 1990; Sheffi and Powell, 1983; Yang, 1995; Yang *et al.*, 1995). However, they require either complicated procedures to evaluate the gradients of the objective functions, or repeated evaluation of the objective function, limiting their potential for practical applications.

In this paper, we will develop and test an algorithm that does not need excessive evaluation of the objective function and the derivatives of the objective function. The two problems will be discussed separately in sections 2 and 3, each of which contains the problem formulation, the algorithm and the test result. Descriptions in section 3 are brief because the structure of the problems and the algorithms are very similar to those in section 2. The paper is summarised in the last section.

2. THE TRIP MATRIX ESTIMATION PROBLEM

2.1. The Problem

The bi-level trip matrix estimation problem is one in which the matrix estimator tries to minimise the matrix estimation error while the link flow pattern adjust itself accordingly to a UE. The problem may be written as (Yang *et al.*, 1992)

$$\underset{t}{\text{Min}} \ Z(\mathbf{t}, \mathbf{V}(\mathbf{t})) = (\mathbf{\bar{t}} - \mathbf{t})^T \mathbf{U}^{-1}(\mathbf{\bar{t}} - \mathbf{t}) + (\mathbf{\bar{v}} - \mathbf{V}(\mathbf{t}))^T \mathbf{W}^{-1}(\mathbf{\bar{v}} - \mathbf{V}(\mathbf{t}))$$

subject to $\qquad\qquad\qquad\qquad \mathbf{t} \geq 0$

where

$\mathbf{t}$ is the vector of the trip matrix to be estimated, $\mathbf{t} = (\dots, t_i, \dots)$, $i \in I$;

I is the set of O-D pairs;

$\mathbf{\bar{t}}$ is the vector of the target matrix, $\mathbf{\bar{t}} = (\dots, \bar{t}_i, \dots)$, $i \in I$;

$\mathbf{\bar{v}}$ is the vector of the observed link flows, $\mathbf{\bar{v}} = (\dots, \bar{v}_a, \dots)$, $a \in \bar{A}$;

$\bar{A}$ is the subset of links with observed link flows, $\bar{A} \subseteq A$;

A is the set of links in the network;

$\mathbf{U}$ and $\mathbf{W}$ are the variance-covariance matrices of $\mathbf{\bar{t}}$ and $\mathbf{\bar{v}}$ respectively;

$\mathbf{V}(\mathbf{t})$ is the UE assignment mapping.

We have used the Generalised Least Square (GLS) estimator for matrix estimation, although other estimators may also be used, such as the Entropy Maximisation estimator. Note that the second term of the objective function is defined only for the subset of links with observed flows.

It is not necessary to have all the links in the network observed. The UE assignment mapping, $V(t)$, is the lower-level problem. It is well-known that the UE assignment problem is equivalent to the following optimisation problem

$$\underset{\mathbf{v}}{\text{Min}} \ z(\mathbf{v}, \mathbf{t}) = \sum_{a \in A} \int_0^{v_a} c_a(x) dx \qquad (2.1a)$$

subject to

$$v_a = \sum_{r \in R} f_r \delta_{ar}, \ f_r \geq 0, \ \sum_{r \in R_i} f_r = t_i \qquad (2.1b)$$

where $\mathbf{v}$ is the vector of the link flows to be estimated, $\mathbf{v} = (\dots, v_a, \dots)$, c_a is the cost on link a, f_r is the flow on route r, $\delta_{ar} = 1$ if link a is in route r, and 0 otherwise, R_i is the set of routes for O-D pair i, and R is the set of all routes in the network.

2.2. The Solution Algorithm

Suppose we have a current solution, $[\mathbf{t}^{(n)}, \ \mathbf{V}(\mathbf{t}^{(n)}), \ \mathbf{P}^{(n)}]$. At each iteration, the upper-level problem is firstly solved to get an auxiliary solution of the trip matrix, $\mathbf{t}^*$, assuming $\mathbf{v} = \mathbf{P}^{(n)}\mathbf{t}$. Then, a UE assignment is performed to find the auxiliary solution of UE link flows, $\mathbf{v}^*$, or $\mathbf{V}(\mathbf{t}^*)$ for $\mathbf{t}^*$. Thus, as the trip matrix varies from $\mathbf{t}^n$ to $\mathbf{t}^*$, the link flow pattern varies from $\mathbf{V}(\mathbf{t}^{(n)})$ to $\mathbf{V}(\mathbf{t}^*)$. Assuming that the variations are linear, we can parameterise the two sets of variables by a single parameter β:

$$\mathbf{t}(\beta) = \mathbf{t}^{(n)} + \beta(\mathbf{t}^* - \mathbf{t}^{(n)}) \qquad (2.2a)$$
$$\mathbf{v}(\beta) = \mathbf{v}^{(n)} + \beta(\mathbf{v}^* - \mathbf{v}^{(n)}) \qquad (2.2b)$$

Then an optimal step length β^* can be found by setting the first derivative of $Z(\mathbf{t}(\beta), \mathbf{v}(\beta))$ with respect to β to be zero. The new solution of trip matrix is then given by (2.2a) with β^*. However, $\mathbf{v}(\beta^*)$ obtained by (2.2b) is only an approximation to $\mathbf{V}(\mathbf{t}(\beta^*))$. Therefore, another UE assignment is performed to find the exact new link flows, $\mathbf{V}(\mathbf{t}(\beta^*))$, for $\mathbf{t}(\beta^*)$. The algorithm can be outlined as follows.

Step 0: Initialise $\mathbf{t}^{(0)}$, $\mathbf{v}^{(0)}$, and $\mathbf{P}^{(0)}$; set $n = 0$.
Step 1: Determine $\mathbf{t}^*$ by, for example, a GLS estimator, assuming $\mathbf{v} = \mathbf{P}^{(n)} \mathbf{t}$.
Step 2: Find $\mathbf{V}(\mathbf{t}^*)$ for $\mathbf{t}^*$ by UE assignment.
Step 3: Find β which minimises $Z(\mathbf{t}(\beta), \mathbf{v}(\beta))$.
Step 4: Set $\mathbf{t}^{(n+1)} = \mathbf{t}^{(n)} + \beta(\mathbf{t}^* - \mathbf{t}^{(n)})$.

Step 5: Find $\mathbf{v}^{(n+1)}$ as well as $\mathbf{P}^{(n+1)}$ for $\mathbf{t}^{(n+1)}$ by UE assignment.

Step 6: If the convergence criterion is met, stop; otherwise, set $n:=n+1$ and go to step 1.

At step 0, the initial trip matrix can normally be set to be the target matrix. The initial UE link flows and link use proportions are obtained by assigning the target matrix to the network. At step 1, the solution to the GLS estimator of the trip matrix is given by (Cascetta, 1984)

$$\mathbf{t}^* = (\mathbf{U}^{-1} + \mathbf{P}^{(n)T}\mathbf{W}^{-1}\mathbf{P}^{(n)})(\mathbf{U}^{-1}\bar{\mathbf{t}} + \mathbf{P}^{(n)T}\mathbf{W}^{-1}\bar{\mathbf{v}})$$

At step 2 and 5, the UE link flows can be found by the well-known Frank-Wolfe algorithm. At step 3, an optimal β can be found by the Newton method. The stopping criterion can be based on the maximum change in the elements of the estimated trip matrix at successive iterations:

$$\text{Max}_i\,(|t_i^{(n+1)} - t_i^{(n)}|/\,t_i^{(n)}) \le \varepsilon$$

where ε is the error tolerance.

Because $\mathbf{t}^*$ is found by solving the upper-level problem (Step 1), the upper-level objective function is reduced from the current solution: $Z(\mathbf{t}^*,\mathbf{v}^{(n)}) \le Z(\mathbf{t}^{(n)},\mathbf{v}^{(n)})$. However, the auxiliary solution $[\mathbf{t}^*,\mathbf{v}^*]$ found at Step 2 does not necessarily point to a descent direction from the current solution $[\mathbf{t}^{(n)},\mathbf{v}^{(n)}]$. Therefore, the search for optimal step length is not limited to be positive in Step 3. The task of objective function reduction is left to the line search instead of finding a descent direction. Although in theory it is possible for an auxiliary solution to point at a direction that is perpendicular to the gradient of the objective function, this is not a problem in the implementation of the algorithm because in numerical calculations it is almost impossible for an auxiliary solution and the current solution to be at *exactly* the same contour line of the objective function. Numerical tests with different networks so far have shown that, in the first few iterations, the auxiliary solution does provide a descent direction. In addition, the first few iterations are most "cost effective"; the solution is close to the optimal one after only a few iterations. This is a desirable feature for practical network estimations where high convergence previsions may not be justified due to errors in the target matrix and the observed link flows.

2.3. Test Results

The Sioux Falls network is used for testing the algorithm. The network has 24 nodes, 76 links, and 528 O-D pairs. This network has been widely used for testing (equilibrium) assignment models. Information in the data includes network characteristics (link-node topology and link performance parameters) and a demand trip matrix. The demand matrix is treated as the true matrix. The target matrix and the observed link flows are generated by (Yang *et al.*, 1992)

$$\bar{t}_i = t_i^+(1.0 - C_{vod}\xi_i)$$
$$\bar{v}_a = v_a^+(1.0 - C_{vlk}\zeta_a)$$

where t_i^+ and v_a^+ are the elements of the true matrix and link flows, ξ_i and ζ_a are randomly generated $N(0,1)$ variables, and C_{vod} and C_{vlk} are the coefficient of variations reflecting the random variations of the target matrix and observation errors in link flows respectively. The true link flows are calculated by UE assigning the true matrix to the network. The variance-covariance matrices, **U** and **W**, are assumed to be the identity matrix. The estimation results for C_{vod}=0.2 and C_{vlk}=0.1 by the new algorithm and the alternate algorithm are compared in Figure 1. It can be seen that the new algorithm converges quickly while the alternate algorithm seems to converge first and then approach some other solution. Similar results were found with different values of C_{vod} and C_{vlk}.

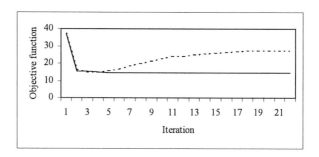

Figure 1. Comparison of the new algorithm and the alternate algorithm for matrix estimation; —— the new algorithm, ---- the alternate algorithm.

3. THE TRAFFIC SIGNAL OPTIMIZATION PROBLEM

3.1. The Problem and the Algorithm

The bi-level signal optimisation/UE assignment problem is one in which a traffic engineer tries to optimise the performance of signals while road users choose their routes in a UE manner. The most commonly used policy for signal optimisation is to minimise the total journey costs in the network. The combined signal optimisation and UE assignment problem has been formulated as the following bi-level programming problem (Fisk, 1984)

$$\operatorname*{Min}_{s} \; Z(\mathbf{s}, \mathbf{V}(\mathbf{s})) = \sum_{a \in A} V_a(\mathbf{s}) \, c_a(V_a(\mathbf{s}), s_a)$$

subject to $s_a^{max} \geq s_a \geq s_a^{min}, \; a \in A; \; \sum_{a \in A_j} s_a = 1, \; A_j \subset A$

Here, s_a is the ratio of green for link a, $\mathbf{s} = (\ldots, s_a, \ldots)$; s_a^{max} and s_a^{min} are maximum and minimum allowable ratio of green for link a, $s_a^{min} > 0$, $s_a^{max} < 1$; A_j is the set of links heading for the jth signal controlled intersection. If link a is not controlled by a signal, then s_a^{max}, s_a, and s_a^{min} will all be equal to 1; $\mathbf{V}(\mathbf{s})$ is the UE assignment mapping for given $\mathbf{s}$. The assignment model has the same form as (2.1a)-(2.1b). Note that the trip matrix is assumed to be fixed in the signal optimisation problem.

The new algorithm proposed here is very much the same as that for trip matrix estimation, with matrix estimation being replaced by signal optimisation with fixed link flows. Given a set of link flows, the problem of signal optimisation is reduced to several sub-problems of determining the optimal ratio of green for each signal controlled intersection. Each of them may be solved by any standard one-dimensional optimisation algorithm, such as the Newton method.

3.2. Test Results

A grid network shown in Figure 2 is used for testing the algorithm. The network has 9 nodes and 24 links. There are 4 centroids (nodes 1, 2, 3, and 4) and 4 O-D pairs (1→3, 2→4, 3→1, 4→2). The demand matrix is fixed: [950 800 900 850]. For simplicity at this stage, the BPR (Bureau of Public Roads) link performance function

$$c_a(v_a) = c_a(0)\left[1 + \alpha \left(\frac{v_a}{q_a s_a}\right)^{\gamma}\right]$$

is used with $\alpha = 2.62$ and $\gamma = 5$. The uncongested link costs $[c_a(0)]$ and link capacity $[q_a]$ are listed in Table 1. There is one node (node 9) that is controlled by traffic signal. The signal optimisation problem in this example reduces to that of a one-dimensional optimisation problem. The optimal solution of the problem can thus be found by direct search and was found to be [0.357 0.643]. The objective function value at the solution is 280836.4.

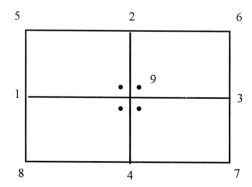

Figure 2. The test network. All links are two-directional.

Table 1. Uncongested link travel costs and link capacities on the testing network.

Link No.	Link nodes	$c_a(0)$	q_a	Link No.	Link nodes	$c_a(0)$	q_a
1	1→5	10	800	13	5→1	10	800
2	1→8	10	800	14	5→2	15	800
3*	1→9	15	500	15	6→2	15	800
4	2→5	15	800	16	6→3	10	800
5	2→6	15	800	17	7→3	10	800
6*	2→9	10	500	18	7→4	15	800
7	3→6	10	800	19	8→1	10	800
8	3→7	10	800	20	8→4	15	800
9*	3→9	15	500	21	9→1	15	500
10	4→7	15	800	22	9→2	10	500
11	4→8	15	800	23	9→3	15	500
12*	4→9	10	500	24	9→4	10	500

Note: *links controlled by traffic signal

Figure 3 shows the comparison of the new algorithm and the conventional alternate procedure. The new algorithm converges to the optimal solution in just a few iterations from different initial conditions. On the other hand, the alternate algorithm converges much more slowly. Also, not surprisingly, it converges to a different solution. Further iterations for the alternate algorithm shown that the solution is [0.224 0.776] and the objective function value at the solution is 282782.3. This is the solution to the Nash noncooperative game, as has been mentioned in the introduction. "Noncooperation" here means that the traffic engineer does not know or consider the route choice behaviour of drivers. Consequently, the total delay in the network of the noncooperative solution is larger than that of the leader-follower solution. The difference between the two solutions is not much in this particular example because there is only one intersection that is controlled by traffic signal. The problem may be more serious if there are more nodes or intersections in the network controlled by traffic signals.

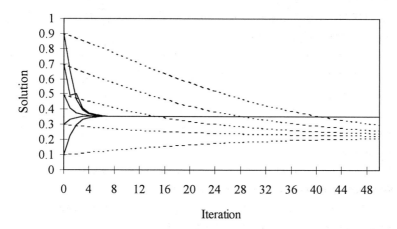

Figure 3. Comparison of the new algorithm and the alternate algorithm for signal optimisation; —— the new algorithm, ---- the alternate algorithm.

4. CONCLUSIONS

The problems of trip matrix estimation and traffic signal optimisation on congested networks have been formulated as bi-level programming problems in the literature. However, the conventional alternate procedures for solving the two problems do not converge to the solution of the bi-level programming problems. In this paper, a new algorithm for solving the two types of bi-level problems has been proposed. In tests on some small problems, it has been shown that the algorithm converges quickly to the optimal solutions. The algorithms proposed here involves only standard routines of trip matrix estimation, UE assignment, and one-dimensional optimisations. This means that the algorithm may be readily programmed, making use of existing programs for the alternate procedures for matrix estimations and signal optimisations.

In signal optimisation algorithm, the BPR link performance function has been used for simplification. This may not be a reasonable function for links controlled by traffic signals. A more realistic function should consider delay caused by the signal, such as the Webster's delay formula. With the Webster's formula, however, the delay tends to infinite as flow tends to the capacity. The Frank-Wolfe algorithm for UE assignment does not treat capacity constraint explicitly. Therefore some modifications are necessary to adapt the algorithm to the delay formula. It is intended to carry out further investigations in this direction.

REFERENCES

Cascetta, E. (1984). Estimation of trip matrices from traffic counts and survey data: a generalised least squares estimator. *Transportation Research*, **18B** (4/5), 289-299.

Fisk, C. S. (1984). Game theory and transportation systems modelling. *Transportation Research*, **18B** (4/5), 301-313.

Hall, M. D., D. Van Vliet, and L. G. Willumsen (1980). SATURN: A simulation assignment model for the evaluation of traffic management schemes. *Traffic Engineering and Control*, **21** (4), 168-176.

Heydecker, B. G. and T. K. Khoo (1990). The equilibrium network design problem. In: *Proceedings of AIRO'90 Conference on Models and Methods for Decision Support*, Sorrento, pp 587-602.

Fisk, C. S. (1988). On combining maximum entropy trip matrix estimation with user optimal assignment. *Transportation Research*, **22B** (1), 69-79.

Sheffi, Y. and W. B. Powell (1983). Optimal signal setting over transportation networks. *Transportation Engineering*, **109** (6), 824-839.

Yang, H. (1995). Heuristic algorithms for the bi-level origin-destination matrix estimation problem. *Transportation Research*, **29B** (3), 1-12.

Yang, H., T. Sasaki, Y. Iida, and Y. Asakura (1992). Estimation of origin-destination matrices from link traffic counts on congested networks. *Transportation Research*, **26B** (6), 417-434.

Yang H. and S. Yagar (1995). Traffic assignment and signal control in saturated road networks. *Transportation Research*, **29A** (2), 125-139.

BACKGROUND OF A TOOL FOR OPTIMISING RESOURCES ON A TRANSPORTATION NETWORK

Karl Jansen, CECIL, Göteborg, Sweden

ABSTRACT

The paper describes the conceptual and mathematical background of a tool for modelling and optimising transportation networks. The tool has been created to model resources and cargo flows in port terminal networks. The concept behind the tool is characterised by three features. Firstly the nodes are non-stationary and can be adapted to accommodate changed requirements in the transportation system. Secondly a high degree of complexity is built into the operational modes of the nodes. Thirdly, the concept incorporates an information network interacting closely with nodes and links. All networks can be optimised employing powerful algorithms and tools from system theory, neural networks and combinatorial graph theory.

1. INTRODUCTION

The network credo "Everything can be modelled as a network" plays a central role in our work. It is with this background the network concept NeuComb and more specifically the tool NeuComb/Port has been developed. A cargo terminal, *e. g.* a port terminal, can be described in several ways but very few of these descriptions are made from a network perspective (Ojala, 1992). Valuable insight into how a cargo terminal works is given by a network perspective on transport. A network approach also makes the concept and model presented fairly easy to understand. Networks are in fact a common metaphor for many activities in modern society (Casti, 1995).

The NeuComb concept and the tool NeuComb/Port, for modelling port terminals, was originally developed for use in the EU project EUROBORDER[1]. NeuComb/Port has been tested and utilised within the project. In this paper mathematical and conceptual backgrounds to the tool are presented in a generic way and are valid for different kinds of terminals. The concept aims at solving problems where nodes and hence also links may be changed and transformed in a dynamic fashion. Basically the concept works with flows on graphs, or specifically networks, where the nodes are characterised by complex node functionalities. Any terminal or enterprise functioning analogously to a port terminal can be modelled using the NeuComb concept. The tool, however, currently only exists for port terminals.

---

[1] EUROBORDER is a project supported by the Commission of the European Communities - DGVII

NeuComb combines two strong mathematical theories combinatorial graph theory and neural networks. Combinatorial graph theory supplies the concept of capacitated links along with strategies for describing and optimising these (Chen, 1990 and Jungnickel, 1994). We have, however, found that the combinatorial graph description of nodes just as junctions between links is insufficient. The nodes themselves have to be given more properties, than just being junctions. In neural network theory on the other hand only the nodes have properties and the links are just connections between nodes. It turns out that a combination of these two theories gives a strong and interesting new concept. It should be noted that the learning storage features of neural networks are not used in NeuComb.

With the capacitated links and complex nodes the framework or an infrastructure for the modelling is created. Onto the infrastructure cargo flows and movements of transportation resources are added. Our aim at CECIL has been and still is to develop strong mathematical concepts and to combine existing ideas so they can be used in modelling transportation. A system background of NeuComb from a port terminal point of view is given in a series of papers by Waidringer and Lumsden (1997a, 1997b and 1998).

2. THEORETICAL FRAME OF REFERENCE

2.1 Infrastructure

The infrastructural network of NeuComb is created using links and nodes. Properties of these elements are defined and described below.

Definition 1. Let G(V,E) be a graph with a set of nodes, V, and a set of links, E={(i,j) i,j ∈V×V}. A function, w, attaching a weight w(e) ∈R to each link e ∈(V∪E) of the graph G(V,E) is introduced. The couple (G,w) is called a network. w(e) can be any relevant quantity associated with the network, e. g the capacity of the link or the time it takes to traverse the link.

1st is standard in combinatorial graph theory (Chen, 1990 and Jungnickel, 1994).

Definition 2. Let every node, k, in the network have a set of in-channels

$$C(k)_{in} = \left\{\alpha_i\right\}_{i=1}^{N} \qquad (2.1)$$

where N is the number of in-channels and a set of out-channels

$$C(k)_{out} = \left\{\beta_j\right\}_{j=1}^{M} \qquad (2.2)$$

where M is the number of out-channels. An example is given in 1st.

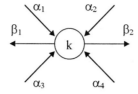

Figure 1. An example node, k, with four in-channels $C(k)_{in} = \{\alpha_i\}_{i=1}^4$

and two out-channels $C(k)_{out} = \{\beta_j\}_{j=1}^2$.

Definition 3. The transmission matrix for a node k is defined as

$$S(k) = \{s_{ij}(k)\}_{1\ 1}^{N\ M} \qquad (2.3)$$

where $s_{ij}(k)=1$ if the channel $a_i \rightarrow b_j$ is open and $s_{ij}(k)=0$ if the channel $a_i \rightarrow b_j$ is closed.

The transmission matrix of 3rd has limited practical value, since only the direct flow from a_i to b_j is considered in a static case. To make the transmission matrix more useful several properties have to be added. Time dependence is one of these properties, since all systems that are modelled with NeuComb are dynamical.

Definition 4. The time dependent transmission matrix for a node k is defined as

$$S(k, t) = \{s_{ij}(k, t)\}_{1\ 1}^{N\ M} \qquad (2.4)$$

where $s_{ij}(k,t)=1$ if the channel $a_i \rightarrow b_j$ is open at time t and $s_{ij}(k,t)=0$ if the channel $a_i \rightarrow b_j$ is closed at time t.

Also the time dependent transmission matrix of 4th has limited practical value, since only the direct flow from a_i to b_j is considered. Several important properties of transmission matrices can, however, be studied using this definition. In a real case the flow may depend on other inputs. These inputs can be either information flow or physical flows. Each of the elements in the transmission matrix has an equation describing it (Andersson *et al.*, 1997). In general the flow trough a node is not the unit flow, but something else. To compensate for this a weight, w, is assigned to the node or some of the channels passing through the node in a fashion much the same as in 1st.

Definition 5. The weighted time dependent transmission matrix for a node k is defined as

$$S(k, t, w) = \left\{ s_{ij}(k, t, w) \right\}_{1\ 1}^{N\,M} \qquad (2.5)$$

where $s_{ij}(k,t,w)=w_{ij}$ if the channel $a_i \rightarrow b_j$ is open at time t and $s_{ij}(k,t,w)=0$ if the channel $a_i \rightarrow b_j$ is closed at time t. w is the weight associated with the node.

2.2. Simplifications in the infrastructure

It is obvious that a system containing complex nodes described above quickly becomes very complex. A general system of this kind is in fact NP-complete (Huckenbeck, 1997). The complexity increases even more when resources and cargo flows are introduced. Given a general system with such high complexity simplifications have to be introduced. Every possible simplification will make the algorithms more efficient but less accurate. Our first simplification is the introduction of layer networks.

Definition 6. Consider a set of nodes arranged as columns, where each column is called a layer. In a general network a node can be connected with all other nodes. However, if the network is layered the nodes can only be connected to nodes that are in the layer before or after the layer which contains the node. An example is given in 2nd. Layered networks are common in the theory of neural networks (e. g. Domany and Meir 1991).

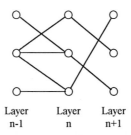

| Layer | Layer | Layer |
| n-1 | n | n+1 |

Figure 2. An example of a layered network.

By layering networks the number of possible links, thus the system complexity, are reduced considerably. Even though this simplification is introduced the system will still be quite complex. Simplifications which are specific for the each application will have to be introduced as well. Knowing *á priori* information about the system make intelligent assumptions possible, but acquiring this information is in many cases hard work.

3. THE NEUCOMB/PORT TOOL

This part of the document concerns the NeuComb/Port tool, which was developed for use in the EU project EUROBODER. We are convinced that the main ideas used in this tool will be helpful in other similar tools.

3.1 Nodes and links in the tool

In the NeuComb/Port tool the infrastructure is made up of nodes and links. The internal nodes used in the tool are of three kinds; action, wait and queue nodes. All types of nodes have a maximum number of cargo units which can reside in the node simultaneously. If the maximum number of cargo units is reached the node is temporarily closed.

Definition 7. In an action node there are c service places where any kind of action can take place. Each of these service places has a service time t_s assigned to them. If the number of cargo units in the node exceeds the maximum number of units that can be served simultaneously the excess units are placed in a queue and are accessed according to the principle first in first out.

The action carried out in an action node can be anything that is found to be relevant. From the tool's point of view, however, all actions that can be carried out are equivalent. Actions only take time no matter what their origin is. In the tool the main parameter is time. It is therefore of interest to calculate the sum of service time and queuing time for a unit entering an action node.

Theorem 1 Consider an action node with c service places, where all service places have the same service time t_s. If there are free service places in the node it is obvious that the sum of service and queuing time for a unit entering the node will be τ. The situation becomes more complicated if all service places are occupied. For the n:th unit entering the queue to the total node time will be:

$$T_n = (k+2)t_s - \tau_\ell \qquad (3.1)$$
$$k = \text{int}(n/c) \qquad (3.2)$$
$$\ell = n - kc \qquad (3.3)$$

where τ_l is the ℓ:th longest time spent in the service places $\ell <= c$.

Definition 8. A wait node is a node where the cargo units are put and accessed in an arbitrary order. In these nodes only the number of cargo units stored have any meaning. If the maximum number of units is reached the node is closed temporarily until there is space available again.

Definition 9. *In a queue node the cargo units are placed and accessed according to the principle first in first out. If the maximum number of units is reached the node is closed temporarily until there is space available again*

The three types of nodes defined above are in the NeuComb tool connected with links. By using specific combinations of nodes and links a terminal is created. Onto this infrastructural network the cargo and resources are added during the running of the tool.

3.2 Parameters in the tool

Times, capacities and costs are all entered into the model as parameters and a specific system depends on these quantities. So before running the model in a specific port environment the relevant parameters have to be entered. The main parameter of the NeuComb tool is time. This means that everything that is possible in the systems is expressed using time. Of course there are other parameters as well but these are subordinated time. In an action node the cargo units and resources have to remain until the time it has to spend there has elapsed. The situation is equivalent in the links. For each resource the length of the link is expressed using time.

3.3 Complexity of systems modelled

A medium sized port terminal which is modelled using the NeuComb/Port tool has about 100 nodes and 300 links where cargo and resources are allowed to travel. In addition to these there are a large number of information relations. This kind of system is highly non-linear, and new optimisation strategies should be considered. The algorithms used in the model should take care of the modelling in a parallel manner, however, the logic for this has not been developed. Many of the existing algorithms are quite time consuming and have to be compressed and smart versions have to be developed. Meanwhile a rational way to work around this problem is to linearise the system. In the NeuComb tool the linearisation is done in a common manner; the calculations are divided into time steps.

4. INFORMATION

It is important to point out that transportation and information are two parts of the same network, since the interaction is very large and important for the system to work. Information is not only generated by data specially prepared for the transport it also originates from the transport system itself. There are several levels of information that are needed when describing and modelling a transportation system. These levels range from static information to dynamic information. Roughly the static information describes the infrastructure and the dynamic information describes cargo and resources in the system. The dynamic information can be divided into two specific categories, internal and external information. This division is defined by the system boundary. Internal information, generated by the activities within the system, is crucial for the system and tells the system itself about its own status. The state of the nodes, including queues in nodes and if the node is temporarily closed, is for instance internal information. Internal information can be viewed as feedback loops, similar to those in control theory. These feedback loops are information links in the equations describing the elements of the transmission matrices of the nodes (Andersson *et al.*, 1997). Links in the tool can be either physical links, where the transportation resources move resources, or information links supplying the system with needed information. Using this definition physical links are also information links, because each transportation resource and cargo unit supplies information to the system. External information, which is added to the system from the outside world, is not crucial for the system to work but has to be included to make the tool of practical use. From a system point of view the external information only slows down the system.

5. OPTIMISATION

One of the main reasons for developing the NeuComb concept is the use in optimisations of transportation systems. There is a large potential for optimisations in these systems. Only the keeping track of resources with modern technology can increase the efficiency considerably (Swahn and Söderberg, 1992). If intelligent strategies for routing resources are combined with the technology the systems will become even more efficient.

Further the optimisation strategy used can in short be described in the following way. First the cargo flows are optimised using free flow. Free flow, in this sense, means that there are no resource problems obstructing the cargo flows. In this step an optimal choice of routes for the cargo flows is made. The second step is adding the constrains of the transportation resources. Unless the number of resources are unlimited these will delay the cargo flows. To minimise the influence of the resources these also have to be optimised.

In the NeuComb/Port tool the only difference between an optimisation and a simulation is the way in which resources are routed. When the program is run in simulation mode, a resource is sent to pick up a new cargo unit as soon as the resource is free. In optimisation mode, the resources have to wait for a new assignment until the program makes an overall assignment. Optimisation are carried out at regular intervals. The algorithm used for the optimal assignment problem is called "The Hungarian Algorithm". Resources are sent to their new locations so that the weight is minimised. We are currently working on improving the optimisation strategies.

6. FUTURE WORK

The NeuComb concept and tools derived from the concept still need further development and documentation. Complex systems like the ones modelled with NeuComb work are still not fully understood. There are a number of questions that still have to be answered. These questions include: Whether reliable predictions of the systems can be made, and defining the limitations of the concept?

Other parts that have to be developed and investigated further are the algorithms which are used. NeuComb includes elements from many different areas. All these elements have to be co-ordinated. The overall goal is to develop a strong concept that can be used to model and optimise transportation systems.

7. CONCLUSIONS

In this paper the background for a concept and tool for modelling transport systems have been presented. One of the main features is the inclusion of complex node functionalities in the networks. This make NeuComb an extension of combinatorial graph theory, which is a powerful theory used in a number of different areas (Jungnickel, 1994). NeuComb can be used, in a quite generic way, to model and optimise a vast number of logistic networks and situations. These situations range from container routing, complex delivery systems to highly optimised logistic schemes. However, it will take time to investigate all the properties of the NeuComb concept, at which work is progressing continuously. The NeuComb/Port tool has shown that the concept works well and inspires further work.

8. REFERENCES

Andersson, S., K. Jansen and J. Waidringer (1997). Optimisation of a generic multi-node transportation and information network. In: *Information Systems in Logistics and Transportation* (B. Tilanus, ed.), Chap. 21, pp 329-333, Pergamon, London.

Casti, J. L. (1995). In: *Networks in Action* (D. Batten, J. Casti and R. Thord, eds.), pp 3-24, Springer Verlag, Berlin.

Chen, W.-K. (1990). *Theory of nets: flows in networks*, John Wiley & Sons, New York.

Domany, E. and R. Meir (1991). In: *Models of Neural Networks* (E. Domany, J. L. Van Hemmen and K. Schulten, eds.), Chap. 9, pp 307-334, Springer Verlag, Berlin.

Huckenbeck, U. (1997). On valve adjustments that interrupt all s-t-paths in a digraph. *Journal of Automata, Languages and Combinatorics*, **2**, 19-45.

Jungnickel, D. (1994). *Graphen, Netzwerke und Algorithmen, 3. Auflag,* B. I. Wissenschaftsverlag, Mannheim.

Ojala, L. (1992). *Modelling approaches in port planning and analysis.* Publications of the Turku school of economics and business administration, Series A-4:1992, Turku

Swahn, M. and P. Söderberg (1992). *Lönsam logistik - med sikte på 2000-talet, Teledok rapport 75,* Teldok, Stockholm.

Waidringer, J. and K. Lumsden (1997a). Modelling a port terminal from a network perspective. 8th IFAC/IFIP/IFORS Symposium on Transportation Systems '97, Crete, Greece.

Waidringer, J and K. Lumsden (1997b). A parametric network model for a port terminal optimisation and simulation, NeuComb/Port. SPC7-The Seventh International Special Conference of IFORS, Göteborg, Sweden.

Waidringer, J and K. Lumsden (1998) Simulation and optimisation of port terminals, using a network concept. 8th World Conference on Transport Research, Antwerp, Belgium.

THE CONE PROJECTION METHOD OF DESIGNING CONTROLS FOR TRANSPORTATION NETWORKS

Andrew Battye, Mike Smith and Yanling Xiang, Department of Mathematics, University of York, Heslington, York, YO1 5DD

ABSTRACT

This paper outlines a steady state multi-modal equilibrium transportation model which contains elastic demands and deterministic route-choices. The model may readily be extended to include some stochastic route-choice or mode choice. Capacity constraints and queueing delays are permitted; and signal green-times and prices are explicitly included. The paper shows that, under natural linearity and monotonicity conditions, for fixed control parameters the set of equilibria is the intersection of convex sets. Using this result the paper outlines a method of designing appropriate values for these control parameters; taking account of travellers' choices by supposing that the network is in equilibrium. The method may be applied to non-linear monotone problems by linearising about a current point. An outline justification of the method is given; a rigorous proof of convergence is as yet missing. Thus the method must now be regarded as a heuristic.

1. CHOOSING THE ROAD AHEAD

Urban Transportation is at a cross-roads; with changing targets ahead, and an expanding plethora of increasingly sophisticated controls to assist in the task of meeting them, the transport planner faces a daunting task.

Currently, computer models of transportation only assess strategies given to them. So the planner is left with complete responsibility for devising strategies likely to be successful when tested (in computers and in reality) against new and changing targets. Thus while vast computational resources are employed for the *assessment* of options for controlling town traffic; the *design* of these options is left to the planner.

Now the design of optimal or near optimal strategies for controlling urban travel is difficult. In general it is far, far more difficult than assessing any given option. Thus at the moment computational assistance is only available for the easier task; by far the harder task is left entirely to the Transport Planner, Local Authority, Central Government or the European Union.

The single most important tool currently entirely absent from the transportation planner's toolbox is a decision-support system based on mathematical optimisation and implemented within helpful and easy-to-use software. This paper outlines a possible approach to the design of such a support system.

Two of the most central control variables in urban networks are traffic signal green-times and prices. We comment briefly on each of these.

2. SIGNAL CONTROLS

Traffic signal controls are already available for controlling traffic in towns; and decisions as to how to set signals are necessarily made very frequently (although often automatically). These decisions have a significant effect on short- and long-run traffic congestion (for example). It would be reasonable, now, to expect signal timings which which seek to reduce future congestion to be routinely implemented. *This is not the case;* mainly because traffic control systems have (for historical reasons) almost always been aimed at (i) short term traffic targets and (ii) management targets.

3. PRICES

Prices are important controls of economic and transportation systems. The decision support tool based on mathematical optimisation whose development is envisaged here should have immediate application to the estimation of approximately optimal prices for transportation taking account of real-life constraints and also taking into account other controls such as traffic signal control settings and bus lane provision. Realistic approximately optimal prices are likely to be very different to the unrealistic marginal cost prices which have so far received most attention, both within EU-funded work and elsewhere.

Marginal cost prices as usually interpreted pay little regard to practical constraints. In order to calculate prices which are confined by realistic constraints it is necessary to solve the problem known as the "second best" problem; that is: what should the prices be set at when some prices cannot be marginal cost prices. To tackle such problems we need a more general optimisation procedure which allows realistic constraints on the prices.

4. THE TRANSPORTATION MODEL PRESENTED HERE

The method of choosing controls outlined here applies to prices and green-times within a model with *inelastic and elastic demand, and deterministic and stochastic route and mode-choice*; and which embodies explicit capacity constraints and explicit queueing delays. However here we illustrate the method on a model without stochastic choice.

The optimal control problem for urban traffic has also been addressed, within rather different models, by Allsop (1974), Chiou (1997), Tobin et al (1988) and Yang et al (1995).

In this paper the equilibrium model will take the form of a monotone complementarity problem.

Monotonicity for fixed controls

The model will be monotone in the flow, delay and cost variables for each fixed control vector p. This will allow a fairly wide range of interactions between flows on one link and costs on other links but will also mean that there is at least one algorithm which is bound to converge to equilibrium for any fixed signal control settings and prices.

5. ACHIEVING THE COMPLEMENTARITY FORMULATION

The notation adopted is shown below; there is a base network and a multi-copy (Charnes and Cooper (1961)) version of this. **Within each copy the travellers or vehicles all have a single destination node**. This network structure is very similar to that in Smith (1996); here we have chosen to give a route-formulation in which each copy has links which comprise *routes* in the base network. In this paper then a link will always be an element of the base network; links in the multi-copy network will here be called routes since they are routes constructed from base network links.

The slightly more general structure in Smith et al (1996) contains the route-flow formulation and the link flow formulation of the assignment problem as special cases.

Route costs on the copies may have components which are sums of costs on base network links (link delays will add in this way along routes) but may also have elements which are non-additive with respect to base network links; so that bus fares and so on may be represented.

The structure also represents multi-mode networks; flow on the routes in a single copy may represent travellers using a single mode; or vehicles of a certain type. Within the framework given here multi-modal effects are most easily represented by thinking of all flows on the multi-copy network in units of passenger car unit equivalents and defining the demand function (W(.)

below) appropriately. All the links in the multi-copy part of the network will have suffix r (as here they are all *routes*) and base network links will have suffix i.

For each node or junction there is also a copy comprising just the base network links terminating at that node: to represent signal stages. Stage k (say) will be a subset of base network links with the same exit node and which may be shown green at the same time. If a junction is not signal controlled then there will be only one stage at that junction and this will be given green all the time.

Notation

Variables which are to be found in the equilibrium problem

X_r =   flow along route r in the multicopy network;

C_n =   least cost of reaching the destination from node n in the multi-copy network, where $C_n = 0$ if n is a destination; and

b_i =   bottleneck delay at the exit of link i in the base network.

Control variables

P_r =   price to be paid for traversing route r; and

Y_k =   proportion of time stage k is green.

Fixed given variables

s_i =   saturation flow at exit of base link i (may be infinite);

S_r =   free-flow travel cost along route r ($S_r > 0$).

Multi-copy network structure

B_{nr} =   1 if node n is the entrance node of route r in the multi-copy network (n Before r) and 0 otherwise;

N_{ik} = s_i if link i is in stage k and 0 otherwise; and

M_{ir} =   1 if route r in the multi-copy network contains link i and 0 otherwise.

Nominal link capacity

y_i = $\sum_k N_{ik} Y_k$, the nominal capacity of link i.

Functions

$W_n(C)$ =   demand at node n if the cost to destination vector is C; and

$g_i(b_i, y_i)$ =   maximum possible average flow when the nominal link capacity is y_i and the bottleneck delay on link i is b_i.

It should be noted that, following Payne and Thompson (1975), the bottleneck delays b_i and here the node costs C_n are regarded as independent variables. Also, destination nodes are slightly special: if n is a destination node then $C_n = 0$ and $W_n(C) = 0$ for all C. (The demand for travel from a node to itself is zero.)

A non-signalised junction will have exactly one "stage" (the first) shown green for all time and so $Y_1 = 1$. Also N_{i1} is (as in the general case above) to be s_i and then the nominal link capacity will be still be $y_i = \Sigma_k N_{ik} Y_k = N_{i1} Y_1 = s_i$. The formula above will then determine the maximum possible flow consistent with this and a bottleneck delay b_i.

<u>Equilibrium, demand and capacity constraints</u>
We use Wardrop's (1952) condition: *more costly routes carry no flow*. But we choose to write this in the following form: *for each route r the (least) cost to the destination from the node B(r) upstream of route r (Before r) is no more than the least cost to the destination via route r, and if it is less then no flow will enter route r*, or

$$w_{1r} = \Sigma_n B^T_m C_n - S_r - P_r - \Sigma_i M^T_{ri} b_i \leq 0, \text{ and}$$
$$\Sigma_n B^T_m C_n - S_r - P_r - \Sigma_i M^T_{ri} b_i < 0 \text{ implies } X_r = 0.$$

Here $\Sigma_n B^T_m C_n$ is just a way of writing $C_{B(r)}$; the cost at the node upstream of route r. This sum comprises just the single cost at that node at the entrance of route r.

The (elastic or inelastic) demand constraints may be written: *the total route-flow $\Sigma_r B_{nr} X_r$ out of node n equals the demand* $W_n(C)$. Rewriting this in a slightly artificial way we obtain:

$$w_{2n} = W_n(C) - \Sigma_r B_{nr} X_r \leq 0, \text{ and}$$
$$W_n(C) - \Sigma_r B_{nr} X_r < 0 \text{ implies } C_n = 0.$$

This is in fact, under natural conditions, equivalent to the italicised condition.

For the capacity constraint condition we suppose here that for any average bottleneck delay b_i and nominal link capacity y_i there is a maximum possible flow $g_i(b_i, y_i)$ consistent with the delay b_i. Then the capacity constraint may be written:

$$w_{3i} = \Sigma_s M_{is} X_s - g_i(b_i, \Sigma_k N_{ik} Y_k) \leq 0 \text{ and}$$
$$\Sigma_s M_{is} X_s - g_i(b_i, \Sigma_k N_{ik} Y_k) < 0 \text{ implies } b_i = 0.$$

As specified here this condition will ensure that congestion costs normally represented by a cost-flow function will in fact occur as "bottleneck" delays b_i which arise from equilibrating via the function g_i. The g_i may be thought of as the inverse of given cost-flow functions. Given a nominal link capacity y_i, g_i delivers a flow x_i compatible with a given cost or delay b_i; instead of delivering a cost b_i for each flow x_i These inverse cost-flow functions g_i may be expected to have numerical advantages if the cost-flow function is steep; because then the g_i are shallow.

<u>The complementarity formulation</u>. Let $x = (X, C, b)^T$, $p = (Y, P)^T$. Also let

$$w = w(x, p) = (w_1, w_2, w_3)^T = (w_{11}, w_{12}, \ldots; w_{21}, w_{22}, \ldots; w_{31}, w_{32} \ldots)^T$$

be the vector of all left hand sides above. Renumber the co-ordinates of w in the natural way and put

$u(x, p) = [w_1(x, p), w_2(x, p), \ldots w_i(x, p), \ldots, w_n(x, p)]^T.$

Then the above conditions become:

x belongs to R_+^n and $u(x, p)$ is normal, at x, to R_+^n; or (1)

x belongs to R_+^n; $u(x, p) \leq 0$; and $u_i(x, p) < 0$ implies $x_i = 0$. (2)

The set of equilibria (for fixed p) as the intersection of convex sets

Let p be a fixed feasible control vector for the moment. The feasibility and equilibrium conditions (1) or (2) are plainly equivalent to:

$x_i \geq 0$ for $i = 1, 2, 3, \ldots, n$; $u_i(x, p) \leq 0$ for $i = 1, 2, \ldots, n$; and $E_2 = -\Sigma x_i u_i(x, p) \leq 0$ (3)

Now suppose that for each fixed $p \geq 0$ the function $-u(\bullet, p)$ is monotone and linear in x. Then

$$-u(x, p) = A_p x + b_p$$

for some square positive semi-definite matrix A_p and some n-vector b_p which both depend on p. Since A_p is positive semidefinite (that is $x^T A_p x \geq 0$ for each n-vector x), $x^T A_p x$ is a convex function of x (the positive semi-definite A_p is the Hessian of this function); and so $-x^T u(x, p)$ is a convex function of x for each fixed p. By linearity each u_i is convex too. Thus each inequality in (3) specifies a convex set.

Therefore, if $-u$ is linear and monotone, for any fixed p the set of equilibria (the set of those x which satisfy all the inequalities (3)) is the intersection of convex sets.

We make the basic assumption here that there is an equilibrium for any feasible p; so that the intersection of convex sets above is non-empty for any feasible p. This will certainly be a reasonable assumption if demand is sufficiently elastic and bottleneck delays are permitted to be arbitrarily large.

Control constraints

Since we wish to vary p we need constraints on the set of possible p values. Let these constraints be $g_j(x, p) \leq 0$; $j = 1, 2, 3, \ldots m$. These constraints must now be added to the constraints (3) and we suppose that the added constraints, like those in (3), are convex in x for fixed p. Suppose that all the constraints, including non-negativity constraints, are now:

$h_k(x, p) \leq 0$ for $k = 1, 2, 3, \ldots, K$ (4)

Of course when p is allowed to vary these sets are no longer convex. If $h_k > 0$ we will say that the constraint $h_k \leq 0$ is *violated*; otherwise it is *satisfied*.

At non-equilibria there are directions which improve all violated constraints

Assume that all the h-constraints in (4) are convex for each p. Although we only have convexity for fixed p we are able nonetheless to utilise this to obtain the above result in (x, p) space.

Let $C_0(x, p)$ be the cone of directions in (x, p)-space which do not cause any satisfied constraint (in (4)) to become violated (constraints defining C_0 are semi-active) and let $C_1(x, p)$ be the cone of those directions at (x, p) along which no violated constraints (in (4)) become more violated (constraints defining C_1 are active). Then $C_1(x, p)$ is the cone of directions δ (in (x, p)-space) such that $\delta \bullet h_k' \leq 0$ if $h_k > 0$. Directions in the interior of $C_1(x, p)$, $\text{int}C_1(x, p)$, reduce the violation of all violated constraints in (4), at (x, p), simultaneously.

Consider any non-equilibrium (x, p) and join this point to any equilibrium point (x^*, p) by a straight line. *Since all the constraints are convex for fixed p;* along this ray all $h_k > 0$ are diminished to be non-positive; and also no satisfied constraint becomes violated. In other words the direction of this ray is in $C_0(x, p) \cap \text{int}C_1(x, p)$ throughout its length (even though p is constant), and so $C_0(x, p) \cap \text{int}C_1(x, p)$ is non-empty.

6. THE CONE-PROJECTION METHOD

Suppose given a smooth objective function $Z = Z(x, p)$; where x is the vector of flows, delays and costs, and p is the vector of controls (signal green-times and road prices). The ideal form of "the cone-projection method" is to follow a "descent direction" D which simultaneously seeks to reduce all $h_k > 0$ to zero while "doing the best for" the given Z. Such a trajectory may (under natural conditions) be expected to converge to equilibrium and a weak variety of local-optimality simultaneously.

A descent direction D arising from this has the following ideal form:
$$D = D(x, p) = d(x, p) + \text{Proj}_{C(x, p)}(-Z'(x, p)) \qquad (5)$$
where the vector $d(x, p)$ is to be a direction in the cone $C_0(x, p) \cap \text{int}C_1(x, p)$ and $\text{Proj}_{C(x, p)}(-Z')$ is the negative gradient of Z projected onto the cone $C(x, p) = C_0(x, p) \cap C_1(x, p)$. The direction $d(x, p)$ is to have the property that it is non-zero if $C_0(x, p) \cap \text{int}C_1(x, p)$ is non-empty.

Consider following D(x, p). How can a trajectory following D(x, p) stop? Only by having $D(x, p) = 0$ (the zero vector).

Plainly $D(x, p) = 0$ implies that both parts of D(x, p) are zero. But $d(x, p) = 0$ implies that $C_0(x, p) \cap \text{int}C_1(x, p)$ is empty; and this only happens if all constraints are satisfied and we are at an equilibrium. $D(x, p) = 0$ also implies that $\text{Proj}_{C(x, p)}(-Z') = 0$ and a Kuhn-Tucker necessary condition holds too. Thus under natural conditions a dynamical system following $D(x, p)$ will only come to rest within the set of equilibria at which a weak necessary-for-optimality condition holds.

The solution method is thus in outline to follow direction $D(x, p)$ at each (x, p). This is intended to be a refinement of the bi-level method proposed in Smith et. al. (1996, 1997 a, b); replacing half-spaces with cones to narrow the search region and reduce numerical/computational problems.

It is natural to choose $d(x, p)$ to be a "centre-line" of the cone $C(x, p)$ obtained by solving a problem $P(x, p)$ which is in a simple case as shown below:

Problem P: Minimise $||\Sigma\lambda_k(-h_k')||$ subject to $\lambda_k \geq 0$, $\Sigma\lambda_k = 1$; and $\lambda_k = 0$ if $h_k \leq 0$ ($=$ 0 by definition if all $h_k \leq 0$).

Now let $d(x, p) = \Sigma\lambda_k(-h_k')$, where λ solves $P(x, p)$.

At the moment we only have an outline justification for the method; we have been unable, so far, to fill in all the details of a rigorous proof. Thus the method is at this stage only a "heuristic". However a fairly rigorous proof is given in Smith et al (1996) where we use a simpler version with half-spaces instead of cones, and Clegg and Smith (1998) give an initial result of the method on a small network.

Implementation and outline justification

To implement the method in Clegg and Smith et al (1998) we used the following two-stage procedure and this does give an outline justification here.

Given a margin of error ε in the required constraints in (4) we consider $h_k(x, p) \leq \varepsilon$. An (x, p) satisfying these constraints will be called an ε-equilibrium.

In the first stage we follow direction $d(x, p)$ or $D(x, p)$ given by (5) repeatedly until we have reached an $\varepsilon/2$-equilibrium. Then we minimise Z subject to the constraint that the flow pattern must be an ε-equilibrium (either by following $D(x, p)$ or by using a standard non-linear optimisation method); to reach a point (x_1, p_1). Now (x_1, p_1) is an ε-equilibrium and (as it minimises Z) the direction $\text{Proj}_{C(x, p)}(-Z')$ must be zero.

Then we repeat these two stages with ε replaced by $\varepsilon/2$ to obtain a point (x_2, p_2). Now (x_2, p_2) is an $\varepsilon/2$-equilibrium and (as it minimises Z) the direction $\text{Proj}_{C(x, p)}(-Z')$ must be zero.

In the nth repetition of this two-stage process we obtain (x_n, p_n). Now (x_n, p_n) is an $\varepsilon/2^{n-1}$-equilibrium and the direction $Proj_{C(x, p)}(-Z')$ must again be zero.

Thus if (x^*, p^*) is any limit point of this sequence (x^*, p^*) must be an equilibrium and must also satisfy an asymptotic Kuhn-Tucker condition: that

$$Proj_{C(x, p)}(-Z') = 0 \text{ for all } (x, p) \text{ in a sequence converging to } (x^*, p^*) \qquad (6)$$

7. CONCLUSION

The paper has outlined a method for calculating signal timings and prices in an urban transportation network which uses directions confined to cones. We have also shown how to express the equilibrium problem with constraints and queueing delays as a problem of determining a point in the intersection of convex sets. The method applies to linear monotone multi-modal deterministic elastic and inelastic problems; and may be extended, by using the work of Akamatsu (1997), Bell (1995) and Fisk (1980), to include stochastic elements.

The linearity may be relaxed by linearising the given problem to establish the non-empty-ness of the cone $C(x, p)$ at each non-equilibrium (x, p). Monotonicity is however essential.

We have given an outline justification of the method; showing that the method yields an equilibrium (x, p) which satisfies an asymptotic Kuhn-Tucker condition. This is a weak necessary condition: a truly optimal (x, p) must satisfy it.

Further work is needed (i) to assess the practical efficiency of the cone projection method, (ii) to convert the "outline justification" given here to a rigorous proof of convergence, and (iii) to relax the conditions specified in this paper.

ACKNOWLEDGEMENTS

We are grateful for financial support from: DETR, EPSRC, ESRC, DGVII (E2). We are also appreciative of constructive criticism and comments on the general equilibrium framework from our LINK partners at UCL and MVA.

REFERENCES

Akamatsu T (1997), Decomposition of Path Choice Entropy in General Transport Networks, *Transportation Science*, **31**, 4, 349 - 362.

Allsop R E (1974), Some possibilities for using traffic control to influence trip distribution and route choice, *Proceedings of the 7th International Symposium on Transportation and Traffic Theory*, 345 - 374.

Bell M G H (1995), Alternatives to Dial's LOGIT Assignment Algorithm. *Transportation Research* 29B, 287 - 296.

Chiou S-W (1997), Optimisation of area traffic control subject to user equilibrium traffic assignment, *Proceedings of the 25th European Transport Forum, Seminar F, Volume II, 53 - 64.*

Charnes A and Cooper W W (1961), Multi-copy traffic network models. *Proceedings of the Symposium on the Theory of Traffic Flow*, held at the General Motors Research Laboratories, 1958 (Editor: R Herman), Elsevier, Amsterdam.

Clegg J and Smith M J (1998), Bilevel optimisation in transportation networks. Paper presented at the IMA Conference on Mathematics in Transport Planning, Cardiff.

Fisk C (1980), Some developments in Equilibrium Traffic Assignment. *Transportation Research*, 14B, 243 – 255.

Payne H J and W A Thompson (1975). Traffic assignment on transportation networks with capacity constraints and queueing. Paper presented at the 47th National ORSA/TIMS North American Meeting.

Smith M J, Xiang Y, Yarrow R and Ghali M O (1996), Bilevel and other modelling approaches to urban traffic management and control, Paper presented at a Symposium at the Centre for Transport Research, University of Montreal, and to appear.

Smith M J, Xiang Y, and Yarrow R (1997a), Bilevel optimisation of signal timings and road prices on urban road networks. Preprints of the IFAC/IFIP/IFORS Symposium, Crete, 628 - 633.

Smith M J, Xiang Y, and Yarrow R (1997b), Descent Methods of Calculating Locally Optimal Signal Controls and Prices in Multi-modal and Dynamic Transportation Networks, Presented at the Fifth Euro Meeting on Transportation, University of Newcastle-on-Tyne, and to appear in the Proceedings.

Tobin R L and Friesz T L (1988), Sensitivity analysis for equilibrium network flow, *Transportation Science*, 22, 242 - 250.

Wardrop J G (1952) Some Theoretical Aspects of Road Traffic Research, Proceedings, Institution of Civil Engineers II, 1, 235-278.

Yang H (1997), Sensitivity analysis for the elastic-demand network equilibrium problem with applications, *Transportation Research*, 31B, 55 - 70.

Incorporating Heteroscedasticity within Individual and between Individuals to Overcome Repeated Measurement Problems

Kang-Soo KIM, Institute for Transport Studies, University of Leeds, UK

Abstract

A method is derived for estimating a discrete choice model incorporating heteroscedasticities to reflect repeated measurement problems. Heterogeneity of each observation is characterised by a specific scale function and individual heterogeneity is introduced in the random utility choice model. This research proves that the unobserved influences affecting a specific individuals' mode choice are correlated from one of his or her selections to the next repeated questions. This research also suggest a strong evidence of learning effect, implying variances would be decrease as the responses faces repeated questions.

1. Introduction

There is considerable confidence in the validity of Stated Preference(SP) technique as a means cost effective of collecting data in surveys handling several alternatives including new or unfamiliar options. SP surveys usually comprise a series of the choice observations from each respondent. However, limitation, known as the " Repeated Measurements Problem " has been identified and generally ignored in practice (Bates(1997)). The problem arises because the assumption that multiple response are independent is not valid.

Recently, the repeated measurement problem has become more widely recognised and some interesting approaches have been proposed to overcome it. The Jack-knife and bootstrap resampling techniques have been applied by Cirillo *et al* (1996) in an effort to obtain an unbiased estimate of a statistic and its variance for coefficients. However, Cirillo *et al* confined the repeated measurement problems to under (or over) estimation of standard error of coefficients.

Outwerslot and Rietveld (1996) have suggested that the total error in a random utility model can be decomposed into two components; an individual specific effect and an observation specific effect. They used the minimum distance method proposed by Chamberlain (1984), and the procedure for estimation of the minimum distance parameter can be found in their paper.

Abdel-Aty, Kitamura and Jovanis (1997) have used mixing distribution to explore the repeated measurement error. They assume that the variation between respondents is random and normally distributed and incorporated this random effect in the standard logit form. The repeated measurement problems in their studies were confined to problems of correlation of the unobserved part of utility. However, this paper suggests that heteroscedasticities of observations should also be considered in the repeated measurement surveys because variances of utility may increase or decrease depending on characteristics of observations such as the number of attributes, the number of alternatives, correlation of attributes and the number of questions.

We define the repeated measurement problem in this paper as the treatment of the combination of observation specific variations and individual specific variations according to repeated observations. In other words, we assume that there will be a observation specific and individual specific variations can increase or decrease as the number of responses increases and that the unobserved factors across responses are correlated in the repeated questions.

The purpose of this paper is to decompose the repeated measurement problem into these two specific variations and then to present methodology to combine these approaches. This will be used to investigate the effect of repeated measurement problems.

2. OBSERVATION SPECIFIC VARIATIONS

The observation variances within an individual can be specified by recognising of role of scale factor in the random utility choice model. Swait and Admowicz (1996) proved that a specific scale function which varies by observations can be incorporated into the variance term of the logit model.

2.1 MNL Model and Scale Function

Suppose that the utility function for i-th alternative in A(n) - all the alternatives of n-th choice set- is additive and the sum of an observable (or systematic) component V_{ipt} plus an error term ε_{ipt}; stochastic component.

$$U_{ipt} = V_{ipt} + \varepsilon_{ipt} \qquad (2.1)$$

where p refers to individual respondent, p=1,2,...P and t refers to each element of responses for individual p, t=1,2,...T$_p$.

If we were to assume that the ε_{ipt} are IID(mutually independent and identically distributed) with a Gumbel distribution and a common scale factor μ, the Multinominal Logit (MNL) model is derived. (Ben-Akiva and Lerman(1985))

$$\text{Pr}_{ipt} = \frac{\exp(\mu V_{ipt})}{\sum_{j \in A(n)} \exp(\mu V_{jpt})} \tag{2.2}$$

The scale parameter μ is related to variance of error term, ε_{ipt} and it is expressed by;

$$\mu_{ipt}^2 = \frac{\pi^2}{6\sigma_{ipt}^2} \tag{2.3}$$

Where σ_{ipt}^2 is the variance of error term of utility of alternative i for person p at time t.

Now, we shall assume that ε_{ipt} is Gumbel distributed, independent across person p and alternative $i \in A(n)$, with scale factors $\mu_{ipt} = \mu_t(D_t)$ so that $\mu_t(D_t) = \frac{}{\sqrt{6\sigma_{ipt}}}$. It is required that $\mu_t(D_t)$ is positive and D_t is certain task demands or whatever factors affect the variances of observations. Therefore, the scale factor, $\mu_t(D_t)$ vary observation specifically as a function of the D_t. If we multiply the equation (2.1) by the scale factor $\mu_t(D_t)$,

$$\mu_t(D_t) \cdot U_{ipt} = \mu_t(D_t) \cdot V_{ipt} + \mu_t(D_t) \cdot \varepsilon_{ipt} \tag{2.4}$$

and

$$\text{Pr}_{ipt} = \text{Pr}\{ \mu_t(D_t) \cdot V_{ipt} + \mu_t(D_t) \cdot \varepsilon_{ipt} > \mu_t(D_t) \cdot V_{jpt} + \mu_t(D_t) \cdot \varepsilon_{jpt}, \forall j \neq i, i, j \in A(n) \}$$

$$= \text{Pr}\{ \mu_t(D_t) \cdot [V_{ipt} - V_{jpt}] > \mu_t(D_t) \cdot [\varepsilon_{jpt} - \varepsilon_{ipt}], \forall j \neq i, i, j \in A(n) \} \tag{2.5}$$

Because of the IID property of the error(see above) difference $\mu_t(D_t) \cdot [\varepsilon_{jpt} - \varepsilon_{ipt}]$, it then follows that choice probability is simply a MNL model,

$$\text{Pr}_{ipt} = \frac{\exp[\mu_t(D_t|\theta) \cdot V_{ipt}(X_{ipt}|\beta)]}{\sum_{j \in A(n)} \exp[\mu_t(D_t|\theta) \cdot V_{jpt}(X_{jpt}|\beta)]} \tag{2.6}$$

where β, θ are parameter vectors, X_{ipt} is vectors of attributes for alternative i and person p. $\mu_t(D_t)$ is the scale function.

Note that the derivation of (2.6) assumes that the $\mu_t(D_t)$ vary only by time(observation) and not by alternative or person. It has the same IIA property as the standard Multinominal Logit Model.

2.2 Specific Scale function($\mu_t(D_t)$)

Before we specify scale function, the following conditions should be considered. Firstly, the scale function $\mu_t(D_t)$ is related inversely with variance, thus it should increase(decrease) as variance decreases(increases). Secondly, scale function $\mu_t(D_t)$ always should have positive value to keep the probability equation (2.5) unchanged.

It is proposed to use the Distance between alternatives in attribute space to represent scale function of each response. It is a measure which gives a clear understanding of the overall choice circumstances in a choice set and the distance measure is related to the correlation of the attributes. In addition, it is varied according to the number of attributes and the number of alternatives.

The proposed approximate measure to represent the observation specific scale function is expressed,

$$\widetilde{D}_t = (\sum_{i,j \in \{COST,TIME,RELI\}} d_{X_i X_j}) \tag{2.7}$$

where, X_i and X_j are vectors of attributes for alternatives. $d_{X_i X_j}$ is the sum of binary cases distance for X_i and X_j.

Based on this measure, we also define proxy for cumulative distance measure to reflect the number of questions which are given to the respondents.

$$\widetilde{\psi}_t = \begin{cases} 0 & for \ t = 1 \\ \sum_{t'=1}^{t-1} \widetilde{D}_{pt'} & for \ t = 2,...,T \end{cases} \tag{2.8}$$

where t refers to replication index, T is the total number of questions given per each respondent and $\widetilde{D}_t$ is the sum of attribute distance of the t-th question.

Now, we need to consider the functional forms for the scale function. We propose the following exponential function because it guarantees that the scale function is always positive.

$$\mu_t(\widetilde{D}_t | \theta) = \exp[\theta_1 \widetilde{D}_t + \theta_2 \widetilde{\psi}_t + \theta_3 (\widetilde{\psi}_t)^2] \tag{2.9}$$

Where θ are the coefficients for estimation.

Note that scale function $\mu_t(D_t)$ is related inversely with the variance. Other functional forms could be adopted. However , the above function is simple and this exponential functional form has been applied in Swait and Admowicz(1996).

In the scale function, we investigate two hypothesis for analysing repeated measurement problems. Firstly Distance($\widetilde{D}_t$) would increase as scale function increases, leading to a decrease of variance of utility , thus θ^1 would be positively significant. Secondly, the variance of utility is a convex function of cumulative Distance($\widetilde{\psi}_t$) implying that the variance would decrease as the respondent faces repeated questions, but beyond some point of cumulative Distance, the variance would be increased. Therefore, we expect that θ^2 , θ^3 would be positively and negatively signed respectively.

3. INDIVIDUAL SPECIFIC VARIATIONS

Abdel-Aty *et al* (1997) postulated that repeated observations for each respondent gives rise to an obvious correlation of the unobserved portion of utility over repeated questions. Based on their postulate, the disturbance, which is variations in the unobserved portion of utility will be more duplicated in repeated observations because the unobserved factors may be invariant across the repeated observations.

In this section, to account for the correlation of the unobserved factors over repeated questions, an individual specific error component method (See Abdel-Aty *et al* (1997)) will be explored. The individual specific error component is incorporated in the unobserved portion of utility with specific scale function which explains observation specific variance.

3.1 Incorporating Individual Specific Error Component

Individual heterogeneity(ξ_p) is introduced into model by assuming that the probabilities are conditional on $\mu_t(D_t) \cdot V_{ipt}$ Equation (3.1) becomes when we introduce individual heterogeneity into equation (2.6) reflecting each person's sequence of observed choice. That is, the unobserved portion of utility of the alternative i, ε_{ipt}, is decomposed into $\xi_p + \varepsilon'_{ipt}$ and the ξ_p is treated explicitly as a separate component of the error. ξ_p is a random term with mean of zero distributed over the population and ε'_{ipt} is random term with mean of zero that is IID over alternatives.

Given the value of ξ_p, the conditional sequence of observed choice is simply logit, since the remaining error, ε'_{ipt} is IID extreme value;

$$\Pr_{ipt} = \prod_{t=1}^{T_p} \frac{\exp[\mu_t(D_t|\theta) \cdot V_{ipt}(X_{ipt}|\beta) + \xi_p]}{\sum_{j \in A(n)} \exp[\mu_t(D_t|\theta) \cdot V_{jpt}(X_{jpt}|\beta) + \xi_p]} \tag{3.1}$$

Since ξ_p is not given, the unconditional probability is the logit formula integrated over all values of ξ_p weighted by the density of ξ_p. If we assume ξ_p follows a normal distribution $N(0,\sigma^2)$, independent of the $\mu_t(D_t)\cdot V_{ipt}$, the probability can be represented by the integral over all possible values of ξ_p.

$$\Pr_{ipt} = \int_{\xi_p=-\infty}^{\xi_p=\infty} \prod_{t=1}^{T_p} \frac{\exp[\mu_t(D_t|\theta)\cdot V_{ipt}(X_{ipt}|\beta)+\xi_p]}{\sum_{j\in A(n)}\exp[\mu_t(D_t|\theta)\cdot V_{jpt}(X_{jpt}|\beta)+\xi_p]} f(\xi_p)d(\xi_p) \tag{3.2}$$

Models of this from are called " Mixed Logit " since the choice probability is a mixture of logits with f as the mixing distribution.

The parameters (θ and β) can be estimated using a maximum likelihood approach, applying the equation (3.2) to the observed choice of each observation within each individual and maximising the sum of the logged probabilities across the responses.

The log-likelihood functions for estimating parameters becomes, thus

$$L = \sum_{p=1}^{P} \ln \int_{\xi_p=-\infty}^{\xi_p=\infty} \prod_{t=1}^{T_p} \frac{\exp[\mu_t(D_t|\theta)\cdot V_{ipt}(X_{ipt}|\beta)+\xi_p]}{\sum_{j\in A(n)}\exp[\mu_t(D_t|\theta)\cdot V_{jpt}(X_{jpt}|\beta)+\xi_p]} f(\xi_p)d(\xi_p) \tag{3.3}$$

Note that observation specific variances within each individual could be tackled by incorporating $\mu_t(D_t)$ into the model and individual specific random term(ξ_p) used to account for the correlation of unobserved errors.

The log likelihood function in equation (3.3) has no closed form expression to calculate the integral function. The evaluation of the integral can be performed using a simulation method or a numerical approximation method. In this paper, the numerical approximation method known as Gaussian Quadrature has been adopted. This procedure can obtain highly accurate estimates of the integration in the likelihood function by evaluating the integration at a relatively small number of support points, thus achieving gains in computational efficiency of several orders of magnitude.

By applying Gaussian Quadrature methods, equation (3.3) is transformed so that Gauss-Hermite Quadrature approximation method could be applied to compute the integral . To do so, we need to define a variable $\frac{1}{\sqrt{2}\sigma}(\xi_p) = u$. Then, $f(\xi_p)d(\xi_p)= \frac{1}{\sqrt{\pi}}e^{-u^2}du$ because

$$f(\xi_p) = \frac{1}{\sigma\sqrt{2\pi}}\exp(-\frac{1}{2\sigma^2}(\xi_p)^2), \quad -\infty < \xi_p < +\infty \tag{3.4}$$

Also define a function K_{pt} as

$$K_{pt}(u) = \prod_{t=1}^{T_p} \frac{\exp[\mu_t(D_t|\theta) \cdot V_{ipt}(X_{ipt}|\beta) + \sqrt{2}\sigma \cdot u]}{\sum_{j \in A(n)} \exp[\mu_t(D_t|\theta) \cdot V_{jpt}(X_{jpt}|\beta) + \sqrt{2}\sigma \cdot u]} \tag{3.5}$$

We can re-write the loglikelihood function as

$$L = \sum_{p=1}^{P} \ln \frac{1}{\sqrt{\pi}} \{ \int_{\xi_p = -\infty}^{\xi_p = \infty} K_{pt}(u) \cdot e^{-u^2} du \} \tag{3.6}$$

The expression within brackets in the above equation can be estimated using the Gaussian Hermite Quadrature approximation method, which replaces the integral by a summation of terms over a certain number of support points. The programming for maximising the loglikelihood function is carried in GAUSS software.

4. ESTIMATION

4.1 Data

ITS (Institute for Transport Studies) have undertaken a study of unitised freight between the UK and Europe (Tweddle *et al* 1995). The study aimed to investigate the behavioural aspects of Freight Transportation actors' decisions as to choice of mode. Within the study, a survey of manufacturers, international hauliers and freight forwarders was undertaken prior to the Channel Tunnel being opened to normal traffic. From that targeted companies, a typical international flow of freight was taken and each respondent was asked to express their preferences by rating for 4 alternatives (Ferry, New Ferry, Le Shuttle and Rail) which consists of three variables (Cost, Arrival Time and Reliability). At least 6 question were asked per respondent, but the number of questions were varied across respondents. For estimation of choice models, we simply assume that the highest ranked option would be chosen.

4.2 Estimation Results

In this section, I will refer to two forms of logit model. Firstly, the Mixed Logit model which is estimated by maximising the loglikelihood function (3.6). Secondly the standard Logit model means simple Multinominal logit model, assuming that all observations are independent. Table 1 compares the standard logit results to the results from the mixed logit model.

Table 1 Estimation Results

	Mixed Logit		Standard Logit	
PARAMETERS	Coefficient	t-stat	Coefficient	t-stat
COST	-0.0215	-1.827	-0.0097	-4.460
TIME	-0.0322	-6.580	-0.0448	-1.605
RELI	0.1284	24.193	0.1350	2.978
ASC(SH)	-0.0049	-0.093	-0.1331	-1.002
ASC(RAIL)	-0.3428	-2.696	-0.1829	-1.183
θ^1	0.6982	7.059	-	-
θ^2	0.5588	8.804	-	-
θ^3	-0.0627	-1.003	-	-
σ	1.1346	21.587	-	-
Statistics				
$l*(0)$	-0.0459545		-1.3862401	
$l*(\theta)$	-0.0352685		-1.35200	
ρ^2	0.233		0.0247	
No. of Obs.	361		361	

Note : σ : *Standard Deviation of* ξ_p, $l*(0)$: *Loglikelihood at Zero*, $l*(\theta)$: *Loglikelihood at convergence*, ρ^2 : *likelihood ratio index (1- $l*(\theta)$ / $l*(0)$)*

We would expect parameters of COST, TIME to be negative and RELI(Reliability) to be positive. This occurs in all models. In the Mixed Logit model, the Cost variable is insignificant, while Time variable is significant at the 95% confidence. RELI are significant at the 95% confidence for all models.

Alternative Specific Constants, ASCs, reflect the tastes of respondents by showing their propensity to switch to mode of transport, to using the Tunnel (Shuttle or Rail) from their base choice, all other things being equal. ASCs can be either positive or negative according to taste of respondents.

Table 1 shows that all ASCs are negative implying respondents have a preference for Ferry (Ferry and New Ferry), thus discount is required to switch existing mode (Note that the survey was performed before opening the Channel Tunnel) to Shuttle and Rail. However, ASC(SH) variables have large variances compared to their coefficient size in the all models. The size of ASC(SH) coefficient is very small(-0.0049) in the Mixed Logit model.

ASC(RAIL) becomes significantly in the Mixed Logit model. The size of ASC(RAIL) in Mixed Logit is bigger than Standard Logit Model. Difference between ASC(SH) and ASC(RAIL) is also larger. We can conclude by interpretation of coefficients that the Mixed Logit model suggest respondents have a strong bias against Rail and would require a large cost discount to encourage switching from their current mode choice (Ferry) to Rail.

We investigate the repeated measurement effect by testing our hypotheses on the coefficients and investigating significance of σ.

We defined a hypothesis that θ^1 would be positive and significant, implying Distance($\tilde{D}_t$) would be increase as scale function increases, leading to a decrease of variance of utility, and , θ^2, θ^3 would be positive and negative, respectively, meaning the variance of utility is convex function of cumulative Distance($\tilde{\psi}_t$).

Table 1 shows that for parameter θ^1, the hypotheses can be rejected as it states that θ^1 equals zero at 95 % confidence. This implies that Distance has a significant impact on the variances of utility. The positive value also shows that as Distance increases, the scale function increases, leading to decreases of variances.

Table 1 also presents the parameter θ^2 can be rejected the hypothesis that θ^2 equals zero for 95 % confidence, implying a strong evidence of learning effect and the variance would be decrease as the responses faces repeated questions.

However, interestingly Table 1 shows that for parameter θ^3 the hypothesis that θ^3 equals zero at 95 % confidence cannot be rejected, though it does not have a negative value as expected. This implies that the variance of utility varies according to the number of questions facing respondents because of combination of repeated measurement effect such as learning and fatigue effect. However, it does raise the question of at which point of cumulative Distance decrease variance because of coexistence of fatigue effect and learning effect in the cumulative response.

The significance of σ illustrates that the unobserved influences affecting a specific individuals' mode choice are correlated from one selection to the next. This suggests that this factor should be incorporated in the estimation model.

5. CONCLUSIONS

The primary conclusion of this research is that repeated measurement effects exist and heteroscedasticities of observations should be considered in the repeated measurement problems along with individual specific variations. The significance of σ proves that the unobserved influences affecting a specific individuals' mode choice are correlated from one of his or her selections to the next repeated questions. In addition, Distance($\tilde{D}_t$) is proved to have a significant impact on the variances of utility. Variances of utility increase as Distance($\tilde{D}_t$) increases indicating that the variance of utility increases as the number of attributes, the number of alternatives and the correlation of attributes increases. Therefore, the repeated measurement effect should be considered in the early stages of SP design.

However, it is difficult to establish the number of observations from each respondent as the repeated measurement problem is not too serious because of interaction of fatigue and learning effect. Furthermore, it is recognised that the results presented in this paper may be dependent on the number of repeated observations and definition of the scale function which is measured by Distance.

Further research should closely examine the impact of the number of observations and the assumption about the scale function.

REFERENCES

Abdel-Aty M.A, Kitamura, R., and Jovanis,P.P.(1997), Using Stated Preference Data for Studying the Effect of Advanced Traffic Information on Drivers' Route Choice, *Transportation Research C*(1), 39-50

Bates, J.J. and Terzis, G.(1997), Stated Preference and the "Ecological Fallacy", *PTRC*

Ben-Akiva, M. and Lerman,S.(1985), *Discrete Choice Analysis : Theory and Application to Travel Demand* , MIT Press, Cambridge,M.A

Bhat, C.R.(1995), A Heteroscedastic Extreme Value Model of Intercity Travel Mode Choice, *Transport Research B*, Vol. **29**, No. 6, 471-483

Bradley, M., Ben-Akiva, M., and Bolduc, D.(1993), Estimation of Travel Choice Models with Randomly Distributed Value of Time, *PTRC*

Chamberlain, G.(1984), *Panel Data*, In Z. Griliches and M.D. Intriligator(eds.), Handbook of Econometrics, 1247-1318, North Holland, Amsterdam

Cirillo, C., Torino, Daly ,A. and Lindveld(1996), Eliminating Bias due to the Repeated Measurements Problem in SP data, *PTRC*

Hensher, D.A., Louviere, J. , and Swait,J.(1997), Combining Sources of Preference Data : The Case of the Lurking λ's, *Working Paper ITS-WP-97-8*, Institute of Transport Studies, The University of Sydney

Ouwersloot H & Rietveld, P. (1996), Stated Choice Experiments with Repeated Observations, *Journal of Transport Economics and Policy*, May, 203-212

Swait,J. and Admowicz, W. (1996), The Effect of Choice Environment and Task Demands on Consumer Behaviour: Discrimming Between Contribution and Confusion, *Staff Paper 96-09*, Dept of Rural Economy, University of Alberta

Tweddle, G., Fowkes, A.S., and Nash, C.A. (1995), Impact of the Channel Tunnel : A Survey of Anglo-European Unitised Freight, Results of the Phase I Interviews, *Working Paper 443*, Institute for Transport Studies, University of Leeds, October

MODELS, FORECASTS, OUTTURN BEHAVIOUR IN THE CONTEXT OF DYNAMIC MARKET CONDITIONS

Austin W. Smyth, R. Jonathan Harron, Transport Research Group, University of Ulster, UK

ABSTRACT

The appropriateness of traditional travel models has recently been questioned on the grounds that behaviour change represents an ongoing process subject potentially to the effect of new policy initiatives in the long term. The travel market is a dynamic function of change in lifecycle stage as well as underlying demographic, economic and level of service effects. Belfast has recently experienced the opening of two new rail projects. The decision to proceed with these projects was informed by two modelling exercises. The paper considers not only the models, forecasts and outturn travel behaviour change but the appropriateness of the models in the context of a dynamic market.

1.   THE MODELLING CONTEXT

1.1   Overview and Modelling

The University of Ulster's Transport Research Group (TRG) is undertaking a low cost monitoring and evaluation study to assess the potential impacts of two rail schemes. Part of the exercise involved attempting to validate the modal forecasts with particular reference to the SP procedures. Two demand forecasting approaches, Revealed Preference(RP) and Stated Preference(SP), were employed in generating patronage projections for both schemes. A model system based on revealed preference (observed travel behaviour) was employed by government appointed consultants, Halcrow Fox Associates (H.F.A.). This suite of models employing a conventional four stage urban travel demand structure, had been developed to undertake the city-wide Belfast Transportation Strategy Review (B.T.S.R.) (Halcrow Fox and Associates,1987). In contrast to the H.F.A. models Northern Ireland Railway's consultants Transecon/Transmark employed a purpose designed series of stated preference modal choice models. In the case of the former the data upon which the models were calibrated was collected in the Autumn of 1985 while for the case of the stated preference models surveys were mounted in early 1986 (Transecon/Transmark,1986). A review of the N.I.R. models is given in Smyth (1989).

1.2   The Revealed Preference Model System

In order to undertake the B.T.S.R. H.F.A. developed a traditional four stage transportation model. The model was calibrated using data collected from household interview surveys, roadside interview surveys and surveys of public transport users (bus and rail). Of particular significance in the assessment of the rail projects are the distribution and modal split models. The former took the form of a classic two person type gravity model as follows:

$$T_{ij}^{n} = G_{i}^{n} A_{j} g_{i}^{n} a_{j} \exp(-\beta^{n} C_{ij}^{n}) \qquad (2.1)$$

where:

n = person type, (i.e. whether a person is a member of a car available or non car available household)

T_{ij}^{n} = trips generated in zone i and attracted to zone j by person of type n

G_{i}^{n} = trips generated in zone i by person of type n

A_{j} = trips attracted to zone j

C_{ij}^{n} = generalised cost of travel from zone i to zone j by person of type n, expressed in equivalent minutes

g_{i}^{n}, a_{j} = the balancing factors to achieve the doubly constrained model such that

$$\sum_{j} T_{ij}^{n} = G_{i}^{n} \qquad \sum_{i} \sum_{n} T_{ij}^{n} = A_{j} \qquad (2.2)$$

β^{n} = the model calibration coefficients to achieve the required trip cost distribution fit to survey results

The modal split model encompassed a range of level of service characteristics at a zonal level. It used the formula

$$P_{pt} = \frac{\exp(utility(pt))}{\exp(utility(pt)) + \exp(utility(car))} \qquad (2.3)$$

where

$$Utility\,(pt) = x_{1} \times income + x_{2} \times car\,ownership + x_{3} \times household\,size \qquad (2.4)$$

$$= y_{1} \times walking\,time + y_{2} \times waiting\,time + y_{3} \times in\,vehicle\,time$$
$$+ y_{4} \times parking\,charges + y_{5} \times fares \qquad = C \qquad (2.5)$$

The coefficients of x and y were calculated using household survey data and other surveys

1.3   The Stated Preference Modal Choice Models

A stated preference (SP) study was conducted in early 1986 with the aim of providing patronage and revenue forecasts for both the Cross Harbour Rail Link and the new railway station at Great Victoria Street. Obtaining patronage forecasts from the SP responses was undertaken using a two stage process. The initial calibration stage involved the estimation of the weights attached to each variable in the experiment. The variables were walking time from the station to Central Belfast, train in-vehicle time, 'rail-link' bus fare from the station to Central Belfast, train fare, train frequency and bus fare and car parking costs (See Table 1). The second stage in the forecasting process involved employing the estimated coefficient to produce demand forecasts. The incremental logit model was used for this purpose. This takes the form

$$P_t' = \frac{P_t \ \exp(V_t' - V_t)}{P_n \ \exp(V_n' - V_n)}$$ (2.6)

where P_t' and P_t are the market share of rail in the after and before situations respectively, (i.e. with GVS and without GVS). V_t' and V_t represent the utilities of rail in the after and before situations where

$$V_t = \Sigma a_i \ LOS_i$$

with a_i = the coefficient obtained from the calibration stage for level of service (LOS) variable *i*.

The logit model used is termed the log-odds model. It involves inputting a probability of travelling by train to the likelihood of use responses. The goodness of fit of the calibrated models were not particularly high. However it was encouraging to find that coefficient estimates were consistent with theory. The walk time coefficient generally exceeded those for train in-vehicle time as expected. Moreover, strong constant effects in favour of car were also found.

Although the process of collecting the SP data was considered satisfactory, the exclusion from the SP models of generated trips, redistributed trips and changes in land use patterns would tend to produce underestimates of the use of the new facilities.

Table 1: Coefficient Estimates and t Ratios for Train In-Vehicle Time and Walk Time (single journey)

	In Vehicle Time		Walk Time	
Antrim	-0.01897	(5.00)	-0.01176	(1.77)
Lisburn/Carrickfergus	-0.00719	(2.13)	-0.01042	(3.04)
Newtownabbey/Dunmurray	-0.01643	(5.37)	-0.01545	(4.98)
Ballymena/Portadown	-0.01046	(4.89)	-0.04801	(12.81)
Londonderry	-0.00527	(1.84)	-0.02836	(5.66)
Bangor	-0.01497	(2.35)	-0.03124	(4.83)
Larne	-0.01182	(5.21)	-0.01341	(3.38)

2. MODEL VALIDATION

Validation in the context of modelling is in practice usually restricted to goodness of fit achieved between predictions for the base year and observed behaviour. While Ortúzar and Willumsen(1994) stress that this is an essential element in model validation they call for before and after studies. A basic distinction can be drawn between cross sectional data, the typical basis for model validation and longitudinal/ time series data including panel surveys. Originally TRG intended to assess the wide reaching impacts of all major transport projects being implemented in the city in the early 1990's. However funding dictated resources be primarily channelled into the assessment of travel behaviour change and traffic/level of service. Fieldworkers undertook household and on board rail surveys before and after the opening of the bridge.

3. FORECASTS

3.1   Demand Forecasting: The Importance of Input and Background Assumptions

Model projections of travel demand are a function not only of the model structure and parameters but a wide variety of assumptions (explicit or implicit) including changes to the level of service offered by competing modes, the competitive response of alternative modes, macro economic performance and demographic change.   It is essential to compare the assumptions made concerning such factors and how accurate those assumptions have proved to be in practice.

3.2   Demand Forecasts based on the Stated Preference Models

Table 2 presents the forecast rail patronage generated by the SP models, in terms of total two-way flows per week. Overall the revised forecasts predicted a slightly lower increase in patronage attributable to Great Victoria Street. Annualising the weekly figures indicated that the additional traffic produced by the Cross Harbour link would be approximately 130,000 (or 235,000-287,000 including cross city traffic) while for Great Victoria Street ridership would rise by 544,000 for the west site (ultimately selected) to 681,000 for the east site.

Table 2: SP Generated Patronage Forecasts - Total of Rail Trips to and from *Central* Belfast per week

	With Cross Harbour Link		Link & GVS (East site)		Link & GVS (West site)
Lisburn	5132	(5132)*	5963	(5988)	5848
Carrickfergus	6393	(7150)	6745	(8501)	6628
Ballymena	3451	(3396)	5046	(4065)	4725
Portadown	10200	(10200)*	14887	(17016)	14140
Newtownabbey	3042	(3459)	3190	(4207)	3115
Dunmurray	2519	(2519)*	3001	(3251)	2937
Bangor	17694	(17694)*	21096	(19354)	20280
Londonderry	5960	(5834)	7348	(7296)	7093
Antrim	1627	(1340)	1718	(1605)	1672
Larne	4715	(4761)	4910	(4976)	4826
Total	60733	(61485)	73903	(76259)	71268

Note: Forecasts obtained previously are given in brackets. Columns may not sum due to rounding.
*   Travel to the centre of Belfast is not affected by the Cross Harbour Link alone.

Table 3: B.T.S.R. Revealed Preference Model Projections for NIR System (2001 AD)

NIR ANNUAL PATRONAGE ('000s)					
Do Minimum	Basic Transport Strategy (BTS)	Cross Harbour Rail Model Runs			Cross Harbour Rail + Great Victoria Street
		A	B	C	D	
6215	5419	6201	6484	6727	7538
		Index	(Do Minimum = 100)		
100	-	100	104	108	121
		Index (B.T.S. = 100)			
-	100	114	120	124	139
DAILY PEAK PATRONAGE					
N/A	4560	4860	5030	6000 N/A	N/A
		Index (B.T.S. = 100)			
-	100	106	110	132 -	-

A = Assumed No modal Shift from Private Transport   B = Provide for Modal Shift from Private Transport
C = As for 'B' plus Assumed Integration of Rail with Bus (as feeder)   D = Preferred B.T.S.R. Strategy (Excluding G.V.S.)

3.3   Demand Forecasts based on the Revealed Preference Models

Table 3 presents a summary of forecasts generated by the B.T.S.R. revealed preference model. It also incorporates indices of change using two 'base' figures, the first reflects the status quo in terms of transport infrastructure in 1986, and the second consistent with a basic transport strategy involving modest change to bus services and substantial highway development notably the Cross Harbour Road bridge. The annual additional ridership for the Cross Harbour link would be in the range 782,000-1,308,000 compared to the situation obtaining under the Basic Transport Strategy of *minus*14,000- *plus* 512,000 compared to the Do minimum approach. In the case of Great Victoria Street the revealed preference models suggest an increase in ridership of 807,000 per annum.

3.4   Adjusting the Demand Forecasts

It is considered appropriate to provide adjusted forecasts of change in demand based on a consideration of the various assumptions referred to.   Table 4 summarises these adjustments for both the revealed preference and stated preference models.

Table 4: Original and Adjusted SP and RP Forecasts

Forecast Base	Cross Harbour Bridge	Cross Harbour Rail + Great Victoria Street
(a) Stated Preference Models	Indicators Exclude Reference to Cross Border and non Belfast traffic	Indicators Exclude Reference to Cross Border and non Belfast traffic
Original Base = 100	$104/108^2$	$117/126^2$
Adjusted[1] Base = 100	$102/105^2$	$116/119^2$
(b) Revealed Preference Model	NIR System Patronage Indicator	NIR System Patronage Indicator
Do-Minimum = 100	100/108	121

1.   *Excluding Patronage Increase attributed to Londonderry line rerouting across Cross Harbour Bridge*
2.   *Include Nominal Increase in Cross City travel after Opening of Cross Harbour Bridge*

Table 5: N.I.R. Annual Patronage

Year	Ridership (millions)	Year	Ridership (millions)
1988/89	5.8	1993/94	5.73
1989/90	5.2	1994/95	6.14
1990/91	5.0	1995/96	6.48
1991/92		1996/97	6.18
1992/93	5.21		

Table 5 summarises N.I.R. patronage over recent years, highlighting the volatility of traffic due to political instability, the IRA cease-fire and the re-emergence in 1996 of community tensions and violence. The most meaningful comparison is between 1993/94 and 1994/95 and with 1995/96. 1993/94 represents the year prior to the opening of the new rail and road bridges and thus is typical of conditions referred to under the B.T.S.R. Do minimum scenario. 1994/95 represented the year in which both Cross Harbour Road and Rail schemes were opened (towards the end of the financial year) while 1995/96 experienced the opening of Great Victoria Street (approximately half way through the year). These point to net increases of some 400,000 passengers between 1993/94 and 1994/95 and a further 330,000 between 1994/95 and 1995/96. In both cases however cross border rail traffic, which was unaffected by either scheme, exhibited considerable volatility in ridership. However a number of points must be remembered: The Cross Harbour Rail bridge is not used by the Londonderry line services; The Belfast city centre 'Rail-link' bus service assumed to cease operation with the reinstatement of Great Victoria Street continued after the latter's opening; Larne line services were rationalised after opening of the Cross Harbour Rail bridge; Under the SP models it was assumed that a rather different timetable would operate; The IRA cease-fire extended from mid 1994 to early 1996 while 1996 witnessed an outbreak of serious community strife.

N.I.R. has itself undertaken a review of Great Victoria Street including a comprehensive programme of passenger surveys (Ulster Marketing Surveys, 1997). This, together with estimates by Ferguson (1996), indicates a net increase of some 11% in patronage to/from central Belfast between 1995 and 1996/97. This would suggest that Great Victoria Street is currently attracting approximately 500,000 additional rail passengers. This represents a net effect. T.R.G. reveal a much more dynamic picture with some two thirds of rail passengers surveyed in 1997 attracted to rail since the Cross Harbour Rail bridge opened, one third during the last year. The ratio of new users of rail attracted after opening of Great Victoria Street to those just before, but after opening of the Cross Harbour Scheme is approximately 2.4:1. (Table 6 and Table 7)

Table 6: Overall Turnover in the Commuter Rail Market (as reported before Implementation of the Rail Projects)

TIMESCALE OF USE	%	TIMESCALE OF USE	%
Less than 1 month	10	1 - 2 years	25
1 - 6 months	19	More than 2 years	41
6 - 12 months	6		

N.B.   Percentages may not add up to 100% due to rounding error

Table 7: Turnover in the Commuter Rail Market by Line (as reported after Implementation of the Rail Projects)

	LINE			
	Bangor	Larne	Lisburn	Portadown
TIMESCALE				
Less than 1 year	24	30	50	33
1 year - GVS Opened (9/95)	14	16	20	33
9/95 - Rail Bridge Opened (11/94)	10	9	5	8
Before Rail Bridge Opened	51	46	25	25

N.B.   *Percentages may not add up to 100% due to rounding error*

Table 8: Previous Stations used by Existing Rail Passengers on the Main Commuter Lines into/out of Belfast after Opening of the Rail Developments

PREVIOUS STATION	*Percentage*	*PREVIOUS STATION*	*Percentage*
Belfast Central	55	City Hospital	2
Yorkgate	19	Adelaide	2
Botanic	34	Other	1

N.B.   *Percentages may not add up to 100% due to rounding error*

Table 9: Previous Mode of New Rail Passengers on the Main Commuter Lines into/out of Belfast after the Opening of the Rail Developments (%)

	LINE			
	Bangor	Larne	Lisburn	Portadown
PREVIOUS MODE				
Car/Van (as driver)	45	22	37	20
Car/Van (as passenger)	14	9	8	20
Walk	9	9	6	
Bus	23	57	31	40
Cycle			2	
Other	9	4	16	20

N.B.   *Percentages may not add up to 100% due to rounding error*

Ongoing analysis of the T.R.G. household panel data and on board surveys is proceeding focusing not only on patronage change and turnover but also movements among long term rail users to alternative city centre stations (Table 8). This analysis is also investigating the previous mode by route and by other key market segments (Table 9).

3.5   Dynamic Effects

Goodwin (1997) has emphasised the lack of stability in everyday travel. However, stability is implied in conventional cross sectional transport models. He stresses the importance of changes in lifecycle. The panel survey revealed that among rail users recently attracted to that mode the majority indicated change of job and home as the most important factors influencing their choice. This adds credence to calls for the development of dynamic techniques or modifications to existing procedures, where possible, to accommodate dynamic effects. The effect of potentially induced traffic effects must also be considered when arriving at an overall conclusion. The Standing Advisory Committee on Trunk Road Assessment (1994), on the basis of an assumed elasticity concluded that an additional 2% of road trips across the city would be induced by the cross harbour road development. This suggests a requirement for further research into locally estimated induced traffic effects.

4.   CONCLUSIONS

Preliminary assessments of outturn performance in comparison with projected demand for rail services of a number of RP and SP models point to an encouraging level of compatibility. The SP model projections are closer to outturn performance both on the basis of the unadjusted as well as the adjusted projections. Both schemes have been implemented recently and, in the case of Great Victoria Street, additional traffic after some 18 months was within approximately 90% of that forecast by the SP models. Such apparent conformity should however, be treated with caution until comprehensive analysis of all the issues raised in this paper are addressed. Nevertheless, the findings do offer encouragement for the use of stated preference techniques including their use in direct forecasting if sufficient attention is paid to the design of the experiment, and it's execution. Techniques, such as those discussed by Ran and Boyce (1996), can be adapted to permit the inclusion of selected dynamic effects.

ACKNOWLEDGEMENTS

The authors would like to thank Mr Norman Maynes, Marketing Development Manager, Translink for some of the information supplied.

REFERENCES

Ferguson,D. (1996) Monitoring the effect of local rail network improvements to plan improved public transport provision for Belfast's two airports, *Proceedings of PTRC European Transport Forum* , London.

Goodwin,P.B., (1997) Solving Congestion, *Inaugural lecture for the professorship of transport policy,* University College, London.

Halcrow Fox and Associates (1987) A review of transportation strategy for Belfast (1986-2001):rail travel; proposals for change, Report Numbers 6,7,8 and 9, *Report submitted to the Department of the Environment (N.I.)*

Moser,C.A., Kalton,G. (1971) Survey methods in social investigation, Heinemann Educational Books, London.

Ortúzar,J de D, Willumsen,L.G. (1994) Modelling transport (second edition), John Wiley and Sons, Chichester

Ran,B, Boyce, D (1996) Modeling dynamic transportation networks (second revised edition), Springer, Berlin.

Smyth,A.W. (1989) The application of statistical modelling to strategic planning in a competitive market: a case study in rail transport, *Proceedings on 1989 I.L.I.A.M. 6 Conference,* University of Ulster, Belfast

The Standing Advisory Committee on Trunk Road Assessment (1994) Trunk Roads and the Generation of Traffic, HMSO, London

Transecon/Transmark (1986) Cross harbour rail link appraisal, *Technical Report 4, Report submitted to Northern Ireland Railways (unpublished)*

Transportation Planning Associates (1988) Technical audit of Great Victoria Street Railway scheme, *Report submitted to Northern Ireland Railways (unpublished)*

Ulster Marketing Surveys Ltd (1997) Great Victoria Street Railway station passenger survey, *Report submitted to Translink* (unpublished)

Using Consensus and Hierarchical Cross Impact Analysis in Modelling Residents' Perceptions in Their Choice of a Dormitory Settlement

Karen S. Donegan[†], Austin W. Smyth[†] and Henry A. Donegan[*]
[†]*Faculty of Engineering - Transport Research Group, University of Ulster*
[*]*Faculty of Informatics - Mathematics Division, University of Ulster*

Abstract

It is a feature of suburbanised towns and villages in the proximity of larger towns or cities, that the former patterns of rural travel behaviour change significantly. The nature of travel behaviour is mutable as a consequence of rural planning policy where small towns and villages are influenced by government area plans. Northern Ireland offers numerous examples of this phenomenon; and, as part of a wider investigation on sustainable development, this paper reports on a prototype study that assesses the pertinent views of residents within one suburbanised village under the remit of a defined area plan. Primarily, it describes the overall modelling strategy with an emphasis on:

- differences, in respect of travel behaviour, between newcomers and established residents, and
- pull factor weightings that illustrate why a suburbanised village is chosen by its residents.

More particularly, the paper discusses the merits and demerits of using a simplified *hierarchical cross impact* technique in modelling the perception of residents in their choice of dormitory settlement. In supporting this technique, the application of the Gaussian *neighbourhood consensus function* (Dodd, 1993) is introduced as an opinion profiling device.

1. Introduction

Setting the Scene

Although the objectives of this research project are, i) to find out why people choose to move to or remain in suburbanised towns or villages and ii) to collect statistically significant indicators that partition the residents' travel behaviour, this paper is directed at the research methodology and not the consequences. In attempting to find out why people choose to live in a certain town their opinions were sampled and a multi-attribute strategy was adopted. It was argued that the latter would reduce the adverse impact of asking direct attitudinal questions about their choice of settlement. Popular reasons for choosing where to live were determined using expert opinion, and through the medium of a questionnaire, these reasons were each accorded a weighting relative to an approved selection of attributes. The resulting matrix analysis produced priority vectors that revealed behavioural differences between the views of *newcomers* and *established residents*. In assessing opinions from the questionnaire, the problem of global consensus emerged. This was resolved using an adaptation of the normal distribution function that produced opinion profiles. These revealed, for each solicited opinion, a focal point of consensus in the opined interval [0, 5] and a corresponding consensus weighting in an interval]0, 1]. Matrix opinion methods and consensus techniques were also used when local transport issues such as route availability, service frequency and travel comfort were sampled.

The Background

More by coincidence than design, the research theme relates to some of the implicit issues in the current UK debate affecting the future balance between brownfield and greenfield housing. It remains to be seen whether the conclusions of this prototype work will contribute usefully to the debate. However the strategies set forth are being implemented in a much broader study that also embraces sustainability as a key attribute and it is expected that the findings will have an impact on rural planning issues, particularly in Northern Ireland; therefore the reader should not loose sight of the fact that this is merely a prototype study. In comparison with its urban counterpart (IATB, 1989; Marshment, 1995; Stopher and Lee-Gosselin, 1997), there is a dearth of published research on rural travel behaviour. With the exception of aspects of Nutley's work (1992, 1995), centred in Northern Ireland, there is little tangible research on the impact of suburbanisation as it relates to the rural travel scene.

To date, the thrust of this work is concentrated in and around an established suburbanised village, and while many locations in Northern Ireland meet the criteria of suburbanisation, Crumlin in Co Antrim was chosen primarily because it was at a well developed stage under the *Antrim Area Plan*, (DOE, 1989). So far the findings point to some characteristics that assist in the description of rural travel behaviour and rural choice; further rural studies are planned.

Today, perhaps as a consequence of the *Antrim Area Plan*, Crumlin has a total population approaching 4000, with marked housing and small industrial developments. Located close to Belfast, Antrim and Lisburn it is well served in terms of public transport. There is both a bus depot and a train station; these along with the local taxi company result in good access throughout the province. The proximity of Belfast International Airport as a major land-use within the Antrim District and a significant employer generates substantial traffic movements.

2. THE PROBLEM AND THE METHODOLOGY

Primary Objective and Hypothesis

A primary objective of the study, given the proximity of Crumlin to major towns is:

To weight and hence rank the reasons that make Crumlin a residential choice.

Secondly, given that travel behaviour is related to both socio-economic demography and transport, there is a fundamental hypothesis that requires investigation, *viz*:

There are clear differences between the travel patterns of established residents and those of newcomers to Crumlin.

Since questionnaires are considered to be the most important primary data source available in Human Geography (Lenon and Cleaves, 1983), the technique is chosen to help resolve the hypothesis and to underpin the primary objective. This provides a *social scientific* basis for replication studies on other metrically remote suburbanised gatherings.

The questionnaire has twenty two questions arranged as: *Section 1* - Personal Details of Householder, *Section 2* - Householder Opinion and *Section 3* - Public Transport. The format is arranged in such a way that respondents merely place a ✓ in the appropriate box ☐. Section one, the nominal part, comprises 13 questions designed to accumulate data in respect of respondents' socio-economic status, while Section two with 3 questions and numerous subparts comprises the ordinal part requesting opined responses (see - Table 1). Edwards *et al.* (1975), have shown that judgements of magnitude, while amenable to semantic scaling, are best made from a numerical scale.

Table 1. Equivalent Likert range

0	1	2	3	4	5
Not important	Very Poor	Poor	Satisfactory	Good	Very good

The purpose of these questions is to arrive at a 'consensus priority listing' amongst the nine chosen pull factors *viz*. HOUSING, PUBLIC TRANSPORT, GOODS/SERVICES, EMPLOYMENT, SECURITY, EDUCATION, RECREATION, PROXIMITY OF AIRPORT AND DEVELOPMENT. Finally Section three contains 6 questions seeking both nominal and opined responses in respect of Crumlin's public transport.

Sampling Strategy

With the population of Crumlin approaching 4,000 it was considered impractical to question every house-hold in the community and a sample size of 100 was chosen in the interests of expediency. Ideally, for the purpose of subsequent generalisation, a representative sample was calculated using the model presented by Hammond and McCullagh (1974) - taking into account a sample distribution of *newcomers* versus *established residents*.

$$\text{Sample size } n = p \cdot q \cdot \left[\frac{z}{d}\right]^2$$

where arising out of a pilot study:

p represents the percentage of newcomers,
q represents the percentage of established residents,
z corresponds to the normal distribution score relative to a specified level of confidence,
d is the tolerable margin of error at the specified level of confidence.

In this case the pilot study revealed 17 newcomers and 13 established residents which together with a 95% confidence level and a 5% margin of error reveals a sample size of 376. Under rigorous conditions, where a postal enquiry would be envisaged this sample size is desirable, but given the nature of this enquiry it was accepted that a higher margin of error, approximately 10% was tolerable. The sample size is proportioned in accordance with the number of dwellings in each sub-area of Crumlin. The sub-areas A, B, C, . . . , V, W, X, Y (defined in Table 2), fit naturally with the nature of the built environment. Table 2 shows the proportions which are deduced by counting the number of dwelling units in each sub-area as a fraction of the sum of the dwellings in all the sub-areas. These fractions, with the implicit assumption of a uniform spread, are used to apportion the sample size of 100 for each sub-area. To simplify the statistical analysis, each of the sub-areas are arranged in groups numbered 1 to 10 in accordance with tenure, type and date of construction. So that bias is reduced to a minimum, a random selection technique is used in distributing the questionnaires.

Table 2. Proportions and Sample Sizes

Sub-area	Code	Units	Fraction	Sample Size	Group	Group Size
Orchard Grove	A	26	0.023	2	1	12
Orchard Hill	B	47	0.041	4		
Lake View	C	70	0.061	6		
The Beeches	D	140	0.121	12	2	25
Langarve Manor	E	146	0.126	13		
Cairn Gdns/Tce	F	62	0.054	5	3	13
Canning Grove	G	40	0.035	4		
Cidercourt Rd	H	24	0.021	2		
Camlin Gdns	I	22	0.019	2		
Hartswood	J	43	0.037	4	4	8
Cidercourt Park	K	47	0.041	4		
Berkley Court	L	19	0.016	2	5	3
Parkley	M	14	0.012	1		
Glendarragh Pk	N	22	0.019	2	6	7
Camlin Park	O	25	0.022	2		
Glenoak Gdns	P	29	0.025	3		
Glenoak Grange	Q	12	0.010	1	7	3
Parkfield	R	18	0.016	2		
Tromery Park	S	94	0.081	8	8	10
Cumbria Lodge	T	20	0.017	2		
Bramble Wood	U	115	0.100	10	9	14
Weavers Meadow	V	50	0.043	4		
Hill Street	W	8	0.007	1	10	7
Main Street	X	52	0.045	5		
Helen Street	Y	10	0.009	1		

3. THEORETICAL CONSIDERATIONS

The Cross Impact Strategy

The style of the questionnaire is somewhat novel for this type of data collection in that it contains an embedded (latent to the respondent) model of an attribute hierarchy as a basis for the multi-criteria evaluation. The latter refers to the prioritisation of those *pull factors* that make Crumlin a desirable place to live and relies on attitudes rather than the mere collection of nominal data that is characteristic of Section One. The questioning strategy is formulated on the basis of an Analytic Hierarchy (Saaty, 1980), but uses the much more amenable method of Cross Impact Analysis (Dodd, 1993) to determine the multi-attribute weightings. In such a hierarchy, using the terminology of Dodd, the uppermost level is the *PRIMARY OBJECTIVE* or *POLICY*. Level two, comprises the *SECONDARY OBJECTIVES*, and these are supported at level three by a series of *TACTICS* which are directly related to the set of alternative *COMPONENTS* at the lowest level (*there can be lower sub-component levels*). The hierarchical model is amenable to either *cross impact* analysis or analysis involving *pairwise comparisons* - methods of analysis that are familiar to multi-criteria decision makers. In either case the aim is the attachment of subjective weights to all of the objectives, tactics and components, so that in vector form, the *COMPONENTS* are prioritised with respect to each *TACTIC*; the *TACTICS* are prioritised with respect to each *OBJECTIVE* and finally *OBJECTIVES* are prioritised with respect to *POLICY*. Ultimately, the decision weight vectors at each level are grouped in matrix form such that, if they are identified as :

C/T the *COMPONENTS* to *TACTICS* matrix,
T/O the *TACTICS* to *OBJECTIVES* matrix and
O/P the *OBJECTIVES* to *POLICY* matrix ,

then the normalised matrix product: **C/T** $\otimes$ **T/O** $\otimes$ **O/P** yields a *COMPONENTS* to *POLICY* vector **C/P** which gives the desired prioritisation weights.

The hierarchy used in this study is illustrated in Figure 1, and bearing in mind the over-riding condition of independence of issues on each level, the final selection was made only after numerous expert consultants gave their opinion as to the credibility of the various issues listed.

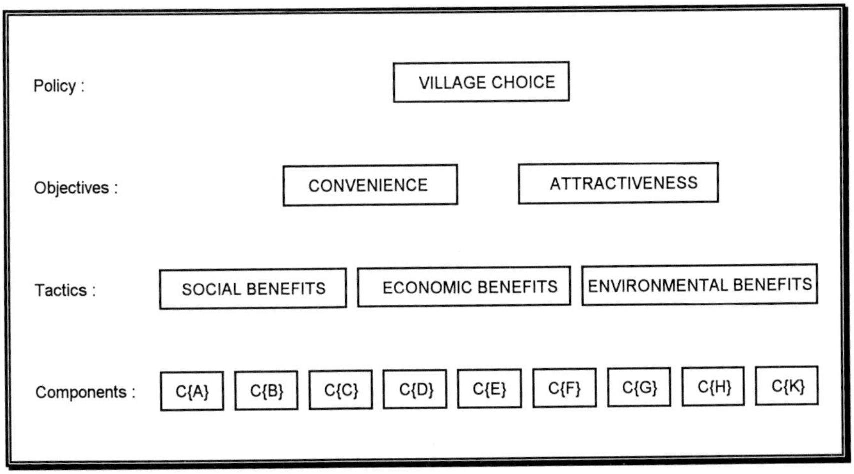

Figure 1. The Analytic Hierarchy

The supporting philosophy for such a strategy as pointed out by Shields *et al.*, (1987) aims at removing the remoteness of the *Policy* from the various *Components* in the decision making process. In simple terms, this means that rather than elicit a prioritisation of the *COMPONENTS* on an *ad-lib* basis directly with respect to *POLICY*, it is better practice to seek incremental priority listings via each *TACTIC*; and similarly to prioritise the *TACTICS* incrementally with respect to each *OBJECTIVE* and so on. The desired priority is then deduced using matrix software by Dodd (1993).

With the principal objective being to assess the relative importance of the above mentioned *pull factors*, the greatest difficulty in the hierarchical design was in obtaining expert agreement on the second level objectives, *viz. CONVENIENCE* and *ATTRACTIVENESS*. Their eventual selection appeared to represent the broadest spectrum of opinion appropriate to ranking the tactics, *viz.* the *SOCIAL, ECONOMIC* and *ENVIRONMENTAL BENEFITS*. There is no real problem in agreeing the independence of the tactics, but at the components level, it is perhaps prudent to agree with Saaty (1980) and work with the assumption of independence. The components are coded as follows:

C{A}= HOUSING   C{D}= EMPLOYMENT   C{G}= RECREATION
C{B}= PUBLIC TRANSPORT   C{E}= SECURITY   C{H}= AIRPORT PROXIMITY
C{C}= GOODS & SERVICES   C{F}= EDUCATION   C{K}= DEVELOPMENT

The unformalised approach used in the Edinburgh model (see Dodd, 1993) is perhaps the most basic form of cross impact assessment. The decision maker, using a Likert scale with values identified as in Table 1 above, assesses the importance of each element on a given hierarchical level relative to each element on the next highest level and produces data corresponding to an opinion matrix such as **C/T** or **T/O** or **O/P**. The simplicity of the cross impact approach has an obvious appeal because of its inherent consistency (Dodd, 1993) as distinct from the transitive consistency check in Saaty's AHP. When more than one decision maker is involved (which is this case for each of the 10 sub-areas of Table 2), a corresponding consensus matrix that summarises all the decisions is calculated using the neighbourhood consensus analyser (Donegan, 1996).

Table 3. COMPONENTS to TACTICS Consensus Results
Established Residents (E/R) and Newcomers (N)

TACTIC →	SOCIAL		ECONOMIC		ENVIRONMENTAL	
COMPONENT ↓	E/R	N	E/R	N	E/R	N
HOUSING	3.7	3.9	3.6	4.1	3.3	4.0
PUBLIC TRANSPORT	2.3	2.0	2.0	2.1	2.3	2.9
GOODS/SERVICES	3.0	3.1	2.9	2.9	2.9	3.0
EMPLOYMENT	2.7	1.0	2.8	1.1	2.8	1.1
SECURITY	2.3	3.0	2.3	3.0	2.4	3.0
EDUCATION	3.5	2.9	3.4	3.2	3.2	3.0
RECREATION	1.5	1.4	1.6	1.7	1.5	1.5
PROXIMITY OF AIRPORT	4.8	4.6	4.7	4.4	4.6	4.3
DEVELOPMENT	3.3	2.9	3.3	3.0	2.7	2.9
MINITAB CODING	C1	C2	C3	C4	C5	C6

Table 3 illustrates the global consensus data for the *components* to *tactics* matrices of both the established residents and the newcomers. The number of rows corresponds to the number of elements on the *given* level and the number of columns to the elements on the level immediately above. The dimensions of all the resulting matrices for the complete hierarchy permit their consecutive multiplication by the matrix product **C/T** $\otimes$ **T/O** $\otimes$ **O/P**. The system facilitates inter-group as well as global ranking comparisons.

Opinion Profiling

The novel concept of neighbourhood consensus (Donegan and Dodd, 1991) is fully addressed in Dodd (1993) which points to its advantages over traditional techniques in the pursuit of voting profiles; of particular novelty is the implicit modelling of each respondent's uncertainty. There are numerous uncertainty functions that can be chosen - in this study the most appropriate is a model based on the normal distribution. The immediate advantage of neighbourhood consensus over other approaches lies with the fact that it gives a pictorial representation, as an opinion profile, of all the respondents' views. The horizontal axis depicts attitude (usually a pseudo Likert scale [0, 5]) and the vertical axis]0, 1] indicates the strength of consensus. For example the ordered pair (2.8, 0.69) represents a *focal point* of consensus at 2.8 in the interval [0, 5] with a 69% consensus weighting. This has been referred to as *definitive consensus* (Donegan and Dodd, 1991). Figure 2 illustrates the screen input in pursuit of a consensus appraisal for the situation involving newcomers expressing their respective views on *SOCIAL ADVANTAGE* with respect to *ATTRACTIVENESS*.

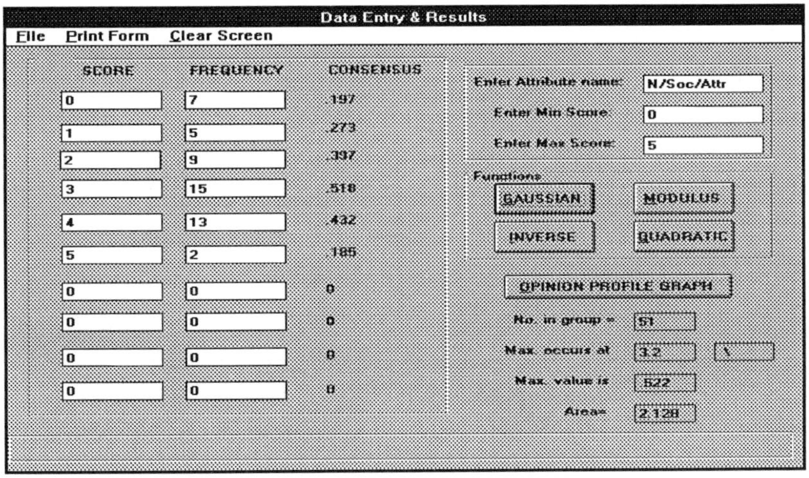

Figure 2. Typical Screen Input to Determine Consensus Output

The results are displayed on the screen after the user of the Neighbourhood Consensus Analyser selects the *function type* button. Figure 3 illustrates the corresponding consensus profile, which is quite useful in a comparative situation particularly in situations where a split consensus exists.

The Neighbourhood Consensus Function

A profiling function that spreads the effect of a score x_i over [0,5] so that its maximum (normalised to 1) occurs at x_i and its value decreases as the numerical distance from x_i increases, is the Gaussian form:

$$g(x,x_i) = \exp\left[- (x - x_i)^2 \, \text{Ln } 2\right].$$

This function, having a variance 1/2Ln2 satisfies two important criteria:

1. the expectation that the distribution of a score x_i should be tolerably uniform over the neighbourhood $]x_i - 0.5, x_i + 0.5[$, (The neighbourhood radius 0.5 was chosen to be consistent with the convention that Likert scales normally comprise ten permitted sub-intervals.) and

2. that its effect should drop significantly (from 1.0 to 0.5 in this case) at one neighbourhood diameter from x_i, fading to insignificance in excess of two neighbourhood diameters.

It is shown in Donegan and Dodd (1991), that the corresponding *neighbourhood consensus* function $f(x)$ for a collection of scores $\{x_i\}_{i \leq n}$ is given by :

$$f(x) \quad = \quad \frac{1}{n} \sum_{i=1}^{n} g(x, x_i).$$

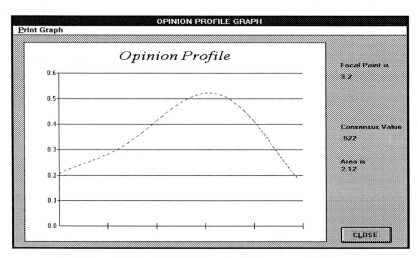

Figure 3. Typical Neighbourhood Consensus Profile

Although a series of statistical experiments would be necessary to clarify the exact nature of the distribution (particularly with regard to (2) above), recent work by Genest and Zhang (1996) has shown that respondents' errors are indeed normally distributed, thereby offering at least a tentative validity to this Gaussian assumption. Clearly, trading against the possibility of a diminishing return on effort, the pursuit of validity must be considered in context; this is particularly so where the variance of subjectivity could far outweigh errors introduced by the adoption of a slightly inaccurate distribution.

Figure 4* goes some way towards demystifying the above idea. It shows four DM profiles $g(x,1.5)$, $g(x,2.1)$, $g(x,2.5)$, and $g(x,2.6)$ together with the corresponding consensus function:

$$f(x) = [g(x,1.5) + g(x,2.1) + g(x,2.5) + g(x,2.6)]/4$$

yielding the maximum amplitude co-ordinates (2.2, 0.884). This means that the focal point of consensus is 2.2 with the degree of consensus weighted as 88.4% for the four DMs. The global profile is by far the most attractive feature of the concept of neighbourhood consensus as it conveys the summary information in the most effective way.

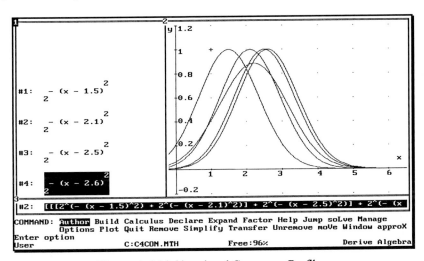

Figure 4. Neighbourhood Consensus Profiles

4. EXAMPLES OF TYPICAL RESULTS

The matrices shown in Figure 5 represent the sample consensus views of newcomers and established residents (the latter in parenthesis) yielding the *pull factor* rankings shown.

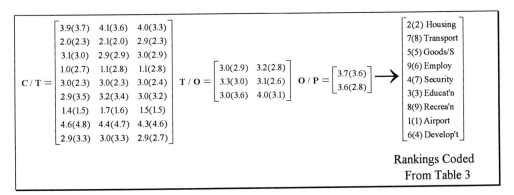

Figure 5. Pull Factor Data and Results for Newcomers and Established Residents

---

* Output from DERIVE A Mathematical Assistant, Version 3.06. Soft Warehouse Inc, 3660 Wail, Waialae Ave, Suite 304, Honolulu, Hawaii, USA.

A fundamental outcome from this enquiry points to a set of key partitioning parameters that distinguishes newcomers and established residents. These are set out in Table 4 together with the corresponding χ^2 data. Using other consensus data from Section Three of the questionnaire a variety of correlation matrices (too extensive to present in this short paper) reveal statistically significant (5%) differences in activity related transport behaviour between established residents and newcomers thereby upholding the stated hypothesis. A fuller set of results may be obtained through communication with the leading author.

Table 4. Significant Partitioning Parameters

PARAMETER	df	χ^2	5% Critical
Work Location of Residents	1	15.984	3.84
Town Choice for Bulk Shopping	2	16.814	5.99
Town Choice for Leisure	1	5.323	3.84
Town Choice for Health Activities	2	6.025	5.99
Town Choice for Commercial Activities	2	18.667	5.99
Frequency of Commercial Activities	1	7.033	3.84

An important conclusion has been the revelation that questionnaires initially designed to elicit expert attitudes in support of prioritisation software are likely to be flawed when used to capture general opinion. Measures are needed to structure questions that mitigate against the selection of "middle of the road" opinions - a common feature of decisions by non experts.

REFERENCES

Dodd, F.J., *Consensus and Prioritisation in Analytic Hierarchies*, DPhil Thesis, University of Ulster, (1993).

DOE (NI), *Antrim Area Plan 1984-2001*, HMSO (1989).

Donegan, H.A. and Dodd, F.J., An Analytical Approach to Consensus. *J. Appl. Math. Lett.* **4**,(2), 21-24, (1991).

Donegan, H.A., A Neighbourhood Consensus Analyser, Development Software, University of Ulster, (1996).

Edwards, W., Guttentag, M. and Snapper, K., "Effective Evaluation: a decision theoretic approach," in *Evaluation and Experiment :* (Editors: Bennett, C.A. and Lumsdaine, A.). New York: Academic, (1975).

Genest, C. and Zhang, S. -S., Hilbert's Metric and the Analytic Hierarchy Process. *Mathl. Comput. Modelling*, Vol. 23, No. 10, pp. 71-86, (1996).

Hammond, R. and McCullagh, P., *Quantitative Techniques in Geography*. Clarendon Press, (1974).

IATB, *Travel Behaviour Research*. Gower, (1989).

Lenon, B.J. and Cleves, P.G., *Techniques and Fieldwork in Geography*. UTP, (1983).

Marshment, R.S., *Urban Travel Characteristics Database*. Inst. of Transport'n Engrs, TCC 6Y-46, (1995).

Nutley, S. and Thomas, C., Mobility in Rural Ulster: Travel Patterns, Car Ownership and Local Services. *Irish Geography*, Vol. 25, No.1, pp. 67-82, (1992).

Nutley, S. and Thomas, C., Spatial Mobility and Social Change: The Mobile and the Immobile. *Sociologia Ruralias*, Vol. XXXV, No. 1, pp. 24-39, (1995).

Saaty, T.L., *The Analytic Hierarchy Process*. McGraw Hill, (1980).

Shields, T.J., Silcock, G.W., Donegan, H.A. and Bell, Y.A., Methodological Problems Associated with the Delphi Technique. *Fire Technology*, Vol. 23, No. 3, pp. 175-185, (1987).

Stopher, P. and Lee-Gosselin, M., *Understanding Travel Behaviour in an Era of Change*. Pergamon, (1997).

THE DELAYING EFFECT OF STOPS ON A CYCLIST AND ITS IMPLICATIONS FOR PLANNING CYCLE ROUTES

Roland Graham, Department of Mathematical Sciences, University of Liverpool, UK

ABSTRACT

Some cycle routes have proved unpopular because at several points a cyclist has to stop or slow very appreciably. Cyclists are discouraged by the resulting extra journey time. Three alternative hypotheses are made for a cyclist's performance in response to an imposed stop. In all he brakes uniformly from a cruising speed. In one, this speed is a personal standard, and he accelerates uniformly to regain it. In the second, because his energy is used in accelerating, he has less to use in cruising. The third hypothesis is that a cyclist's power output in accelerating is the same as in cruising, except at very low speeds, so that he attains his normal cruising speed asymptotically.

Trials were conducted on a lightly trafficked 2.5-km circuit with seven roundabouts. Cyclists made at least two circuits each, one non-stop and one stopping at roundabouts. Pairs of times for each cyclist were compared with derivations from each hypothesis. It was found that the estimates derived from the second hypothesis were much closer to the trial times than those given by the others. The conclusion is that making an adult cyclist stop is equivalent to extending the journey time by as much as an extra 50 m, approximately, would take. In reckoning the lengths of routes in order to select one for development, an equivalent distance of about 50 m should be added for each stop or near stop, unless they are close together.

1. CYCLE ROUTES AND THEIR ATTRACTIONS

The two major purposes for the establishment of urban cycle routes in Britain have been to improve the safety of cycling and to attract more people to cycle. Some routes have been created by adopting alignments along minor roads and paths, with facilities to cross major roads. A major factor in improving the safety of such a route is to attract people who previously cycled. This depends on various characteristics such as scenic views, qualities of riding surface, and, importantly, the time taken to traverse the route compared with that for a previous route.

The time difference has been more commonly expressed in terms of distances. In the 1980s a common rule of thumb was that a designated cycle route should not be more than 10% longer than an alternative route. More recently this has been replaced by the rule that cyclists should go as small a distance as possible (see, e.g., C.R.O.W, 1993).

Both of these rules are naive in that, whenever there is a choice of two routes, some travellers can be expected to take one and the others to take the other. If a route can be varied according to a parameter, the proportions are expected to depend on this logistically (see, e.g., Hensher and Johnson, 1981).

The specific purpose of this paper is to analyse the time taken to traverse a route, which is a major factor in its attraction. It is assumed that the route is flat, so that the time taken depends primarily on the length of the route. Wind is ignored, but air resistance determines a cyclist's cruising speed. The time also depends on traffic conditions, in particular the need to pause, i.e. stop or virtually stop, where other traffic has priority. It further depends on the alignment of the route, especially at corners where a cyclist's speed is all but reduced to zero. The effect on the time due to these pauses is the subject of the analysis.

That such pauses have an effect on the attractions of a route has been shown through responses to a questionnaire of Liverpool student cyclists (Graham, 1988).

2. CRUISING, BRAKING, AND ACCELERATING

When a cyclist has conditions that enable him to maintain a constant speed, he will normally do so. This normal cruising speed is determined by his power output and the air resistance. In contrast to constant cruising speeds, cyclists may brake and accelerate in many ways. Braking to a foreseen stop normally has a larger magnitude than acceleration, and takes little time; but three basic rules of acceleration from rest can be envisaged.

The first rule is simple: the cyclist maintains constant acceleration till he has reached normal cruising speed. As a consequence of this, a typical cyclist uses more power in accelerating than in cruising. If he attains his normal cruising speed, his average power output would be higher, which is likely to produce discomfort either because of the high demand on his system or because he gets too hot for his clothing. The likely consequence is that his cruising speed is reduced below normal. This leads to the second rule for acceleration: it is constant till the cyclist attains a cruising speed below his normal cruising speed. The third possible rule for acceleration is that the cyclist's power output is the same as when cruising; in this case his speed tends asymptotically to normal cruising speed.

The air resistance experienced by a cyclist is proportional to the square of his speed v, as verified by Whitt and Wilson (1982). Thus his power output is

$$p = mv\frac{dv}{dt} + cv^3 = \frac{m}{3}\frac{dv^3}{ds} + cv^3,$$

(2.1)

where m is the total mass of the cyclist and bicycle, c is a constant related to them, and s is the distance travelled in time t. Hence in cruising at constant speed v_0 the power output is

$$p_0 = cv_0^3. \tag{2.2}$$

According to Whitt and Wilson, typically $c = 0.27$ kg/m. The value depends on the cyclist; this one is consistent with a frontal area of 0.5 m^2. In equation (2.1) the ratio m/c is significant. Taking 84 kg as a typical value of m, a typical value for m/c is 312 m.

More significantly for this paper, we take a typical braking rate of 3 m/s^2, and a typical acceleration of 1.5 m/s^2. These values will provide a guide to which terms may be neglected.

3. THE TIME LOST IN PAUSES

Braking to a pause is straightforward. Although it is unlikely to be perfectly regular, we suppose that the cyclist's retardation in braking is a constant r, i.e. the acceleration is $-r$, produced partly by braking and partly by air resistance. Then the time taken to stop from speed v is v/r, and the distance travelled is $v^2/2r$.

Suppose that a cyclist's acceleration follows the first rule given in Section 2: the cyclist maintains a constant acceleration until the normal cruising speed, v_0, is reached. The results are similar to braking: the time taken is v_0/a; and the distance travelled is $v_0^2/2a$, which would only take a time $v_0/2a$ at the cruising speed v_0. Thus if the cyclist only pauses momentarily, the extra time taken by the pause, through both braking and accelerating, is

$$\frac{v_0}{2a} + \frac{v_0}{2r}. \tag{3.1}$$

Now suppose that a cyclist's acceleration accords with the second rule envisaged in Section 2, so that his maximum speed $v_m < v_0$. Further suppose that, as with the first rule, he maintains a constant acceleration a from a standing start till his speed is v_m. Then, from equation (2.1), the energy expended in accelerating is

$$\int_0^{v_m/a} p\,dt = \int_0^{v_m/a} \left(mv\frac{dv}{dt} + cv^3 \right) dt = \frac{mv_m^2}{2} + \frac{cv_m^4}{4a}. \tag{3.2}$$

The distance spent in accelerating is $v_m^2/2a$, and that in braking $v_m^2/2r$, so in covering a distance s_1, greater than the sum of these, from start to stop, we find the total energy expended, using equation (3.2), is

$$cv_m^2\left(s_1 - \frac{v_m^2}{2a} - \frac{v_m^2}{2r} \right) + \frac{mv_m^2}{2} + \frac{cv_m^4}{4a} = \frac{v_m^2}{2}\left[2cs_1 - \frac{cv_m^2}{2}\left(\frac{1}{a} + \frac{2}{r}\right) + m \right]. \tag{3.3}$$

The total time taken is similarly

$$\frac{1}{v_m}\left(s_1 - \frac{v_m^2}{2a} - \frac{v_m^2}{2r} \right) + \frac{v_m}{a} + \frac{v_m}{r} = \frac{1}{v_m}\left[s_1 + \frac{v_m^2}{2}\left(\frac{1}{a} + \frac{1}{r}\right) \right]. \tag{3.4}$$

Hence if the journey is made up of sections of lengths s_i, $i=1(1)n$, between stops, with $\sum_{i=1}^{n} s_i = s$, the total energy expended in the journey is

$$\frac{v_m^2}{2}\left[2cs-\frac{ncv_m^2}{2}\left(\frac{1}{a}+\frac{2}{r}\right)+nm\right],$$

and the total time taken is

$$\frac{1}{v_m}\left[s+\frac{nv_m^2}{2}\left(\frac{1}{a}+\frac{1}{r}\right)\right]. \tag{3.5}$$

Assuming that the average power output during the journey is the same as if the cyclist maintained his normal cruising speed throughout, with power output p_0 as given by equation (2.2),

$$\frac{v_m^3}{2}\frac{\left[2cs-\frac{ncv_m^2}{2}\left(\frac{1}{a}+\frac{2}{r}\right)+nm\right]}{\left[s+\frac{nv_m^2}{2}\left(\frac{1}{a}+\frac{1}{r}\right)\right]}=cv_0^3,$$

whence

$$\frac{v_0^3}{v_m^3}=\frac{1-\frac{nv_m^2}{4s}\left(\frac{1}{a}+\frac{2}{r}\right)+\frac{nm}{2cs}}{1+\frac{nv_m^2}{2s}\left(\frac{1}{a}+\frac{1}{r}\right)}.$$

Both the numerator and denominator of this expression are dominated by 1, so a power series expansion may be made. Taking the typical values for acceleration and deceleration given in Section 2, and a value for nm/cs from the trials described in the next section, that seems typical of urban cycle routes, it is appropriate to expand to first order in nv_m^2/sa and nv_m^2/sr, and to second order in nm/cs. This achieves an approximation to within the order of one per cent, which is adequate for the purpose of this paper. It is found that

$$\frac{v_0}{v_m}=1-\frac{nv_0^2}{12s}\left(\frac{3}{a}+\frac{4}{r}\right)+\frac{nm}{6cs}-\frac{n^2m^2}{36c^2s^2};$$

here v_0^2 has been written in place of v_m^2, which is consistent to the approximation used. Hence, from equation (3.4), the total time taken, to the same approximation, is

$$\frac{s}{v_0}\left[1+\frac{nv_0^2}{12s}\left(\frac{3}{a}+\frac{2}{r}\right)+\frac{nm}{6cs}-\frac{n^2m^2}{36c^2s^2}\right]. \tag{3.6}$$

If the cyclist's acceleration accords with the third rule of Section 2, so that his power output is as in normal cruising, given by equation (2.2), then from equation (2.1)

$$\frac{dv^3}{ds}+\frac{3cv^3}{m}-\frac{3p_0}{m}=0. \tag{3.7}$$

This equation shows that as $v\to0$, $dv/ds\to\infty$, so it can not be followed absolutely. Using the typical values given in Section 2, $dv/dt=5$ m/s^2 when $v=0.14$ m/s. If this speed is attained from rest with constant acceleration 5 m/s^2, the time taken is 0.028s and the distance covered is 2 mm. The equation gives smaller values and the differences between these and those with constant acceleration are negligible. So, using equation (3.6), if the cyclist starts from rest with $v=0$ at $s=0$,

$$v^3=v_0^3(1-e^{-3cs/m}).$$

Since $v=ds/dt$, the time taken to reach a distance s_0 is

$$t = \int_0^{s_0} \frac{ds}{v_0(1 - e^{-3cs/m})^{1/3}}.$$

Thus, for large s_0 the extra time taken to cover the distance due to the need to accelerate is approximately

$$\frac{1}{v_0} \int_0^\infty \left[\frac{1}{(1 - e^{-3cs/m})^{1/3}} - 1 \right] ds = \frac{I\,m}{3cv_0}, \tag{3.8}$$

where

$$I = \int_0^\infty [(1 - e^{-x})^{-1/3} - 1]dx \approx 0.75.$$

For a distance of about 82% of m/c, the extra time is about 96% of this, so for the accuracy attained in this paper the approximation is adequate for distances of over about 260 m.

4. TRIALS OF CYCLISTS

To test the hypotheses, some trials of cyclists were conducted. A level circuit was used in order to reduce the effects of hills, which would be difficult to estimate, and of traffic, which would require a large number of trials to produce a reliable average. The circuit was 2.5 km in length in New Brighton, Wirral, and included seven roundabouts. At the times of the trials it was lightly trafficked.

Cyclists were timed going round the circuit in two modes of operation, each cyclist going singly at their own speed and completing circuits in each mode. In one mode, a cyclist stopped only in completing a circuit, unless held up by traffic. In the other, he paused at six or all seven roundabouts. In both modes the start was from rest. Most cyclists made two or more journeys in each mode.

If a cyclist got held up by traffic at a roundabout, he estimated the time and reported it to the timekeeper. The estimated delay was then subtracted from his time for the circuit. This occurred in only one case, and the time involved was only a second, which was small compared with the differences in times by one cyclist completing two circuits in the same mode.

Though they were not told enough to appreciate the differences between the hypotheses, the cyclists were aware of the general purpose of the trials, and were told that accuracy of timings according to their normal behaviour was important for the scientific nature of the investigation. Twelve cyclists took part in trials. The times in each mode of one of them were very inconsistent. Another was a racing cyclist whose average extra time with pauses was only 17 s; those expected by all three hypotheses were over 27 s. Possibly he easily reached maximum speed, limited by a low top gear; but also his times with pauses are suspect. So the times of these two cyclists are ignored in the analysis that follows.

Table 1 has a row for each rider. One column is for the average time without a planned pause and one for the difference between this and the average time with pauses. Two columns give the average values of n, equal to the number of pauses plus 1, in the two cases. Three columns give the estimates of the normal cruising speed v_0 according to the three hypotheses for acceleration, using the values for a, r, and m/c given in Section 2 and equations (3.1), (3.6) and (3.8) with the lower average value of n. Each estimate was used to find the extra time that would be taken according to the respective hypothesis, using the corresponding equation with the higher average value of n; the results are given in the next three columns.

TABLE 1. AVERAGE TIMES OF RIDERS AND PREDICTIONS ACCORDING TO THE THREE HYPOTHESES

Rider	Average time with few pauses (s)	Actual average extra time (s)	Average n with few and several pauses	v_0 according to hypothesis			Extra time expected by hypothesis		
				1 (m/s)	2 (m/s)	3 (m/s)	1 (s)	2 (s)	3 (s)
1	500	64	1	8	5.03	5.12	5.16	17.6	65.7	111.8
2	345	45	1	8	7.33	7.43	7.50	25.6	51.4	81.8
3	438	64	1	7.5	5.75	5.84	5.90	18.7	56.0	92.3
4	333	72	1	8	7.60	7.70	7.77	26.6	50.4	79.6
5	366	60	1	7	6.89	6.99	7.06	20.7	46.5	73.3
6	400	45	1.5	7	6.33	6.48	6.58	19.0	44.3	71.2
7	448	58	1	7	5.61	5.70	5.76	16.8	53.2	87.0
8	436	40	1	7	5.77	5.87	5.93	17.3	52.1	84.8
9	349	52	1	8	7.24	7.34	7.41	21.7	51.7	82.3
10	324	40	1	7	7.80	7.90	7.98	23.4	43.4	66.6

Though it would have been theoretically simpler and more accurate to estimate a rider's cruising speed from a circuit made with a flying start and a flying finish, this would have been more difficult for the timer and for a rider making several circuits. The loss of accuracy, due mainly to the assumed values of a and r, was small, because these quantities have little effect on the estimate. The circuits made by each rider were completed in close succession, so that the cyclist was in fairly consistent physical condition; where four circuits were made, the sequence was NPPN or PNNP, where N = no pause, P = with pauses. Eight riders made six circuits, with the sequence NPPNNP. It was apparent that some of them needed time to warm up, and others flagged at the end. For each the most consistent four consecutive times were taken; in two cases the selected sequence was PPNN.

Being warmed up helps to produce a consistent physical condition, but it is not the usual condition of cyclists commuting over a short distance. It is more important, however, to find the difference between the two times than to obtain an accurate estimate of either in ideal or normal commuting conditions.

The values for the extra time expected by the first hypothesis were much smaller than the actual ones. Much lower values of a and r would be needed to explain the observations, a possibility which can be excluded. Thus we conclude that the first hypothesis is false; and a cyclist's cruising speed is affected by the need to accelerate.

The third hypothesis predicted extra times all of which were larger than the actual ones, by factors between 1.1 and 2.1. So we may also reject this hypothesis.

Two of the extra times predicted by the second hypothesis were within a second of the actual ones; of the remainder, four were less and four more. The second hypothesis thus appears to be a good working hypothesis, though individual predictions are scattered on both sides of it. The average time difference (actual   hypothesis 2) was less than 3 s; and the root mean square error in the trial time, estimated from the 17 pairs of times, each in one mode for one cyclist, was over 7 s. Thus the time difference was not greater than expected at the 10% level of significance; but it could also be explained by an average pause of less than 0.5 s, or errors in the assumed values of a, r and m/c. The variance in the average time differences, allowing for the mean square error in trial times, implies a root mean square error in time differences of 6 s: this could be due to personal differences in coasting, braking, pausing and accelerating, but possibly some other personal factor was also involved. So hypothesis 2 gives a good estimate of average extra time; but personal differences, for whatever reasons, also play a small part.

It is notable that the prediction of the extra time per pause according to the second hypothesis is approximately $m/6cv_0$. If a cyclist were able to go at normal cruising speed throughout, then in the extra time taken per pause he would be able to go a distance $m/6c$, i.e., with the value of m/c given in Section 2, 52 m. This distance is independent of the power output of the cyclist. A similar prediction results from the third hypothesis, with an equivalent distance per pause of $m/4cv_0$.

5. DISCUSSION

An effect that has not been taken into account is that of the duration of the journey. It has been assumed that this does not affect a cyclist's power output, but this assumption is not perfectly true. According to a graph produced by NASA (1964), when a duration of 10 minutes is increased by 25-30%, there is a decrease in the power output of a "healthy man" of about 3%; thus there is a decrease in cruising speed of about 1%. This is small compared with the accuracy achieved here, and so there is little point in pursuing this aspect further.

It is likely that none of the hypotheses very accurately reflects cyclists' performances. The second hypothesis is likely to be deficient because acceleration is not uniform. This, however, does not greatly affect the prediction, because it is not very sensitive to the magnitude of the acceleration, but dominated by the effect of producing the extra power required for it. Also in the second hypothesis, the supposition that the average power production is the same as if the cyclist maintained cruising speed throughout is unlikely to be perfect. For instance, if the pauses occur in one half of a journey, a more accurate prediction might be obtained by supposing that in each half journey the average power output is as in normal cruising, so that normal cruising speed is attained in the half without a pause; for it is unlikely that a cyclist could comfortably expend energy in repeatedly accelerating during the first half of the journey though it would be recovered later, or similarly store it up during the first half for later expenditure.

Again, if two pauses occur so close together that between them a cyclist could only achieve a low speed before he has to brake, he might improve his performance by saving his energy and making his maximum speed even lower. This emphasises the inadequacy of the hypotheses. There are two reasons. One is that if two sections have different lengths, the two cruising speeds required for optimal performance are not the same, as had been assumed. Optimal performance may be regarded as minimizing the total energy expenditure for a given total time. It may be found by minimizing the sum of the energies as in equation (3.3), subject to the sum of times as in equation (3.4) being constant. If the lengths are large, the difference in cruising speeds is small. Strictly, however, the result obtained for the second hypothesis is only true if the sections are of equal length. The other reason is that a cyclist's performance in one section, as stated for the second hypothesis, is not optimal. The optimal performance may be found by using optimal control (see, e.g., Leitmann, 1966) and consists of four stages. In the first stage, the cyclist maintains maximum acceleration, which may depend on speed. In the second, he maintains a constant cruising speed. In the third, he coasts or freewheels, expending no energy and slowed by air resistance. In the fourth stage, he brakes to a halt.

The time spent in coasting is $m/\sqrt{3}cv_m$, typically about 30 s. This is very different from the practice of all the experienced cyclists who took part in the trials: four said they did not coast, five for 1 - 2 s, and three for up to 5 s. Thus it appears that cyclists are not able to store up appreciable amounts of energy for long periods, and the theoretically finer results to be obtained from the optimal-control solution, and differences in cruising speeds with the second hypothesis or the optimal-control solution, can not be realised in practice.

The realisation that cyclists do not redistribute their power output significantly over more than about 15 s puts the trial results and the second and third hypotheses in a new light. In the third hypothesis, a cyclist does not redistribute his power output from the braking phase, when it is zero. In the second hypothesis, the redistribution takes place to the acceleration phase from both the braking phase and the cruising phase, which may well last for half a minute or more. Thus many cyclists may perform in a period around a pause according to the second hypothesis; but if they have a long period without a pause following a period with several close pauses, they may then attain a higher cruising speed gradually, as in the third hypothesis. It seems that an improved model would combine these hypotheses in some way, taking into account how well a cyclist redistributes power output over time.

Finally, we should consider the stops that make a cyclist wait before restarting, because of other traffic or a signal. If the wait is longer than about ten seconds, the performance of cyclists described in the last paragraph suggests that very little of the energy that would have been expended after this can be used to increase power output in cruising. For the first few seconds of a wait, however, he may be able correspondingly to increase his power output in cruising. This is proportional to the cube of the speed; it follows that the time recovered will only be about a third of that spent in waiting. Thus the time that can be recovered from a wait is at most a third of the first few seconds of the wait; the remainder is lost. In these cases the time is the same for all cyclists, as opposed to the distance for having to pause.

6. CONCLUSIONS FOR PLANNING CYCLE ROUTES

To make a cycle route effective in attracting cyclists, the time taken to traverse it is an important factor. A route may be planned with several points at which a cyclist has to pause, i.e., stop and wait, stop momentarily, or almost stop, which add appreciably to the travel time. This paper has dealt primarily with pauses that do not involve waits, and found that the extra time is more than that which would be required if the cyclist simply braked from normal cruising speed and then recovered it with normal acceleration. Trials of cyclists gave support for the nature of two hypotheses based on the power outputs of individual cyclists in accelerating and cruising. Both of these predicted that in the extra time taken by an individual cyclist due to a pause, he could travel an extra distance which depends little on his power output. According to the second hypothesis, which was well supported by the trials, the distance is about 52 m plus a small distance which depends on the individual cyclist's cruising speed, acceleration and braking; according to equation (3.3) this is about

$$v_0\left(\frac{1}{4a} + \frac{1}{6r}\right),$$

which, for the typical cyclist of Section 2, is 3 m. Again according to equation (3.5) there is a small reduction per pause, which is proportional to their density; for an average distance between pauses of 520 m, the reduction is 5.2 m. Thus most support is given to an equivalent distance of 55 m per pause, reducing to 50 m per pause for a spacing of 520 m.

If two pauses occur in close succession, it is unlikely that a cyclist will attain between them the cruising speed attained elsewhere on the journey. In this case the combined effect of the pauses will be appreciably less than twice the effect of one pause. An extreme example of this occurs where cyclists have to cross a busy road in two stages, having two pauses to cross streams of traffic, with the second of the order of 10 m after the first. This should be treated as one pause, with an extra time required to move from one stage to the next.

If a cyclist has not only to pause but to wait, then the journey time will be increased by most of the waiting time; a third of the first few seconds of it may be recovered by cruising faster.

7. ACKNOWLEDGEMENTS

The author is indebted to Mr H. McClintock for advice and to members of Merseyside Cycling Campaign and cyclists who came to a special event in National Bike Week 1998 for participating in the trials.

REFERENCES

C.R.O.W. (Centre for Research and Contract Standardization of Civil and Traffic Engineering The Netherlands) (1993). In: *Sign up for the bike*, Chap. 3, pp. 44 - 66. C.R.O.W, Ede.

Graham, R. (1988). *Liverpool's first cycle routes: development, experience and recommendations.* Merseyside Cycling Campaign, Liverpool.

Hensher, D. A. and L. W. Johnson (1981). *Applied Discrete-Choice Modelling.* Croom Helm, London.

Leitmann, G. (1966). *An introduction to optimal control.* McGraw-Hill, London.

NASA (National Aeronautics and Space Administration) (1964). *Bioastronautics Data Book,* document SP-3006.

Whitt, F. R. and D. G. Wilson (1982). *Bicycling Science.* M.I.T. Press, London.

AIR QUALITY MANAGEMENT AND THE ROLE OF TRANSPORT PLANNING AND CONTROL

Bernard Fisher, School of Earth & Environmental Sciences, University of Greenwich, Medway Campus, Chatham Maritime, Kent ME4 4TB, UK

ABSTRACT

This study contains an assessment of air pollution levels in Trafalgar Road, Greenwich. This is a congested road on a main route into central London, which has achieved notoriety since its residents took legal action in an attempt to restrict traffic. The four types of dispersion model used in the assessment are briefly described. All four models show that there is a likelihood that air quality standards for a number of pollutants, particularly PM_{10}, will be exceeded in the vicinity of busy congested roads in London. Zones where standards are exceeded are restricted to regions within 10m or so from the road. Any assessment has to take account of concentrations on a very fine scale (at distances of 10m from a road).

1. INTRODUCTION TO AIR QUALITY MANAGEMENT

In the UK Government's National Air Quality Strategy eight air pollutants were included. These are ozone, benzene, 1,3 butadiene (two carcinogenic volatile organic compounds), SO_2, CO, NO_2, particles, and lead. The 1995 Environment Act requires local authorities to undertake reviews of the air quality in their areas. Following a review local authorities may designate an area an "air quality management area", representing an area with poor air quality. The authority is then required to bring forward plans to improve the air quality. It is envisaged that parts of urban regions will be designated air quality management areas, because of the effect of road transport emissions. The future powers of local authorities could range from measures to ensure road vehicles comply with emissions standards, to the closing of roads to traffic and the use of land-use planning measures within the authority's power.

The first step in this process is an air quality review. The present analysis makes use of a desktop study, using generally available information. Later it may be necessary to undertake more detailed assessments involving monitoring, and more complex modelling. The air quality review should consider the future air quality in the year 2005.

This study contains an assessment of air pollution levels in Trafalgar Road, Greenwich. This is a congested road on a main route into central London (Fig 1), which has achieved notoriety since its residents took legal action in an attempt to restrict traffic.

Trafalgar Road is a two-lane road on the Inner London cordon surrounding central London about 9km from the Houses of Parliament. Radial two-way traffic flows into London at this distance from the centre are about 2 million vehicles per day. Trafalgar Road has a traffic flow of about 20,000 vehicles per day which is characteristic of many urban streets which are major routes (Department of Transport, 1993 and 1995). About 50% of the major roads in London carry traffic flows of between 15,000 and 25,000 vehicles per day.

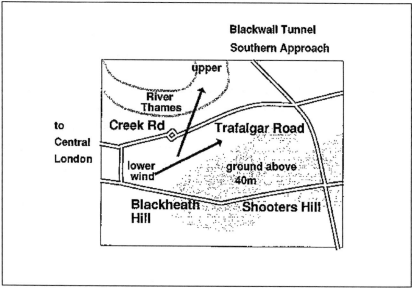

Figure 1. Sketch of major roads near Trafalgar Road, Greenwich

2. AIR QUALITY MODELLING

In an air quality assessment of roads within urban areas the direct effect of emissions from traffic on the road and the combined effect of emissions from other traffic, industrial, commercial and domestic sources within the urban area, which contribute to the urban background concentration, are important. The direct influence of emissions from traffic on a road is significant close to the road. At greater distances from the road the influence is reduced until it cannot be distinguished from the many influences that produce the urban background concentration.

The urban background concentration will be high, more or less everywhere in the urban area, in certain weather episodes. The weather conditions when these occur are associated with persistent low wind speeds. Low winds speeds are generally associated with anticyclonic conditions. In winter particularly, the episode may be associated with very stable air in which vertical motion is suppressed and the cold dense air near the ground tends to follow the lie of the land. The latter condition is associated with low-level inversions, which restrict vertical mixing and increase pollution levels. In these conditions pollution will be trapped within a shallow layer next to the ground, as little as 100m deep, so that as air travels across sources in the urban area, pollution levels will gradually build up with limited opportunity for dilution.

Key Receptor Sites. This assessment has been undertaken for a number of roads in Greenwich in the vicinity of Trafalgar Road (see Fig 2):
(1) Trafalgar Road (20,000 vehicles per day, 10m from road centreline)
(2) Blackwall Tunnel Southern Approach (60,000 vehicles per day, 30m from centreline)
(3) Shooters Hill (23,000 vehicles per day, 40m from road centreline)
(4) Blackheath Hill (23,000 vehicles per day, 10m from road centreline)
(5) Creek Road (20,000 vehicles per day, 10m from road centreline).

Receptor sites at points A to E have been chosen to be representative of sensitive locations, general houses, on each of these roads. Concentrations are calculated at receptor sites (see Fig 2). Locations at which traffic counts are available are shown as points 1 to 5 on Fig 2.

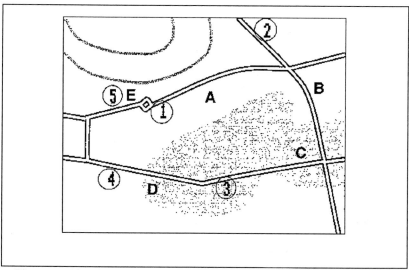

Figure 2. Sketch of road system showing positions of traffic counts and key receptor points

3. AIR POLLUTION ASSESSMENT

At the core of the National Air Quality Strategy are the air quality standards listed in Table 1.

Table 1. Standards and Specific Objectives in the National Air Quality Strategy

POLLUTANT	STANDARD		SPECIFIC OBJECTIVE
	concentration	measured as	to be achieved by 2005
Benzene	5 ppb	running annual mean	the air quality standard
1,3 Butadiene	1 ppb	running annual mean	the air quality standard
Carbon monoxide	10 ppm	running 8-hour mean	the air quality standard
Lead	0.5 μg/m^3	annual mean	the air quality standard
Nitrogen dioxide	150 ppb	1 hour mean	150 ppb hourly maximum
	21 ppb	annual mean	21ppb annual mean
Ozone	50 ppb	running 8-hour mean	50 ppb measured as the 97 percentile to be achieved by 2005
Particles PM$_{10}$	50 μg/m^3	running 24-hour mean	50 μg/m^3 measured as the 99 percentile to be achieved by 2005
Sulphur dioxide	100 ppb	15 minute mean	100 ppb measured as the 99.9 percentile to be achieved by 2005

ppm = parts per million, ppb = parts per billion, μg/m^3 = microgrammes per cubic metre

The standards are based on an assessment of the effects of each of the eight most important air pollutants which affect public health. In setting standards the judgements of the Expert Panel on Air Quality Standards have been used. The standards have been set at levels at which there would be an extremely small or no risk to human health. In addition to the standards, the Strategy sets air quality objectives. The objectives represent the progress which can be made in a cost-effective manner towards the air quality standards by 2005.

These standards involve a complex mixture of concentration levels and averaging times. Some of the standards refer to long-term averages (annual means), while others refer to short-term average concentrations over 8 hours or 1 hour. Measures to reduce emissions of air pollution during episodes would tend to improve compliance with the CO, NO$_2$ and particulate standards in Table 1. Measures taken during episodes would have little effect on annual averages and hence do little to improve compliance with the annual average benzene standard in Table 1.

The types of dispersion model used in the assessment are briefly described below.

Design Manual for Roads and Bridges. The Department of Transport published in 1994 a revision of their Design Manual for Roads and Bridges (Dept of Transport, 1994). This includes a method of comparing the air pollution effects of proposed road schemes.

The DMRB method is a screening method to identify locations where vehicle exhaust concentrations may exceed air quality standards. It is based on knowledge of the traffic flow along roads, distance from a chosen receptor to the centre-line of roads out to distances of 200m from the centre-line of the road, the percentage of heavy goods vehicles travelling on roads, vehicle speeds and the type of area (rural, urban, suburban) under consideration. The background concentrations assumed in the DMRB are typical values for urban areas and are not specific to London.

The DMRB method does not take into account specific local features, such as local buildings or topography. The structure of buildings along the side of a road can influence dispersion. If a street has a series of more-or-less attached tall buildings (relative to the width of the road) on either side of the road, then there is a tendency for pollution released in the road to be trapped in the street "canyon".

CAR Model. The CAR (Calculation of Air Pollution from Road Traffic) model is a dispersion model developed in the Netherlands for assessing air pollution levels in streets in built-up areas (Eerens *et al*, 1993). The CAR model may be used to calculate concentrations of air pollutants close to streets, at or near the kerb side. The version used here has been adjusted so that emission factors appropriate to the UK apply.

The main factors that the CAR model takes into account are the traffic flow along a street, the fraction of heavy goods vehicles, the distance from the road centre-line to the exposure point, the speed of the traffic and the type of road. The latter factor takes account of buildings close to one side or other of the street. Assumptions have to be made regarding the background concentrations within the urban area. These were based on measurements made at the monitoring site at Bloomsbury in Central London, and the site at Eltham in Greenwich, interpolating where appropriate (Beevers *et al*, 1995; Bower *et al*, 1995).

The results from the DMRB and CAR assessments are shown in Table 2 with the exceedences of Air Quality Standards shown in bold.

Table 2. Results from the DMRB and CAR Models at Receptor Points on Roads near Trafalgar Road

Concentrations	CO (ppm) max 8h mean		Benzene (ppb) annual average	
Receptor site	DMRB	CAR	DMRB	CAR
A	**14.1**	**11.4**	**6.3**	3.6
B	**18.1**	**12**	**8.6**	3.0
C	5.7	6.7	2.3	1.5
D	**17.5**	**17.6**	**7.9**	**5.6**
E	**12.1**	9.4	**5.5**	2.2

CALINE4 Dispersion Model**CALINE4 Dispersion Model**. The CALINE4 dispersion model has been developed and validated in the USA and has been widely used in the UK for predicting

the dispersion of pollution near roads. A feature of the CALINE4 model is that it may be used to estimate concentrations in specific weather conditions.

The CALINE4 model has been used to look at CO and PM_{10} concentrations from traffic emissions in Trafalgar Road in meteorological conditions, which are associated with a high background concentration over London. These were conditions, which would (1) tend to produce a winter episode i.e. low wind speeds (1m/s), stable air (category F) and a low mixing depth (100m), and (2) tend to produce a summer episode i.e. low wind speeds (1m/s), unstable air (category A) and a strong persistent inversion, although the mixing depth might be quite deep (1000m). The wind direction was taken to be parallel to the road.

The maximum 1 hour maximum CO at the central London site in Bloomsbury in 1992 and 1993 was about 5.5ppm (Bower *et al*, 1995). At monitoring sites in London which are subject to a greater influence from vehicle emissions, concentrations are higher. The background concentrations of carbon monoxide was assumed to be 5.5 ppm.

Calculations have been made of the concentrations of fine particles during summer and winter episodes. Beevers *et al* (1995) report on PM_{10} measurements made at Eltham during a summer episode (12 July 1994) and a winter episode (23 December 1994). During 24 hours on the 11-12 July the maximum mixing depth was estimated to be 1000m, the wind speed was 2m/s and variable and the highest 1-hour PM_{10} concentration was about 60µg/m³ at Eltham. On the 23 December the mixing depth was estimated to be 100m, the windspeed 1m/s and highest 1-hour PM_{10} concentration was about 60µg/m³ at Eltham. Taking the concentration in background air to be 60µg/m³, the 1-hour average concentrations of PM_{10} were estimated in typical summer and winter episode conditions. The results of the calculation of CO and PM_{10} are shown in Table 3.

Numbers printed in bold in Table 3 are ones where the standards are likely to be exceeded. In brackets the percentage contribution from the background air is given. Clearly the influence of the background increases at greater distances from the road.

Table 3. 1-hour Average Concentrations from the CALINE4 Model in Summer and Winter Episode Conditions at Different Distances from the Centre-line of Trafalgar Road

Distance (m)	CO summer episode(ppm)	CO winter episode(ppm)	PM_{10} summer episode µg/m³	PM_{10} winter episode µg/m³
10	**13.9** (40)	**15.6** (35)	**210**(29)	**240**(25)
20	8.7 (63)	10.2 (54)	117(51)	144(42)
30	7.6 (72)	9.0 (61)	98(61)	123(49)
40	7.1 (77)	8.4 (65)	89(67)	111(54)
50	6.8 (81)	8.0 (69)	83(72)	105(57)

The results in Table 3 are broadly in agreement with the calculations using the DMRB and CAR models in Tables 2. Concentrations near the roadside for some pollutants are likely to exceed the standards. These exceedences occur during meteorological episodes associated with high background pollution. However the exceedence is caused by traffic in the road. 50m from the road where the majority of the CO and PM_{10} is from the background air no exceedence occurs.

GRAM Dispersion Model. The GRAM model has been developed by the University of Greenwich to permit simple estimates of CO, benzene, NO_2 and PM_{10} concentrations combining features of the DMRB and CALINE4 models. The intention is to have a simple approach which starting from the input data used by the DMRB produces concentrations in a form that can be directly compared with the air quality standards. Results from the GRAM model are shown in Table 4.

Table 4. Results from the GRAM Model at Receptor Sites Compared with Air Quality Standards

Receptor	Max 1-hour NO_2 ppb	Annual mean NO_2 ppb	Max 8-h CO ppm	Annual mean benzene ppb	Max 24-h PM_{10} $\mu g/m^3$
A	121	38	15.3	3.7	140
B	135	41	18.4	4.1	160
C	109	34	13.2	3.3	113
D	124	39	16.6	3.9	139
E	112	35	13.9	3.5	117

Table 4 shows that the exceedence of air quality standards occurs at all the receptor sites. Measurements made in Creek Road (20,000 vehicles/day) which is a continuation of the same route into London on which Trafalgar Road lies, but about 2km nearer to central London, indicates typical 8h average PM_{10} concentrations at the kerbside of over 100 $\mu g/m^3$ on a number of 1995-6 winter days. Measurements from an automatic continuous analyser (Donaghue and Ma, 1996) indicate a maximum NO_2 concentration of 200ppb and an annual mean of 38.5ppb. A local authority might have to consider designating an area adjacent to Trafalgar Road, and other roads in the area, as Air Quality Management Areas.

Improvement in Concentrations by National Measures. By 2005 the introduction of exhaust emission technology to new petrol driven cars is likely to lead to reductions per vehicle of 80% of CO, NOx and volatile organic compound emissions under ideal operating conditions. Regardless of the mix of vehicles in the motor fleet by 2005 it is likely that air quality standards for CO and benzene will not be exceeded. It appears probable that the annual mean NO_2 limit value will still be exceeded.

National inventories of PM_{10} emissions can be somewhat misleading (EPAQS, 1995), since the 19% of the national total from diesel and the 5% of the national total from petrol vehicles will lead to much greater exposures in urban areas compared with emissions from fossil-fuelled power stations and mining and quarrying. 86% by weight of primary PM_{10} emissions in Greater London are derived from vehicle exhausts.

Reductions in national PM_{10} emissions in the future will arise from the significantly lower emissions from heavy goods vehicles. Reductions will also arise from the increasing number of cars fitted with emission control equipment which run on unleaded petrol eliminating the release of lead-rich particles and reductions in the sulphur content of diesel will reduce emissions per vehicle. Although predictions of future national emissions are inevitably uncertain, they (QUARG, 1993; Royal Commission on Environmental Pollution, 1994) all indicate significant reductions by 2005 of national PM_{10} emissions from road transport of up to 50%.

Some of the PM_{10} particles are sulphate and nitrate aerosols, produced in the rural atmosphere from distant sources elsewhere in the UK or abroad and are not related to urban emissions in London. National measures would reduce **primary** PM_{10} concentrations, mostly derived from transport sources in London, below the air quality standard in most areas of London during most episodes. In the immediate vicinity of Trafalgar Road national measures would produce significant improvements but not necessarily ensure the standards were not exceeded.

Improvements in Concentrations by Local Measures. Traffic on radial routes into Inner London have shown a small overall increase between 1972 and 1993 (about 6%). A rather larger increase in cars has been offset by a fall in heavy goods vehicles (Department of Transport, 1993 and 1995). PM_{10} emissions from an average heavy duty vehicle is much greater than the PM_{10} emissions from an average light duty vehicle (a factor of roughly 30). Along Trafalgar Road about 20% of the traffic used to be heavy duty vehicles before the introduction of a lorry ban in central Greenwich, so action to restrict emissions from heavy duty vehicles would have the most dramatic immediate effect. However this action would not eliminate PM_{10} episodes along the Trafalgar Road, since high concentrations over London would still occur on the 20 to 30 days of episodic conditions. The diversion of traffic away from Trafalgar Road would reduce the air pollution exposures along other roads in the centre of Greenwich. However it has already been seen that traffic would be diverted to other roads which are already polluted.

The results of the London Congestion Charging Research Programme (MVA Consultancy, 1995) suggests that the introduction of the most stringent measure considered in the report, which is based on charging vehicles coming into London, would reduce traffic flows (measured in vehicle kilometres) by about 10%. This is insufficient on its own to eliminate exceedences of the PM_{10} air quality standard.

Effect of Land Use Planning Measures. Land use planning policies can influence travel demand only in the long-term. The average rate of turnover of the built environment is small, of the order of one or two per cent per annum. One conclusion of the Royal Commission on Environmental Pollution (1994) report is that there is sufficient uncertainty to make the search for an 'ideal' development pattern unproductive. In any case land use policies must start with the physical infrastructure that already exists and the choice of development patterns and location policies must be flexible in response to local circumstances. From an air quality point of view it is likely that zones immediately adjacent to busy streets may currently exceed standards and continue to exceed standards in the future. Land use planning has a role here by seeking to avoid the close proximity of busy roads with residential housing.

4. CONCLUSIONS

(1) All four models used in the assessment show that some air quality standards will be exceeded in the vicinity of a busy congested roads in London. The exceedence of standards is restricted to regions within 10m or so of the road centreline. Any assessment has to take account of concentrations on a very fine scale (at distances of 10m or so from a road).

(2) Measures to restrict vehicle emissions in Trafalgar Road permanently would be effective in reducing concentrations of carbon monoxide, nitrogen dioxide, benzene and fine particulates close to the road. Further from the road the levels of pollution are increasingly determined by pollution in the background urban air and not by the road.

(3) High concentrations in episodes would still remain, because of elevated levels in background air. One would wish to discourage traffic during episodes rather than forcing traffic to divert to other roads.

(4) When considering alternative traffic routes, one should take into account the number of exposed properties very close to the road. Land-use planning may have a role by discouraging residential housing in the neighbourhood of busy roads.

REFERENCES

Beevers, S., S. Bell, M. Brangwyn, G. Fuller, T. Laing-Morton and J. Rice (1995). *Air Quality in London*, Second Report of the London Air Quality Network, SEIPH.

Bower J. S., G. F. J. Broughton, P. G. Willis and H. Clark (1995). Pollution in the UK: 1993/4, AEA Technology, National Environmental Technology Centre Report AEA/CSR1033/C.

Department of Transport (1993).Transport Statistics for London: 1993, HMSO, London.

Department of Transport (1994). *Design Manual for Roads and Bridges* (DMRB) Volume 11, HMSO.

Department of Transport (1995). *London Transport Monitoring Report: 1995*, HMSO London.

Donaghue S. and M. Ma (1996). NOx concentrations at Creek Road Deptford during 1994-5, private communication.

Eerens H. C., C. J. Sliggers and K. D. van den Hout (1993). The CAR model: the Dutch method to determine city street air quality, *Atmospheric Environment*, 27B, 389-399.

Expert Panel on Air Quality Standards (1995). Particles, HMSO, London.

MVA Consultancy (1995). The London Congestion Charging Research Programme - Principal Findings, Government Office for London, HMSO London.

Quality of Urban Air Review Group (1993). *Diesel vehicle emissions and urban air quality*, Second Report of the Quality of Urban Air Review Group, Department of the Environment.

Royal Commission on Environmental Pollution (1994). *Eighteenth Report*, HMSO London Cm 2674.

MAPPING ACCIDENT CASUALTIES USING SOCIAL AND ECONOMIC INDICATORS

Ibrahim M. Abdalla, Department of Mathematics, Napier University, Edinburgh

ABSTRACT

Geographic information for the home address of the accident casualty is obtained from the home-address post-code for each casualty. This allows the STATS 19 data base, the UK police system for reporting accidents, for the former Lothian Region in Scotland, 1990 to 1992, to be linked to social and economic indicators in the 1991 UK census and to the corresponding digitised boundaries at the smallest census geographical level (Output Areas, OAs) and post-code sector level in Scotland. For each post-code sector Standardised Casualty Ratio (SCR) which is commonly used in epidemiology to study rare diseases is calculated from the ratio of the number of casualties observed to that expected in the area. Adjusted SCRs are calculated, they are the ratios of the numbers of casualties predicted by social and economic factors that are measured at the census using Poisson regression to the expected numbers. Empirical Bayes Estimates (EMEs) are applied to prevent the results from areas with small populations being shown as too extreme. Results from the analysis indicate that accident risk to residents from deprived areas is high compared with those from affluent areas. Finally maps that can be used to identify areas in Lothian where there is relatively high SCRs are presented.

1. INTRODUCTION

Previous results in Abdalla et al. (1996) indicate that the rate (per resident population) and the type of road accident casualty varies between different neighbourhoods. Casualty rates, per population and per registered vehicles in the area, (Abdalla, 1997a) are generally higher amongst people from poorer neighbourhoods. The overall rate of pedestrian casualties (per population) from deprived areas is 4 times the rate of their counterparts from affluent areas. Car driver casualties have similar rates amongst the deprived and the affluent residents despite the low level of car ownership amongst residents from the deprived areas (see details in Abdalla et al., 1997b and c).

This paper has two objectives. The first one is to investigate the risk of involvement of residents from deprived areas in road accidents outside the area (OA) where they live. This investigation is carried out in order to examine the hypothesis that casualties from deprived areas carry their accident risk with them when they move outside their resident boundaries. The second objective is to utilise Standardised Casualty Ratios (SCRs) adjusted for the effect of social and economic factors in the area to predict areas (post-code sectors) with high concentrations of casualties. The effect of small populations is reduced by implementing Empirical Bayes methods. Maps that depict casualty concentrations in different areas are included in the paper.

The distribution of casualties across the 5774 OAs (an OA covers 2 to 3 postal addresses and 50 household on average) in Lothian is very sparse. Many OAs (29%) had no casualties during the period from 1990 to 1992, 31% had 1 casualty each, 20% had 2 casualties each and the rest had 3 or more casualties, therefore it would not be appropriate to employ such units in statistical modelling. Post-code sectors (total 130 in Lothian) which are aggregates of OAs provide a fine level of spatial resolution and can be utilised as units of analysis in this development.

In Section 2 the paper discusses the risk of accidents outside the area of residence. In Section 3, Standardised Casualty Ratios (SCRs) are presented and maps that display the distribution of casualties in Lothian are provided. Poisson regression and SCRs adjusted for the effect of social and economic factors are discussed in Section 4. Conclusions are given in Section 5.

2. RISK OF ACCIDENTS OUTSIDE THE AREA OF RESIDENCE

A weighted deprivation index comprising eight deprivation indicators has been utilised (see Abdalla et al., 1997b). This allows establishing a single score for each post-code OA in the Lothian Region and consequently it was possible to establish both the 15% OAs with highest deprivation and the 15% with lowest deprivation (affluent OAs). The connection between the casualties living in the 15% most deprived areas who are involved in road accidents outside their area boundary and the risk of road accidents is now discussed. The data for this analysis which cover all OAs in Lothian are presented in Table 1. This is a one-response one-factor table, where the response is a dichotomous variable with values either a resident who comes from the "15% most deprived OAs" or "not from the 15% most deprived OAs", and the factor has two levels corresponding to sub-populations of road accident casualties and those who were not involved in road accidents in the corresponding area.

Table 1: Average (1990-1992) Casualties Involved (and not Involved) in Road Accidents Outside their OAs in Lothian Region.

Casualties:	From the 15% most deprived	Not from the 15% most deprived	Total
ALL:			
Involved in accidents	509=a	2345=c	2854
Not involved in accidents	98110=b	614845=d	712955
Pedestrians:			
Involved in accidents	177=a	568=c	745
Not involved in accidents	98442=b	616622=d	715064
Car drivers:			
Involved in accidents	107=a	755=c	862
Not involved in accidents	98512=b	616435=d	714947

The *odds ratio* estimate for a casualty being from the 15% most deprived OAs involved in a road accident outside his OA boundary relative to not being involved is calculated as

$$odds \quad ratio = \frac{ad}{bc} \qquad (2.1)$$

The resulting *odds ratios* and the corresponding significance and 95% confidence intervals are displayed in Table 2 for all casualties, pedestrians and car drivers, suggesting that there is some evidence that being a resident of a deprived area (15% most deprived OAs) increases the risk of having a road accident outside the resident area. The odds of residents from the 15% most deprived OAs being involved in road accidents outside their areas are estimated to be 1.4 times (or 40%) as large as the odds of not being involved. Pedestrian risk is double. However, car drivers from the 15% most deprived OAs have almost similar accident risk compared with those from areas outside the 15% most deprived OAs boundaries.

Table 2: Tests of Significance and 95% Confidence Intervals.

Sub-group	*odds ratio*	Sig. test (p-value)	95% C.I.
All casualties	1.4	0.00	1.24 to 1.50
Pedestrians	2.0	0.00	1.65 to 2.31
Car drivers	0.9	0.89	0.72 to 1.09

Similar analysis is carried out to estimate the *odds ratio* for casualties from the 15% most affluent OAs involved in road accidents outside their OA boundary relative to not being involved. The results indicate that residents from affluent neighbourhoods have less risk of involvement in road accidents outside their area boundary. The estimated *odds ratios* for all casualties, pedestrians and car drivers are 0.7, 0.4 and 1.0 respectively, all not significantly different from 1. This suggests that the odds of a resident from the 15% most affluent OAs being involved in accident is 30% smaller than not being involved in an accident and 60% smaller than not being involved in a pedestrian accident.

These differences in risk between residents from the highest and the lowest deprived areas appear to reflect the high level of exposure together with possibly a lack of care and positive attitude towards safety measures amongst residents from the deprived areas. Better provision of road safety education with particular initiatives that target the disadvantaged population should have a significant effect in reducing casualty numbers.

3. STANDARDISED CASUALTY RATIOS (SCRs)

Standardised casualty ratios (SCRs), are calculated for each post-code sector in Scotland. This involves calculating the number of expected casualties for each post-code sector. The ratio of the casualty numbers (C_i) to this expected number (E_i) is the SCR. SCR of 1.0 represents the average casualty rate in Lothian. The post-code sectors differ in size so the expected numbers of casualties in each post-code sector has a wide range (207 casualties).

To correct for bias resulting from small expected numbers, the method of "shrunken" estimates which employs a Bayesian technique is used. Where the population is small the ratio (SCR) is shrunk towards the overall casualty rate in Lothian and where the population is relatively large the ratio is less shrunk towards the overall rate. The shrunken estimates are given by

$$SCR_i = \frac{C_i + \hat{v}}{E_i + \hat{\alpha}} \qquad (3.1)$$

The parameters v and α are to be estimated by using an iterative maximum likelihood approach (Langford, 1994).

Figure 1 displays shrunken SCRs for all casualty data in Lothian, 1990 - 1992. The highest SCRs appear to be clustered in Edinburgh District and part of West-Lothian District, with East-Lothian shown as the least affected part of the whole Region.

Figure 1: Standardised Casualty Ratios (SCRs) for Lothian, 1990 - 1992.

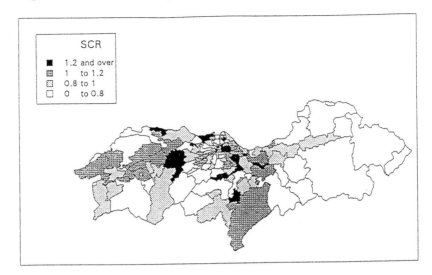

4. ACCIDENT CASUALTIES ADJUSTED FOR SOCIAL FACTORS

Poisson regression was used to analyse the casualty data. The Poisson model uses the number of casualties at the post-code sector as the dependent variable which is assumed to be Poisson distributed about a mean of μ, which in turn is assumed to be proportional to the expected number of casualties (E) for the area. The ratio of the mean (μ) to its expected value (E) is then predicted from the explanatory social and economic variables $\underline{x}$ through the log link function:

$$\log(\mu / E) = \eta = \underline{\beta}^T \underline{x} \qquad (4.1)$$

where η is known as the linear predictor and the vector $\underline{\beta}$ contains the parameters which are to be estimated. The vector $\underline{x}$ contains the values of the explanatory variables.

Maximum likelihood estimates for the parameters of the Poisson regression model can be obtained using an iteratively reweighted least squares procedure in the SAS GENMOD procedure. The quality of the fit between the observed values "number of casualties" and the fitted values $\hat{\mu}$ is measured by the scaled deviance (SD).

Census variables described in Table 3 were used to calculate derived variables that would characterise different aspects of the post-code sector. Exploratory analyses were carried out to determine which measures, from the list given, had the clearest relationships with casualty rates.

Table 3: Variables Selected for Analysis.

Variable	Definition	Mean	Std	Q1*	Q3**
Unemployment (unp)	proportion of economically active adults 16+ seeking work	0.08	0.05	0.05	0.11
Social class 4 & 5 (sc45)	proportion of heads of households in social class 4 & 5 (semi/unskilled)	0.17	0.11	0.10	0.24
No car (nocar)	proportion of residents with no car	0.31	0.17	0.17	0.43
Overcrowding (overc)	proportion of residents in households below the occupancy norm	0.04	0.04	0.02	0.06
Pensioners (pen)	proportion of residents of pensionable age	0.17	0.06	0.14	0.20
Large households (lhose)	proportion of households with 4 or more dependants	0.01	0.01	0.00	0.01
Children in non-earners households (nonch)	proportion of children in households which contains no economically adults	0.16	0.12	0.06	0.22

*Q1: Lower quartile,   **Q3: Upper quartile

A Poisson model containing only an intercept was fitted to the data in order to provide a measure of the variation in the dependent variable (number of casualties in the post-code sector) around its mean. Then different univariate models, each containing an intercept and each of the independent variables (from Table 3) were formed. Those independent variables which produced a statistically significant reduction (at the 5% level) in deviance are displayed in Table 4. For all casualty data, the overcrowding variable (overc) produced the largest reduction in deviance, suggesting that 8.5% of the variation in all casualty accidents in post-code sectors can be predicted by the proportion of overcrowded households. The results also suggest that the rate of all casualties increases in post-code sectors where there is an increase in the rate of unemployment, people in social class 4 and 5 and dependent children in non-earners households and a decrease in the level of car ownership (Table 4).

For pedestrian and PSV casualties, most of the variation is explained by the low level of car ownership and the high proportion of overcrowded households and dependent children in non-earners households (Table 4).

Table 4: Estimated Parameters from Poisson Regression, Univariate Variable Models.

Model		unp	sc45	nocar	overc	pen	lhose	nonch
					Univariate models			
All casualties	est.	2.6	1.4	0.9	4.8	-	-	1.3
	Std err	0.9	0.5	0.3	1.4	-	-	0.4
	change in deviance	151.4	162.4	177.3	220.9	-	-	196.9
	% change in deviance	5.8%	6.2%	6.8%	8.5%	-	-	7.6%
Pedestrians	est.	4.9	2.4	2.3	9.6	-	9.9	2.6
	Std err	0.8	0.5	0.3	1.4	-	8.4	0.4
	change in deviance	175.2	130.8	319.1	254.4	-	10.7	246.8
	% change in deviance	17.5%	13.1%	31.8%	25.4%		1.1%	24.6%
PSV casualties	est.	4.1	1.7	2.3	9.5	5.6	-	2.4
	Std err	1.3	0.8	0.4	2.0	1.4	-	0.6
	change in deviance	49.7	29.9	132.1	106.3	94.3	-	88.8
	%change in deviance	6.5%	3.8%	17.3%	13.9%	12.3%	-	11.6%

Combining all the variables in Table 3 together in one Poisson model in the case of all casualty, pedestrian and PSV casualty data, diminishes the significance of the variables in explaining variations in casualty rates. This result implies that there is some correlation between the different independent variables.

However, when social classes 4 and 5 and the overcrowding variables were combined together in one model for all casualty data, both were statistically significant at the 5% level and the model predicted 10.9% of the variation in all casualty rate (Table 5). The casualty rate positively increases as the result of increasing the proportion of residents in social classes 4 and 5 and the proportion of overcrowded households in the area.

Pedestrian casualties were best predicted by combining unemployment and overcrowding variables, where the PSV casualties were best predicted by joining together the overcrowding and dependency variables (Table 5).

Table 5: Estimated Parameters from Combined Variables Poisson Regression Models.

Model		est.	Std err	change in deviance	% change in deviance
			Models for combined variables		
All casualties	sc45	0.9	0.5	283.1	10.9%
	overc	3.9	1.5		
Pedestrians	unp	2.5	0.9	293.1	29.2%
	overc	7.7	1.6		
PSV casualties	overc	6.8	2.6	120.2	15.7%
	nonch	1.2	0.7		

The contribution of each of the socio-economic variables in explaining variation in casualty rates in post-code sectors is expressed as the inter-quartile ratio. This is the ratio of the SCR for the upper quartile of the distribution of the variable, to that of the lower-quartile, Tables 6 and 7. For example, for the social class variable (sc45) in Table 6 the inter-quartile ratio when using all casualties data is 1.22, it corresponds to the ratio by which SCR for all casualties will increase as the proportion of residents in social classes 4 and 5 changes from 0.10 to 0.23. Ratios above 1.00 suggest that the rates of accidents casualties increase with the increase in the level of area disadvantages.

Table 6: Inter-quartile Ratios - Univariate Variable Models.

| | Inter-quartile ratio | | |
| | Single socio-economic variables | | |
Variable	All casualties	Pedestrian	PSV casualties
unp	1.17	1.34	1.28
sc45	1.22	1.40	1.27
nocar	1.26	1.82	1.82
overc	1.21	1.47	1.46
pen	-	-	1.40
lhose	-	1.10	-
nonch	1.23	1.52	1.47

Table 7: Inter-quartile Ratios - Combined Variables Models.

| | Inter-quartile ratio | | |
| | Combined socio-economic variables | | |
Variable	All casualties	Pedestrian	PSV casualties
unp	-	1.16	-
sc45	1.13	-	-
nocar	-	-	-
overc	1.17	1.36	1.31
pen	-	-	-
lhose	-		-
nonch	-	-	1.37
Full predictor	1.33	1.58	1.59

The results from Tables 4 and 6 suggest car ownership as the strongest predictor of accident casualties in post-code sectors, particularly amongst pedestrians and PSV casualties. The absence of this indicator in the full predictor models in Table 5 suggests that car ownership is strongly correlated with factors that describe income. Such factors include employment levels and other income surrogates such as social class and dependent children living with non-earning adults.

The level of car ownership has been reported as having a direct influence on casualty rates in terms of more exposure to pedestrian travel, particularly young children, in the absence of other alternatives (Harland et al., 1996). Abdalla (1997a) indicates that residents from deprived areas have less access to a car, and are more likely to travel to work by public service vehicle, bike or walking.

Results from a survey (Abdalla, 1997a) indicate that residents from deprived areas are more exposed to accidents risk than their counterparts from the affluent areas. They are more likely to walk long distances and cross busy roads more often than those from the affluent areas.

Overcrowding is linked to the effect of the structure of the area (Preston, 1972). Overcrowded households are more likely to be in overcrowded streets associated with lack of play space for children inside and outside. Within the city centre overcrowded households are more exposed to high traffic density. Within the outskirts of the city, the speed of traffic might also be related to accident rates which could be increased by the increase in the density of the population in the area.

Figure 2 displays SCRs adjusted for social and economic factors, obtained from the combined effect models (Table 5). The highest adjusted SCRs for all casualty data are in the Edinburgh area and the pattern of their distribution can be linked to the pattern shown in Figure 1. This confirms the results from this analysis that areas with high concentration of accident casualties can be linked with social and economic disadvantages.

Figure 2: Standardised Casualty Ratios (SCRs) Adjusted for Socio-economic Factors, Lothian, 1990 - 1992.

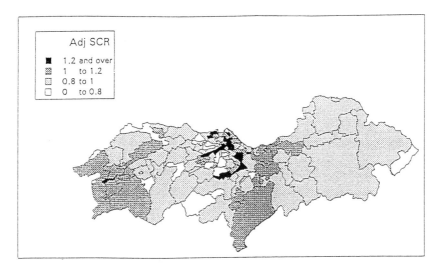

5. CONCLUSIONS

This paper has investigated links between social and economic indicators of the population and road accident casualties. Results from the analysis indicate that the risk to residents from deprived areas outside their resident areas is high compared with those from affluent areas. Models and maps that link accident casualties to the social and the economic indicators from the census were constructed and the effect of these indicators were estimated. An increase in area disadvantages is linked with an increase in road accident casualty rate. Pedestrians are the most affected group.

6. REFERENCES

Abdalla, I. M, R. Raeside, D. Barker, and D. McGuigan (1997b). An Investigation into the Relationships between Area Social Characteristics and Road Accident Casualties, *Accident Analysis and Prevention, Vol. 29, No. 5,*

Abdalla, I. M. (1997a). Statistical Investigation and Modelling of the Relationships between Road Accidents and Social Characteristics, *PhD Thesis, Napier University, Edinburgh.*

Abdalla, I. M, R. Raeside, D. Barker (1997c). Road Accident Characteristics and Socio-economic Deprivation, *Traffic Engineering and Control, Vol. 38, No. 12.*

Abdalla, I. M, R. Raeside and D. Barker (1996). Linking Road Traffic Accident Statistics to Census Data in Lothian, *The Scottish Office Central Research Unit,* Edinburgh.

Bailey, T. and A. Gatrell (1995). Interactive spatial data analysis, *Longman,* Essex, England.

Clayton, D. and J. Kaldor (1987). Empirical Bayes Estimates of Age-Standardised Relative Risks for Use in Disease Mapping, *Biometrics, Vol. 43,* 671-681.

Elliott, P., J. Cuzick, and others (1992). Geographical Environmental Epidemiology, *Oxford University Press,* Oxford.

Harland, D. G., K. Bryan-Brown, and N. Christie (1996). The Pedestrian Casualty Problem in Scotland: Why so Many? *The Scottish Office Central Research Unit,* Edinburgh.

Langford, I. (1994). Using Empirical Bayes Estimates in the Geographical Analysis of Disease Risk, *Area, Vol. 26, No. 2,* 142-149.

Marshal, R. (1991). Mapping disease and mortality rates using empirical Bayes estimators, *Applied Statistics, Vol. 40,* 283-294.

Preston, B. (1972). Statistical Analysis of Child Pedestrian Accidents in Manchester and Salford, *Accident Analysis and Prevention, Vol. 4.*

CONCEPTUAL DATA STRUCTURES AND THE STATISTICAL MODELLING OF ROAD ACCIDENTS

Ken Lupton, Mike Wing and Chris Wright, Transport Management Research Centre, Middlesex University, The Burroughs, London NW4 4BT

ABSTRACT

The aim of this paper is to review the structure of road accident data, the database framework within which it is stored, and the potential for exploiting a hierarchical structure using multilevel statistical models. Most national accident databases regard accidents as the primary units of observation, with other characteristics stored as attributes. But it is more natural to picture the network, accidents, and other variables as a collection of related objects within a hierarchical system, which can be achieved using object-oriented database technology within a Geographical Information System (GIS) framework. This would permit more efficient data capture and storage, facilitate analysis of accident frequencies as a function of road layout, and facilitate the development of multilevel statistical models.

1. THE CONCEPTUAL STRUCTURE OF ROAD ACCIDENT DATA

While the UK national accident database is acknowledged as better than most, its structure was evolved at a time when computer software was limited in scope and data sources were fragmented. It regards accidents as the primary units of observation, with their characteristics stored as attributes. For example, road layout information is stored as part of the accident description and hence limited to what the police can record at the time, and it may be described differently in the records of different accidents occurring at the same site. Other potentially useful sources are not exploited, and sites where there are no accidents do not appear in the database, so that comparisons cannot easily be made between sites having good and bad accident records. The variables that are routinely collected and incorporated within the UK national database are summarised in Table 1, which is intended to give a rough guide rather than a complete description.

We are concerned with two main questions: how are the causal factors linked, and is there a 'natural' structure for road accident data that reflects those links? Mapping the causal 'network' is quite a difficult task, for three reasons. First, the elements don't have equal status and their roles differ in nature. Second, we don't have a definitive list of all the factors involved, and even if we did it would be too cumbersome to be of much use. Third, even with a reduced list, all the elements could conceivably influence each other as well as influencing the outcome (accidents):

the resulting diagram could be very complicated. Fortunately, in some cases, we can discount a causal link altogether, and in others, plausibly assume that the link is unidirectional.

Figure 1 shows one representation, in which the flow of cause and effect runs almost entirely from top to bottom. The location variables and ambient conditions are independent elements so they appear near the top, as do vehicles and drivers. The accident itself is an outcome resulting from the interaction between the other elements, and is best described as an *event*. It appears near the bottom, where it resolves finally into the injuries sustained by the victims together with material damage both to the vehicles involved and to the physical structure of the road environment (these last two are not shown in the diagram).

Table 1: Data fields in the UK national road accident database

Category	Data field
ACCIDENT	Accident reference number
	Time, date
	Map grid reference
	Road number(s) and class(es)
	District
ROAD USERS INVOLVED	Casualty severity
	Age, gender
	Breath test
ROAD ENVIRONMENT	Speed limit
	Pedestrian crossing
	Carriageway type
	Junction control
	Junction detail
	Weather, lighting, road surface conditions
VEHICLES INVOLVED	Vehicle type(s), registration number(s)
	Manoeuvre(s)

Note that unlike the other elements, the characteristics of a vehicle contribute to the outcome indirectly as well as directly, because (a) they mediate the driver's perception of the road environment and the choice of manoeuvres available, (b) they determine whether the vehicle will respond to whatever the driver actually decides to do, and (c) they modify the nature of any impact and its effect on its occupants. Also, injuries are simultaneously influenced by the personal characteristics of their 'owners' and the nature of the impact, which appears explicitly here although only the sketchiest of details are available for real accidents.

When developing a database, we aim for a data structure whose elements are linked in a 'natural' way. If possible, it should also reflect the pattern of causal interactions involved. Here, the elements have a natural hierarchy, of which the most obvious is the spatial nesting of road infrastructure. The road network is a physical *object*: particular sections of road (or sites such as junctions) where accidents occur are part of the road network and lower-level objects in their own right. In turn, physical obstructions and hazards are features of the site having the status of

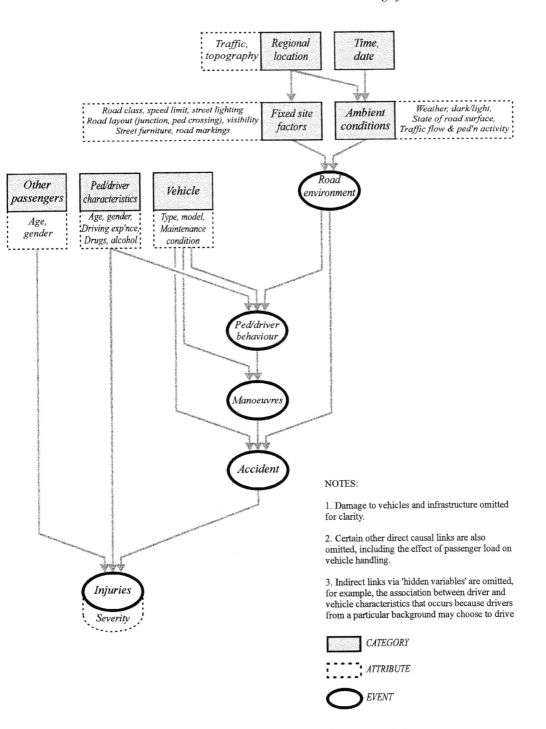

NOTES:

1. Damage to vehicles and infrastructure omitted for clarity.

2. Certain other direct causal links are also omitted, including the effect of passenger load on vehicle handling.

3. Indirect links via 'hidden variables' are omitted, for example, the association between driver and vehicle characteristics that occurs because drivers from a particular background may choose to drive

☐ CATEGORY

⌐⎯⌐ ATTRIBUTE

◯ EVENT

Figure 1: One way of representing the causal factors involved in road accidents

objects. In short, the geographical region 'owns' the road, the roads 'owns' sites, and the sites 'own' their physical features. The status of the geometrical layout is less easy to define: it might be pictured as an *attribute* of the site as opposed to a separate object. The same is true of the speed limit together with ambient conditions such as fog or darkness.

Road users and their vehicles exist independently of the road network, so they are given the status of objects, but they have a special relationship with one another: occupants (including riders) 'belong to' vehicles, but to complicate matters, vehicles also 'belong to' their drivers and riders. For present purposes, it is convenient to represent pedestrians as 'drivers' of 'virtual vehicles', although their status on the road differs in practice from that of drivers and riders. Passengers are subsidiary to vehicles. Manoeuvres are *events* defined in terms of the vehicles that make them and the road environment within which they take place. Injuries are pictured as events that 'happen to' occupants, whereas injury severity and personal characteristics (gender, age, driving experience and so on) are attributes. Logically, the accidents themselves are events or outcomes defined by all the other entities but they have to be represented explicitly because of the attributes attached to them: the identification code, date, time and so on. The complete hierarchy is shown Figure 2.

No representation of this type is unique: for example, we have chosen to represent ambient conditions and other features associated with the accident site as attributes, whereas some could equally well be represented as objects. However, a hierarchical representation is easy to understand, and it reflects at least some of the causal interactions in road accidents.

2. A DATABASE SPECIFICATION

The main problem with the existing UK system is that the primary units are the accidents themselves: starting with an accident record, one is obliged to piece together the circumstances in which the accident took place from whatever information the investigator was able to note at the time, as opposed to viewing the accident as an outcome of the circumstances, many of which are common from one accident to the next. For example, road layout is treated as an attribute and is not always described consistently between the records of different accidents happening in the same place (Lupton, Jarrett and Wright, 1997). Nor does an accident-centred database deliver all the information that we need. It tells us nothing about parts of the network where accidents don't happen, whereas we may have as much to learn from the 'safe' locations as we do from the high-risk ones. Nor can it easily be linked to other sources: for example, many highway authorities now have digitally encoded information about highway geometry and street furniture in the form of GIS data files that could play a useful role in many different types of investigation, a point that we shall return to later. Finally, the road users and vehicles are also treated as attributes within the national database and information that could identify them omitted, so it would be difficult to link the involvement of a particular road user or vehicle with involvement in previous accidents even if identification codes were retained.

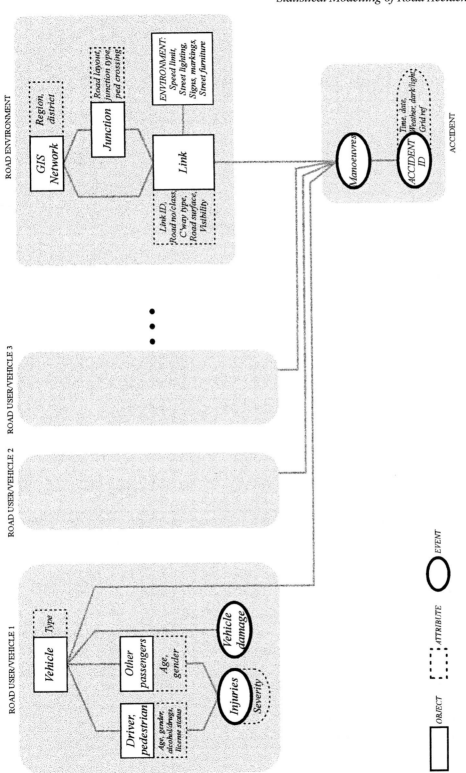

Figure 2: An object-oriented road accident database structure

All these considerations point to the desirability of introducing current database technology in the form of a relational system or an object-oriented system. Either can take advantage of systematic design procedures tailored to the application, such as the Structured Systems Analysis and Design Method (SSADM: see Goodland and Slater, 1995). However, an object-oriented system has three additional advantages.

Firstly, the way the information is structured can be important in terms of data integrity. In an objected-oriented system, objects are defined in such a way that irrelevant attributes cannot be attached to them: a car cannot jack-knife. Furthermore, the links between objects are maintained automatically by the system without the user having to define and input the key codes that are normally used to link files within a relational system. In other words, the natural hierarchy is much easier to articulate - we can design input forms that are easier to fill in and encourage error correction on site rather than in retrospect (Lupton, 1997).

The second has to do with facilitating output: it is easier to extract data in a form suitable for statistical modelling if the data is structured in a way that reflects the causal relationships that we want to estimate.

Finally, an object-oriented system can be fitted more easily into a GIS framework compared with a conventional system based on records and fields arranged sequentially in files. If the location information is based on a GIS representation of the road network, other information such as speed limits, lighting, signage, maintenance, and so on can be tied into it from other sources. The authors are currently using an adaptation of the SSADM method to construct a GIS-based system as part of a longer-term project.

3. MULTILEVEL STATISTICAL MODELLING

Modelling poses a challenge because of the number of factors involved in road accidents and the complex nature of the interactions between them. Not all the relevant data may be available even with a GIS-based system. For example, details of the geometrical layout for road construction schemes are generated during the design stage, but they are rarely accessible to local authority investigators in later years when accident problems emerge. Fortunately, it seems possible to recreate estimates of curvature from digitised road centreline data, and one of the authors is currently developing a demonstration tool for doing this as part of a related project (Zhang and Lupton, 1998). At a more fundamental level, if we are not clear about the causal processes involved, we can be misled by the results of, for example, contingency tables that ignore the influence of intermediate variables. This point is made clear in Tunaru and Jarrett (1998) and also in a parallel paper at this conference by the same authors. Here, we are concerned with the implications of a hierarchical structure among the variables, which again may invalidate the results yielded by conventional models. Hence, we shall concentrate on a model that takes hierarchical structure into account.

Multilevel modelling is an extension of regression modelling in which the explanatory variables operate on different levels (Goldstein, 1995). Its development in the UK has largely been driven by the needs of educational research. A typical problem scenario in road safety research might be as follows. The basic unit of analysis is the high-risk accident site, and the researcher wishes to investigate the relationship between two variables describing the basic unit: say, percentage accident reduction y as the response variable and expenditure x on engineering treatment as the explanatory variable. Treatment sites are grouped into different highway authorities whose treatment policy and methods vary. It is helpful to model the influence of the highway authority at a higher level in order to explain some of the variability in treatment effectiveness at the lower (site) level. As a rough guide, Paterson and Goldstein (1991) have suggested that a minimum sample size of 25 higher-level units (preferably 100) is needed to obtain useful results.

The alternatives would be (a) to estimate the regression coefficients separately for each highway authority using dummy variables, (b) to estimate them using analysis of covariance, or (c) to represent the relationship between treatment effectiveness and expenditure in terms of group means. The first approach does not make efficient use of the data, the second is cumbersome if the number of higher level units is large or the hierarchy extends over three or more levels, and the third overlooks within-group phenomena that might be important.

Table 2: Possible multilevel relationships among factors affecting road accidents

Unit of analysis	Alternative higher level groupings	Response variable	Explanatory variables
Accidents	Geographical area	Severity Pedestrian involvement No of vehicles	Collision manoeuvres Pedestrian involvement Driver/rider characteristics Vehicle type Road environment features Weather Traffic speed, flow
Accident-involved vehicle passengers	Vehicle	Severity	Personal characteristics Vehicle type Speed limit Road environment features
Individual road users	(a) Geographical area of residence (b) Place of employment	Accident involvements	Gender, socio-economic status Alcohol, drug usage Vehicle type Age, driving experience, exposure
Individual vehicles	(a) Geographical area of residence (b) Company fleet	Accident involvements	Make/model of vehicle Driver/rider characteristics Age, accumulated mileage, exposure
Individual accident locations	(a) Length of road (b) Geographical area	Accident frequency	Road user category Vehicle type Road environment features Weather, dark/light Traffic speed, flow

In principle, multilevel models can also be applied to repeated measurements of accident frequencies (considered as the basic units of observation) on, for example, individual traffic sites or larger units (the higher-level grouping). This would allow for variations in trends between different types of site or area.

Other types of road accident data can also be expressed in multilevel form. The most obvious ones involve the grouping of accidents, road users, or vehicles by geographical area. The occupants of any one vehicle involved in an accident also constitute a natural group, as do vehicles or drivers employed within a company fleet. Some of the possibilities for multilevel modelling are listed in Table 2. All the models are discrete response models, with potentially a mix of categorical and continuous explanatory variables.

4. EXAMPLE

The authors have carried out an exploratory analysis of accident counts for 95 road links in Kent over the period 1984-91, using the multilevel software package MLwiN developed at the Institute of Education in London (Goldstein *et al*, 1998; see Kreft and De Leeuw, 1998, for a useful review of alternatives). The aim was to explore variability among the accident trends from link to link.

In principle, the software can handle Poisson counts but it was decided to work initially in terms of the transformed response variable *$\log_e$(annual accidents per 100 000 veh-km)* using link length and two-way traffic flow data supplied by the County Council. The slope of a graph of the response variable plotted for consecutive years could then be interpreted as the annual proportional growth in accident rate per 100 000 veh-km. The values were approximately Normally distributed, but all the links for which zero accidents occurred in one or more years had to be eliminated from the sample. In addition, not all the accident counts were available for all the years for all of the remaining 65 links: those for years in which some form of engineering treatment had been carried out had been removed from the database previously. (The software does not require that the time patterns of observations for individuals in the sample be the same.) Year numbers were converted to index numbers starting with 1984 = 1.

Although it is not easy to follow, we shall use the same form of notation that appears in the software manuals and related textbooks. The response variable for observation i on year j is denoted by *logaccrate$_{ij}$*, while *year$_{ij}$* is the corresponding year number, and β_0 and β_1 are the (constant) intercept and slope coefficients for the linear regression of *logaccrate* on *year*. The random elements of the model operate on two levels. At the lower level, e_{0ij} is the residual of *logaccrate$_{ij}$*, while at the upper level, u_{0j} is the residual of the intercept for link j, and u_{1j} is the residual of the slope coefficient for link j. The variables *dual$_{ij}$* and *speedlt60$_{ij}$* are 0-1 flags indicating whether the link was dualled, and whether it was subject to a speed limit of less than 60 mph, respectively.

Several models were fitted of progressively increasing complexity. Estimates of the model coefficients for four of them are summarised in Table 3 together with their standard errors.

1. Variance components model: $\quad y_{ij} = \beta_0 + u_{0j} + e_{0ij}$

2: Fixed slope and variable intercept: $y_{ij} = \left(\beta_0 + u_{0j} + e_{0ij} \right) + \beta_1 year_{ij}$

3: Variable slope and intercept: $\quad y_{ij} = \left(\beta_0 + u_{0j} + e_{0ij} \right) + \left(\beta_1 + u_{1j} \right) year_{ij}$

4. As 3 but with additional explanatory variables at the higher level:

$$y_{ij} = \left(\beta_0 + u_{0j} + e_{0ij} \right) + \left(\beta_1 + u_{1j} \right) year_{ij} + \beta_2 dual_{ij} + \beta_3 speedlt60_{ij}.$$

The main results are as follows. First, not surprisingly, the accident rate varies considerably from link to link, more so than it varies from year to year within a given link (model 1). Hence it would be misleading to fit a single regression model to all the data. However, adding a linear trend term with an intercept that varies between links (model 2) explains a significant proportion of the total variation in comparison with the simple model. The multilevel formulation is efficient and parsimonious compared to the alternative of fitting separate regression models to all 65 links.

Secondly, allowing the slope coefficient to vary random between links does not appreciably improve the explanatory power of the model (model 3). Interestingly, however, the intercept and slope residuals are negatively correlated: plots show that the annual percentage decline in accidents for links with high accident rates per veh-km is greater than that for links with low ones. We shall remark on this later.

Finally, the additional explanatory variables in model 4 absorb an appreciable proportion of the remaining variability in the data and also account for some of the previously mentioned correlation between slopes and intercepts. The don't, however, affect the value of the slope coefficient β_1.

Table 3: Multilevel models for accidents on 65 road links in Kent, 1984-91

Parameter		*Model 1*	*Model 2*	*Model 3*	*Model 4*
Fixed part	β_0	1.544 (0.077)	1.740 (0.090)	1.735 (0.098)	2.345 (0.105)
	β_1	--	-0.045 (0.011)	-0.044 (0.012)	-0.044 (0.012)
	β_2				-0.717 (0.160)
	β_3				-0.774 (0.112)
Variable part	$var(u_{0j})$	0.340 (0.067)	0.348 (0.068)	0.454 (0.111)	0.166 (0.060)
	$var(u_{1j})$	--	--	0.002 (0.002)	0.002 (0.002)
	$covar(u_{0j}, u_{1j})$	--	--	-0.017 (0.011)	-0.008 (0.009)
	$var(e_{0ij})$	0.253 (0.019)	0.241 (0.018)	0.228 (0.019)	0.227 (0.019)
-2log(likelihood)		752.3	735.1	732.2	681.2

Not shown in the table are the results for two further models in which engineering treatment was added as an explanatory variable ('treatment' included routine resurfacing work and street lighting as well as targetted remedial schemes). Adding a treatment variable did not appreciably improve the model fit, nor did it account for the faster annual decline in accidents per veh-km on links with high accident rates noted earlier. Other possible causes are increasing congestion on major roads, or a regression-to-mean effect. Further modelling work is now being carried out using a log-linear formulation with Poisson error structure, on an extended range of data.

5. CONCLUSION

We have put forward an outline proposal for a hierarchical database structure that is believed to be more robust, more flexible, and more efficient than the existing UK system. In addition, it more closely reflects the causal interactions between the elements and factors involved in road accidents, and in this sense it is better geared to the needs of statistical analysis, where the hierarchical data structure maps naturally into various forms of statistical model, especially multilevel models. However, an object-oriented database structure is not *essential* for statistical modelling, and its main advantage may lie elsewhere. By eliminating data redundancy it reduces the frequency of errors and facilitates data preparation, which may account for a large proportion of the time and effort involved in a statistical modelling project.

Our review suggests that multilevel software has potential applications in many areas of road safety research. A preliminary exercise has demonstrated the effectiveness of the user interface for the particular software package used, and revealed interesting facets of a sample of link accident data from an earlier study. Work continues with log-linear models applied to an extended range of data.

ACKNOWLEDGEMENTS

The authors are grateful to Margaret Thompson and David Jarrett for helpful advice. The research was supported by the Engineering and Physical Sciences Research Council and the Department of Transport, the Environment and Regions.

REFERENCES

Goldstein H. (1995, 2nd ed) *Multilevel statistical models.* London: Arnold.

Goldstein H., J. Rasbash, I. Plewis, D. Draper, W. Browne, M. Yang, G. Woodhouse and M. Healy (1998) *A users's guide to MLwiN.* London: Institute of Education.

Goodland M. and C. Slater (1995) *SSADM version 4.0: a practical approach.* London: McGraw-Hill.

Kreft I. and J. de Leeuw (1998) *Introducing multilevel modelling.* London: Sage.

Lupton K., D. F. Jarrett and C.C.Wright (1997) *The consistency of road accident variables in Great Britain, 1995*. Technical Report no 1997/6, Transport Management Research Centre, Middlesex University (unpublished).

Lupton K. (1997) Accident databases: design concepts. *Highways & Transportation*, 44 (10), 21-22.

Paterson L. and H. Goldstein (1991) New statistical methods for analysing social structures: an introduction to multilevel models. *British Educational Res. J.*, 17 (4), 387-393.

Tunaru R. and D. F. Jarrett (1998) Graphical models for road accident data. In *Proc Universities Transport Study Group Annual Conference, Trinity College Dublin, 2-4 January 1998* (unpublished).

Zhang P. and K. Lupton (1998) Relating road accidents to horizontal curvature within a GIS. In *Proc Universities Transport Study Group Annual Conference, Trinity College Dublin, 2-4 January 1998* (unpublished).

An Analysis of Causality for Road Accident Data Using Graphical Models

R.S. Tunaru and D.F. Jarrett, Transport Management Research Centre, Middlesex University, London

Abstract

The technique of *graphical modelling* (Whittaker, 1990) can be used to identify the dependence relationships between variables representing characteristics of recorded road accidents. It allows large multi-dimensional tables to be analysed by looking for conditional independence relationships among the variables. The variables under study can often be divided into groups that are ordered in time or by a hypothesised causal assumption. For these situations *graphical chain models* (Whittaker, 1990) are used to explore causal relationships between the variables. Some examples are given for a six-dimensional and a ten-dimensional contingency table.

1. Introduction

A national road accident database will contain a large number of variables representing characteristics of the recorded road accidents. An important problem is then to identify the dependence relationships between the variables. For statistical analysis, the data can be summarised in a multi-dimensional contingency table cross-classified by the variables under study. Because of the Yule-Simpson paradox (Simpson, 1951), the analysis of marginal tables, involving only two or three variables at a time, can be very misleading. Graphical modelling (Agresti, 1990, Edwards, 1995, Whittaker, 1990) is therefore an appropriate approach. This allows the full multi-dimensional table to be analysed and looks for conditional independence relationships among the variables. The statistical models, called *graphical models*, are a sub-class of the well-known log-linear models for contingency tables. Graphical models represent the factors in log-linear models by the vertices of a graph and the edges in this graph correspond to two-factor effects. These models can help to reduce the number of variables sufficient to explain and model the data. The statistical software MIM, developed by Edwards (1995), makes the approach easy to use with a large number of variables.

The variables under study can often be divided into groups that are ordered in time or by a hypothesised causal model. For these situations graphical chain models can be used to explore causal relationships between the variables. These are also based on graphical models but the graph has directed edges between nodes in different groups, with the arrows pointing towards response variables.

Examples are given below for a six-dimensional and a ten-dimensional contingency table using British STATS19 data for 1995. Causal models will be proposed for response variables such as the Number of Vehicles involved in the accident, the Number of Casualties and the Accident Severity. Contingency tables summarising road accident data are sparse because of the large number of variables under study and the nature of the data. The standard methods of model fitting and testing can then be misleading. A comparison between these methods and methods based on *exact Monte Carlo sampling* is discussed.

2. GRAPHICAL MODELS

Let $X = (X_1, ..., X_k)$ denote the vector of variables of interest. The *conditional independence graph* (for short the *independence graph*) is an undirected graph $G = (K, E)$, where $K = \{1, ..., k\}$ is the set of vertices corresponding to the set of variables under study, and where (i, j) is not in the edge set E if the variables X_i and X_j are independent given the remaining variables $X_{\{K \setminus \{i,j\}\}}$.

In general we denote the fact that the random variables (or subsets of variables) X_a and X_c are conditionally independent given X_b by $X_a \perp\!\!\!\perp X_c \mid X_b$, or just $a \perp\!\!\!\perp c \mid b$.

The fundamental result, called the *global Markov property*, is a separation theorem proved, for instance, in Whittaker (1990, pp 63-67).

Theorem 1. *Let* X_a, X_b *and* X_c *be disjoint subsets of variables from* X. *In the independence graph of* X, *suppose that the subsets of vertices* a *and* c *are separated by the subset* b, *in the sense that every path connecting any vertex from* a *with any vertex from* c *has to intersect* b. *Then*

$$X_a \perp\!\!\!\perp X_c \mid X_b.$$

This separation theorem gives the basic rule for reading the conditional independence relationships among the variables of X directly from its independence graph.

Graphical log-linear models form a subclass of the class of hierarchical log-linear models. They are built using the same log-linear expansion as log-linear models. However, the rule concerning the interaction terms in the expansion of graphical models is, in a sense, opposite to the rule for hierarchical log-linear models. Using only three variables X_1, X_2 and X_3 for simplicity, the saturated log-linear model (which is hierarchical and graphical) is

$$\log p_{ijk} = \lambda + \lambda_i^1 + \lambda_j^2 + \lambda_k^3 + \lambda_{ij}^{12} + \lambda_{ik}^{13} + \lambda_{jk}^{23} + \lambda_{ijk}^{123}.$$

A hierarchical log-linear model requires that if, for instance, λ_{ij}^{12} is set to zero then t any higher-order interaction term with superscripts containing both indices 1 and 2 need also be set to zero. Thus

$$\log p_{ijk} = \lambda + \lambda_i^1 + \lambda_j^2 + \lambda_k^3 + \lambda_{ik}^{13} + \lambda_{jk}^{23}$$

is a hierarchical log-linear model, but

$$\log p_{ijk} = \lambda + \lambda_i^1 + \lambda_j^2 + \lambda_k^3 + \lambda_{ik}^{13} + \lambda_{jk}^{23} + \lambda_{ijk}^{123}$$

is not hierarchical because λ^{123} is included but λ^{12} is missing. So the removal of an interaction term in an hierarchical model requires the removal of the corresponding higher interaction terms. A graphical model requires that, whenever the model contains all two-factor interaction terms generated by a higher order interaction term, the model has the higher order interaction term too. Thus,

$$\log p_{ijk} = \lambda + \lambda_i^1 + \lambda_j^2 + \lambda_k^3 + \lambda_{ij}^{12} + \lambda_{ik}^{13} + \lambda_{jk}^{23}$$

is not graphical because the inclusion of $\lambda_{ij}^{12}, \lambda_{ik}^{13}$ and λ_{jk}^{23} requires the interaction term λ_{ijk}^{123} too. The conditional independence graph for hierarchical log-linear models is the same as the interaction graph. The latter is an undirected graph, with the set of vertices corresponding to the variables under study and with the pair (i, j) in the edge set E if and only if there is an interaction term λ^a in the log-linear expansion such that $\{i, j\} \subseteq a$. Different hierarchical log-linear models can have the same interaction graph. However, given an undirected graph, there is only one graphical model corresponding to this graph.

Statistical inference can be based on the (scaled) *deviance*, which is a generalised likelihood ratio. If we denote our current model by M_0 and the saturated model by M_s, the deviance $dev(M_0)$ is twice the difference between the maximised log-likelihood function under the saturated model M_s and the maximised log-likelihood function under the model M_0. This statistic is asymptotically distributed as chi-squared with degrees of freedom equal to the number of free parameters. Thus the overall deviance can be used as a measure of goodness-of-fit. For testing nested models $M_0 \subseteq M_1$ we use the deviance difference $d = dev(M_0) - dev(M_1)$: under the hypothesis that M_0 is true, d has an asymptotic chi-squared distribution with degrees of freedom equal to the difference in the number of free parameters between M_0 and M_1. It is always better to use the deviance difference than the overall deviance because the asymptotic test is more reliable. The χ^2 distribution is then used to calculate the P-value, the probability of obtaining the observed or a larger deviance.

Edwards' program MIM includes several methods of model selection and methods for estimation and testing. The procedure of backward elimination starts from the saturated model and at each step removes the edge for which the deviance difference test for edge removal has the largest P-value greater than or equal to a specified significance level α. The edges that are significant (with P-values smaller than α) at one stage of the analysis are not tested again at further stages but always retained in the graph. In the end, when no further edge can be deleted, the corresponding model should fit the data well. Furthermore, all the conditional independences can be read directly from the graph. The backward elimination procedure is to be preferred to a forward inclusion procedure since it is passes through a sequence of models, all of which fit the data, and the models become simpler at each step.

A possible complication is that contingency tables based on road accident data can be expected to be sparse, with many very small cell frequencies. The asymptotic tests based on the deviance are then not very reliable. The asymptotic P-values in the case of large sparse tables tend to underestimate the real P-values. Exact tests are required to overcome this difficulty and MIM provides options for them. Consider, for instance, a 3-dimensional table of counts. To test the hypothesis $H_0 : i \perp\!\!\!\perp j \mid k$ exact tests are constructed by conditioning on the marginal totals. Denote by Ψ the sample space of all possible 3-dimensional tables $\mathbf{n} = \left[n_{ijk} \right]$ with the same fixed margins as the table of observed counts. Then the P-value for the test criterion T is

$$
\begin{aligned}
P_{obs} \quad &= \quad \Pr(T \ge T_{obs} \mid H_0) \\
&= \quad \sum_{\{\mathbf{n} \in \Psi : T(\mathbf{n}) \ge T_{obs}\}} \Pr(\mathbf{n} \mid H_0)
\end{aligned}
$$

where

$$
\Pr(\mathbf{n} \mid H_0) = \prod_k \left\{ \frac{\prod_i n_{i+k}! \prod_j n_{+jk}!}{n_{++k}! \prod_i \prod_j n_{ijk}!} \right\}.
$$

This approach is easily generalised to any higher dimensional table.

The *exhaustive enumeration* method, calculating $T(\mathbf{n})$ and $\Pr(\mathbf{n} \mid H_0)$ for each table $\mathbf{n}$ from Ψ, is not always feasible. The alternative is to use *Monte Carlo sampling*. Following the algorithm in Patefield (1981), N random tables are sampled from Ψ such that the probability of sampling a table $\mathbf{n}$ is $\Pr(\mathbf{n} \mid H_0)$. For the table $\mathbf{n}_r$, define z_r to be 1 if $T(\mathbf{n}_r) \ge T_{obs}$ and to be 0 otherwise; then estimate P_{obs} by $\hat{P}_{obs} = \sum_{r=1}^{r=N} z_r / N$. As these estimated exact P-values are unbiased it is better to use exact conditional tests whatever test criterion T is used, as was pointed out by Kreiner (1987).

The whole methodology can be illustrated using a 6-dimensional contingency table. The data considered is extracted from the STATS19 database for 1995, and consists of all accidents in the county of Bedfordshire. The following variables are considered:

A = Accident Severity (fatal, serious, slight),

L = Light Conditions (daylight, darkness),

N = Number of Vehicles involved in the accident (one, two, three, four or more),

R = Road Surface conditions (dry, wet-damp, snow-ice-frost-flood),

T = Road Type (major roads, minor roads, where major roads are motorways and
A roads, and minor roads are B, C and unclassified roads), and

S = Speed Limit ($\leq$ 40 mph, > 40 mph).

The contingency table is too big to be reproduced here. The variables are treated on an equal footing, that is, all are considered as response variables. This might seem to be a strong assumption, but the purpose here is to show how the technique works. In addition, the conditional independence relationships among the variables can be investigated in an exploratory manner, with the possible benefit of finding an initial model that can be investigated further using more sophisticated techniques.

Applying the above procedures to the data leads to the model represented in Figure 1. Grouping the variables as $a = \{A\}$, $b = \{N, S\}$ and $c = \{L, R, T\}$ it is easy to verify the conditions stated in Theorem 1. Therefore we can read directly from the independence graph that, given the Number of Vehicles and the Speed Limit, Accident Severity is independent of Light Conditions, Road Surface and Road Type. Thus, the important variables for explaining the Accident Severity seem to be the Speed Limit and the Number of Vehicles.

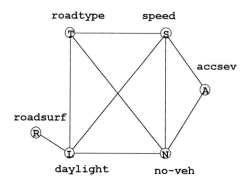

Figure 1. The final model for Bedfordshire data

3. GRAPHICAL CHAIN MODELS

Suppose that, based on external information, the variables $X = (X_1, ..., X_k)$ can be divided into disjoint groups $X = S_1 \cup S_2 \cup ... \cup S_q$, in such a way that any S_i is causally prior to S_{i+1} for all $i = 1, \cdots, q-1$. The subsets S_i are called *chain blocks*. The association structure is assumed to be such that all edges between vertices in the same chain block are undirected, and each edge connecting vertices in different blocks is directed, pointing towards the vertex in the higher-numbered block. The pairwise Markov property on which a chain graph is defined states that there is no edge between two vertices u, v from the same block S_j, or there is no arrow from $u \in S_i$ to $v \in S_j$ for $i < j$, if

$$u \perp\!\!\!\perp v \mid S_1 \cup S_2 \cup \ldots \cup S_j \backslash \{u, v\}.$$

A graph with directed and undirected edges is a chain graph if every cycle consists only of undirected edges. The chain blocks of such a graph are the connected blocks of the graph obtained from the initial graph by removing all directed edges. Graphical chain models are multivariate response models for S_i given $S_1 \cup \ldots \cup S_{i-1}$. The joint density can be factorised as

$$f(x_1, \cdots, x_k) = f(S_1) f(S_2 | S_1) \ldots f(S_q | S_{q-1} \cup \ldots \cup S_1).$$

The case of just two blocks is generic. We consider that the first block S_1 is a set of covariates $X_1, \ldots, X_p$ and the second block S_2 is a set of response variables $Y_1, \ldots, Y_r$. This dichotomy may appear to complicate the modelling process. However, it can be shown (Whittaker, 1990) that the number of models that have to be considered decrease from $2^{\binom{p}{2} + pr + \binom{r}{2}}$ to $2^{\binom{r}{2}} + 2^{\binom{r}{2}}$, which is a considerable improvement. This is related to the important result (Whittaker, 1990, p 304-305) described in the following theorem.

Theorem 2. *The conditional independence graph for a model with the conditional distribution of $S_2 | S_1$ is the same as the conditional independence graph for the model with the joint distribution of S_1 and S_2 where the subgraph corresponding to S_1 is complete. Moreover, the graph has the global Markov property with respect to the conditional distribution of $S_2 | S_1$.*

Therefore the modelling process can be carried out sequentially. At each step, the current block of variables is considered to consist of response variables and all the previous blocks are considered explanatory. The conditional model can be fitted in the joint framework by making sure that the subgraph of the explanatory variables is complete. In this way, we can use all the methods of estimation and inference available for graphical models.

4. APPLICATIONS TO ROAD ACCIDENT DATA

4.1 Bedfordshire Data

First we consider the Bedfordshire data again. The 6 variables can be partitioned into 3 ordered blocks: $S_1 = \{L, R, S, T\}$, $S_2 = \{N\}$ and $S_3 = \{A\}$. The first block contains purely explanatory variables. The independence graph for this block may or may not be of interest. However, it was decided to investigate the conditional independence relationships among the variables in this block. There are two edges missing, between R and T and between R and S. This means that Road Surface is independent of Road Type and Speed Limit, given Daylight Conditions.

The first step in building the graphical chain model is to fit the conditional model for the first two blocks. The subgraph defined by $\{L, R, S, T\}$ is assumed complete; then there is only one missing arrow, between R and N. The next step is to consider Accident Severity, the single variable of the third block, as a response, and to keep fixed the complete subgraph defined by the variables in the first two blocks. There are three arrows missing, between R and A, between L and A and between T and A. The resulting chain graph is shown in Figure 2.

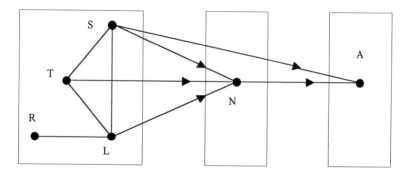

Figure 2. Graphical chain model for Bedfordshire data with 6 variables.
(A = accident severity, N = number of vehicles, S = speed limit, T = road type,
L = daylight/dark, R = road surface condition)

This model has an obvious causal interpretation. The Speed Limit, Road Type and Daylight Conditions all directly influence the Number of Vehicles. Road Surface influences the Number of Vehicles but only indirectly, through Daylight Conditions. Finally, Accident Severity is influenced directly only by the Speed Limit and the Number of Vehicles. The conditional independence relationships can be read from the graph obtained by replacing the directed edges with undirected edges. Care must be taken when reading conditional independences in this way, since certain conditions about the graph need to be verified, see Whittaker (1990, pp 75-80). These conditions hold in this example. The graph shows that $A \perp\!\!\!\perp \{L, R, T\} \mid \{N, S\}$, which means that Accident Severity is independent of Daylight Conditions, Road Type and Road Surface given the Number of Vehicles and the Speed Limit. This is the same conclusion as before. Using these conditional independences, the model is given by the following factorisation of the joint density function

$$f(a,l,n,r,s,t) \quad = \quad f(l,r,s,t) \; f(n|l,r,s,t) \; f(a|l,n,r,s,t)$$
$$= \quad f(l,r,s,t) \; f(n|l,s,t) \; f(a|n,s) \; .$$

For comparing the severity of accidents in urban and rural areas, the estimated probabilities have to be calculated for each level of the variable Number of Vehicles. Trying to use the full dimensional table is not always feasible because of the number of variables involved, and it is advantageous if the table can be collapsed onto an appropriate marginal table involving fewer variables. However, this can be misleading because of Simpson's paradox. Conditions for collapsibility, in the context of graphical models, have been studied by Asmussen and Edwards (1983). Using their results, it can be shown (Tunaru and Jarrett, 1998) that it is possible to calculate the estimated probabilities for Accident Severity in the marginal table defined solely by Accident Severity, Speed Limit and the Number of Vehicles.

In this analysis, the exact Monte Carlo sampling method was used to avoid the problem of sparseness. It turns out, however, that the results are the same using the likelihood ratio test and the stepwise backward elimination procedure.

It is possible to consider a larger number of variables. The table will then be more sparse and the use of exact conditional methods becomes essential. For the same county Bedfordshire we can consider another four variables:

C = Number of Casualties in the accident (1, 2, 3 or more),

D = Day of the Week (Sunday, Monday-to-Thursday, Friday, Saturday),

H = Hour of the Accident (0-6, 7-9, 10-14, 15-18, 19-23), and

P = Pedestrian Crossing within 50m of the place of the accident (no, yes).

We cannot say that all 10 variables can be viewed in a symmetric way. The possible history of the accident gives us a clue about how to partition the variables into recursive blocks. The first block contains the variables $\{D,H,L,P,R,S,T\}$; the reason for choosing these is that their values are related to a site of the road network, and are established well in advance of the occurrence of the accident. The Number of Vehicles is the only variable in the second block, and the last block is $\{A,C\}$; we are able to know the values of these last two variables only after the accident happens. Applying the same method as before, with exact Monte-Carlo sampling, gives the chain graph of Figure 3.

The graphical model for the first block of variables may be of interest or not, but directly from the graph we can read that

$$P \perp\!\!\!\perp \{D,H,L,R\} \mid \{S,T\}$$
$$\{S,T\} \perp\!\!\!\perp \{R,D\} \mid \{H,L\}.$$

Modelling the Number of Vehicles as a response variable, from the chain graph we get that

$$N \perp\!\!\!\perp \{D, L, P, R\} \mid \{H, S, T\}.$$

Finally for the Accident Severity and the Number of Casualties

$$\{A, C\} \perp\!\!\!\perp \{P, R, T\} \mid \{D, H, L, N, S\}$$

$$C \perp\!\!\!\perp \{L, P, R, T\} \mid \{A, D, H, N, S\}.$$

These relationships can help us understand the factors which influence either the Accident Severity or other related variables such as the Number of Vehicles involved and the Number of Casualties in the accident.

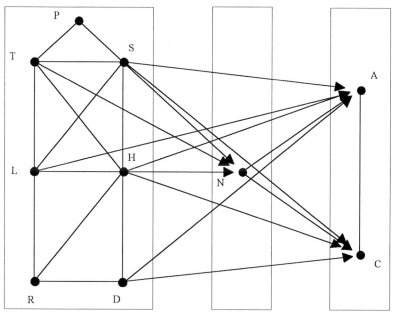

Figure 3. Graphical chain model for Bedfordshire data with 10 variables
(A = accident severity, C = casualties, N = number of vehicles, S = speed limit,
H = hour of day, D = day of week, P = pedestrian crossing, T = road type,
L = daylight/dark, R = road surface condition)

4.2 Bedfordshire and Hampshire data

An interesting question concerns what happens when more data are available. It may be thought that there is no need for exact conditional tests and Monte Carlo methods as there are data available for other counties as well. The contingency table will then cross-classify a larger and larger number of cases while keeping the number of cells fixed. However, this is not necessarily the case. Considering the data from STATS19 for 1995, but for two counties, Bedfordshire and Hampshire, cross-classified by the same 10 variables as before, we get a table which is still sparse. This is due to the nature of the data and is not to do with the sampling method. We expect the table to have small frequencies in the cells corresponding to fatal accidents and large numbers in the cells corresponding to slight accidents, for example.

Applying the same methodology as before results in the graphical chain model of Figure 4. The subgraph corresponding to the first block of variables $\{D,H,L,P,R,S,T\}$ should be considered complete; for clarity, the edges between these vertices are not shown. There are some interesting causal relationships revealed by the chain graph. The presence of a Pedestrian Crossing does not affect the Number of Vehicles involved in the accident, Accident Severity or the Number of Casualties. The Day of the Week influences directly the Number of Vehicles, the Accident Severity and the Number of Casualties. The Accident Severity and the Number of Casualties are directly connected, suggesting that a multivariate regression model (with joint response variables) may be more appropriate than ordinary regression models.

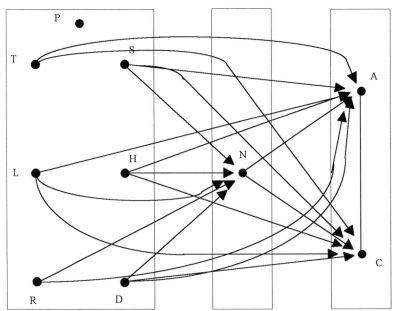

Figure 4. Graphical chain model for Bedfordshire and Hampshire data
(A = accident severity, C = casualties, N = number of vehicles, S = speed limit,
H = hour of day, D = day of week, P = pedestrian crossing, T = road type,
L = daylight/dark, R = road surface condition)

Following the modelling process step by step, it can be informative to describe the conditional independence relationships. We read from the chain graph that

$$N \perp\!\!\!\perp \{P,T\} \mid \{D,H,L,R,S\},$$

and, at the second stage,

$$\{A,C\} \perp\!\!\!\perp P \mid \{D,H,L,N,R,S,T\}.$$
$$C \perp\!\!\!\perp \{P,R\} \mid \{A,D,H,L,N,S,T\}$$

These conditional independence relationships suggest that safety measures, aiming at a reduction in accident severity and the number of casualties, should not consider primarily the presence of pedestrian crossings. The variables in the conditioning set are those that should be targeted because they influence directly the variables of interest, Accident Severity and the Number of Casualties. However, it should be noted that this analysis considers all accidents. It can be shown that a different model might be reached if the analysis is restricted to accidents involving pedestrians.

5. CONCLUSION

Graphical chain models provide a useful exploratory technique for disentangling the potential factors which influence variables such as accident severity or the number of casualties. However, some care needs to be taken in the choice of statistical test used to select a well-fitting model. Using the same 10 variables, the graphical chain models for Bedfordshire, and for Bedfordshire and Hampshire together, are different. This is not surprising since the second model was based on combined data. For the data for Bedfordshire alone, when just 6 variables are used, the graphical chain models obtained using different methods of testing and model selection are the same. However, for the 10-variable table, a different final models is obtained if asymptotic (chi-squared) methods of testing are used instead of the exact Monte-Carlo method used here. As the contingency tables become larger and more sparse, the classical tests are not reliable and the use of exact tests and Monte Carlo simulation procedures become essential.

It would be of interest to compare the conditional independence relationships in different counties and then to test the common ones, if any, in the combined data for the whole of Great Britain. The authors hope to report on such investigations in the future.

ACKNOWLEDGEMENTS

The authors would like to thank Chris Wright and an anonymous referee for helpful comments and suggestions for improving this paper.

REFERENCES

Agresti, A. (1990). *Categorical Data Analysis*. John Wiley, New York.

Asmussen, S. and Edwards, D. (1983). Collapsibility and response variables in contingency tables. *Biometrika*, **70**, 3, 567-578.

Edwards, D. (1995). *Introduction to Graphical Modelling.* Springer, New York.

Kreiner, S. (1987). Analysis of multidimensional contingency tables by exact conditional tests: techniques and strategies. *Scand. J. Statist.*, **14**, 97-112.

Patefield, W.M. (1981). An efficient method of generating random $R \times C$ tables with given row and column totals. *Appl. Statist.*, **30**, 91-97.

Simpson, C.H. (1951). The interpretation of interaction in contingency tables. *J. Roy. Statist. Soc (B).*, **13**, 238-241.

Tunaru, R. and Jarrett, D. (1998). Graphical models for road accident data. *Universities Transport Study Group 30th Annual Conference*, Dublin (unpublished).

Whittaker, J. (1990). *Graphical Models in Applied Multivariate Statistics.* John Wiley, London.

MODELLING OF CELL-BASED AVL FOR IMPROVING BUS SERVICES

M. Law, J.D. Nelson, M.G.H. Bell, Transport Operations Research Group, University of Newcastle, UK

ABSTRACT

Increasing growth in car ownership and private car usage is leading to high levels of congestion and pollution in many busy towns and cities. A reliable and efficient public transport service is one of the solutions for future transport provision. The development of a novel cell-based Automatic Vehicle Locationing system for public transport application is presented in this paper. The technique uses the existing AVL system and combines it with cellular technologies. Little modification is required to turn the existing AVL sequential polling system to the simultaneous cell-based AVL polling system. The fundamental advantage of the cell-based approach to AVL is that direct communication between the vehicles and the cell-based station enables the vehicle fleet to be polled simultaneously. Cell-based AVL System offers possibility for improved public transport service and public transport priority through frequent update of vehicle position and vehicle information. Modelling of cell-based AVL systems using a simulation method is given in this paper. The results show that the polling time for a bus fleet using the cell-based technique is considerably lower than the conventional sequential polling AVL system.

1. INTRODUCTION

The use of Automatic Vehicle Locationing Systems in the public transport sector is increasing. AVL Systems are used for bus fleet control, passenger information systems and for priority of public transport vehicles at signal-controlled junctions. At present, most bus services suffer increased inefficiencies due to congestion, except where sufficient road space allows the segregation of buses using bus lanes or busways. Apart from the segregated facilities, bus priority at signalised junctions is one of the most important facilities available to maintain an efficient bus service. A variety of bus priority techniques (Hounsell, 1994) are available for public transport priority at isolated signals and in network-wide co-ordinated systems. The use of AVL to provide bus detection/location for bus priority is particularly attractive as dedicated transponders and bus detection systems are avoided. The usefulness of the current AVL systems is constrained by the ability to locate the vehicles in real-time, and to predict the arrival time of vehicles at junctions is not accurate enough for use in a dynamic priority system and urban signal control. The polling frequency for most of the AVL systems in UK are of the order of 30 seconds or longer and depend on the size of bus fleet (e.g. London Transport; BUSNET system in Nottingham; and ASIS AVL system in Merseyside). Cell-based AVL Systems offer the possibility for improving the conventional AVL system through a frequent update of vehicle position and vehicle information. The target is to update the vehicle position at a frequency of every few seconds.

2. APPLICATION OF CELLULAR COMMUNICATION TECHNOLOGY TO AVL

There are three major technologies (Scorer, 1993) for AVL, namely beacon-based, satellite-based (GPS), and terrestrial-based systems. The beacon-based AVL systems are similar to the dead-reckoning systems used in dynamic route guidance (Anagnostopoulos *et al.*, 1992). With fixed route transit, distance travelled is sufficient for vehicle locationing. The beacon-based AVL systems are able to provide accuracy to around 10m. The Global Positioning System (GPS) allows determination of vehicle location from the transmission of long-range radio waves from satellites. For the GPS system, an accuracy of the order of a hundred metres is offered (Stewart, 1993). The accuracy of the system can be increased by using differential techniques (Cheong, 1992). The terrestrial radio-based systems make use of the dedicated ground networks by comparing low frequency transmissions received by the vehicle to calculate the position of the vehicle with respect to the radio navigation network. The system is thought to have an accuracy of the order of a few thousand metres, although this is significantly increased when using differential techniques. Due to its high accuracy, the conventional beacon-based system has been selected for the cell-based AVL application.

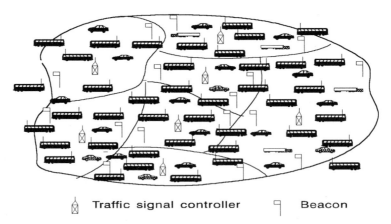

  ⊠ Traffic signal controller   ⊓ Beacon

Figure 1 Cellular Communication Concept to AVL

The basic concepts behind cellular radio are not complex and not new (Bell, 1979; Walker, 1990).

To apply the cellular technology to support conventional AVL systems, it is proposed that the service area is split up into a number of cells each with its own cell identity code (Figure 1). Each cell uses the traffic signal controller as a cell-base-station. Simultaneous interchange of data between the vehicles and traffic signal controllers, traffic signal controllers and public transport control centre forms the basic function of the cell-based AVL system.

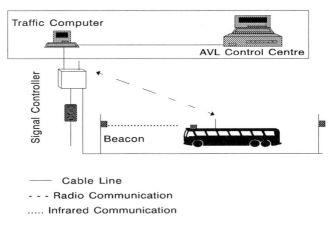

Figure 2. Cell-based AVL System Architecture

Figure 2 depicts the system architecture of the generic Cell-based AVL System. Typically, the system consists of an on-board processing unit (OBU), an odometer, and a cell-base-station communication system. When the bus leaves the depot, the OBU is initialised with the relevant data such as cell_ID, bus_ID, route_ID, and the odometer is set to zero. As the bus travels along its route, it transmits (via radio data communication) a "telegram" to the cell-base-station which contains the cell_ID, bus_ID, the route_ID, Beacon_ID, and the odometer reading. From this information the location of the bus can be calculated. Beacons located along the route are used for correcting errors accumulated in the odometer.

Since all cell-base-stations and vehicles are operated in the same radio frequency it is essential that each cell-base-station has a unique cell_ID, and each bus has a unique bus_ID. To avoid co-channel interference, limited radio transmission power and the time divided technique are used. The short telegram sent from a cell-base-station is received by all the AVL equipped vehicles in that cell, but only the vehicle which has the same bus_ID specified by the cell-base-station's telegram responses to the message. The cell-base-stations transmit the updated information received from the vehicles to the traffic control centre and the data are used for applications such as traffic light priority, passenger information systems and improvement of traffic signal control.

3. MODELLING OF THE GENERIC CELL-BASED AVL

3.1. The SIGSIM Microscopic Traffic Simulator

The SIGSIM microscopic simulator (Law and Crosta, 1998) has been developed jointly between the Transport Operations Research Group at the University of Newcastle upon Tyne and the Centre for Transport Studies at University College London. The modelling of vehicles is at the individual vehicle level, with each vehicle having its own properties: type, length, maximum acceleration, desired speed etc. Vehicle movements are determined according to the route assigned by the user and the Gipps car following model (Gipps, 1981), in which the vehicle acceleration is adjusted according to the relative speed and distance of the vehicle in front. Vehicles are able to change lanes (Gipps, 1986), in order to follow the assigned route, or to gain speed advantage. The five types of vehicles modelled are generated at the start of the route and at intervals according to a shifted negative exponential distribution, the mean arrival rate being user defined.

In SIGSIM, geometrical components layout of the network are required as input to model, such as the length of each link, the number of the controller to which the link is associated, and the number of lanes per link. Detector positions on lanes are user defined, as are lane exit and entry positions, vehicular routes, flows and proportions (by vehicle type). The signal control policy of each junction can be fixed, vehicle actuated system-D, discrete time approach (Bell and Brookes, 1992), or SCOOT-like (Smith *et al.*, 1994a).

Public transport has been included and buses equipped with AVL, bus stops and passengers are modelled. Both the global positioning system (GPS) and the beacon-based approach have been implemented. SIGSIM supports public transport priority at signalised junctions and information from the AVL equipped buses have been used for supporting various traffic control measures.

3.2. Modelling of Cell-based AVL Systems

SIGSIM has been enhanced to represent the proposed cell-based AVL. The modelling of generic cell-based AVL systems in SIGSIM was carried out in four parts: (i) the modelling of cell and cell-based station and communication between vehicle and the cell-base-station; (ii) roadside beacons, and the acceptance of beacon codes by the bus on-board unit; (iii) the modelling of the AVL equipment e.g. polling and odometer counts; and (iv) the modelling of errors which will effect the accuracy of the odometer readings.

A cell consists a set of links and each cell has an identifier. There is one cell-base-station in each cell and they are defined to have position and transmission power. The links for the cells, the cell identification code, position of the cell-base-stations, and the cell-base-station transmission power are defined by the user.

The beacons can be placed anywhere within the network of intersections being modelled. The positioning, number and identification code of the beacons is defined by the user. The simulation model updates the position of each vehicle every 2/3 seconds, corresponding to the reaction time of drivers (Gipps, 1981). The simulation model provides a routine for buses equipped with AVL equipment to check if a beacon has been passed during the update of the bus position. If a beacon lies between the updated and the previous bus position, then the individual beacon code is stored and the odometer reading is reset to zero. Beacons are used for compensating errors which accumulate in the odometer measurements.

The vehicle position is tracked by measuring distance travelled, using the pulses generated from a precalibrated odometer. As the bus follows known routes, distance travelled is sufficient for continuous locationing. Odometer reading are modelled by measuring the accurate distance travelled since the last beacon passed, and then errors in odometer measurements are modelled and added to the distance travelled. The odometer reading is then rounded down to give an integer number of pulses. The number of pulses per metre is defined by the user. Two "forms" of errors accumulate in the odometer measurement (Smith *et al.*, 1994b): (i) consistent or "one-off" errors, related to odometer calibration, the value of which remain constant for a particular vehicle throughout the simulated run, and (ii) inconsistent errors, related to vehicular manoeuvres e.g. overtaking, wheelspin etc, which are generated every time the bus position is updated. The error can be formulated by:

$$y(t) = x(t)(1 + e_1) + e_2(t)$$

where $y(t)$ distance measured by vehicle equipment

$x(t)$ actual distance travelled

e_1 constant error, randomly selected once (when vehicle is generated in simulation model) from a normal distribution.

$e_2(t)$ an error randomly selected from a normal distribution, every time a vehicle position is updated.

The odometer reading is then taken as the highest integer value of $y(t)$ multiplied by number of pulses per metre.

In the communication between vehicle and cell-base-station, the cell-base-station polls a different bus every 0.2 seconds, requesting information concerning that particular bus and a bus telegram consists of the information: a unique vehicle serial number; cell-id, the route number; the time of polling; the code of the last beacon passed; and the odometer reading of distance travelled since the last beacon passed. From this information, SIGSIM is capable of estimating various traffic parameters. An example is given in Table 1 of a bus telegram extracted from SIGSIM simulator.

4. RESULT OF THE SIMULATION

The layout of the network (Figure 3) being modelled is one of the input parameters of the SIGSIM simulator. In this example, the SCOOT-like controlled junction is surrounded by four junctions running appropriate fixed time plans, to ensure that traffic arrives in platoons. Detectors are placed at 240m upstream of the junction on links of 250 m in length.

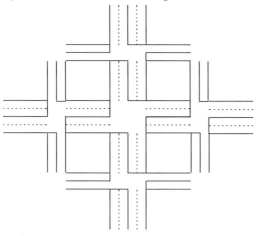

Figure 3. The star Network

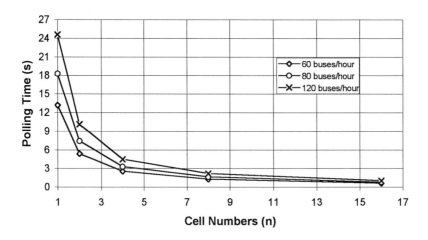

Figure 4. Polling Time vs Cell Numbers

Three different bus densities are simulated: bus entering the network at every 30, 45, and 60 seconds. The network is divided into 1, 2, 4, 8 and 16 cells for use in the simulations. The large number of cells used for a small network is mainly to demonstrate the relationship between poll time and the cell numbers. The results in Figure 4 confirmed that, Cell-based AVL has opened up new possibilities for dramatically increasing the capacity of the data communication between AVL equipped-buses to the traffic signal controller. Real-time information extracted from cell-based AVL can be used to update various traffic parameters hence improve urban traffic signal control, vehicle monitoring, and support public transport priority efficiently at signalised junctions.

Table 1. Example of Bus Telegram

Cell	Lane	Bus ID	R	Polling Time (s)	Distance Traveled from last Beacon	Beacon ID	Odometer Reading	Estimated Distance to Stop Line (m)
3	506n	349	2	443.4000	-1.500000e+00	3	1.480000e+02	-2.000000e+00
3	506n	349	2	445.2000	-1.500000e+00	3	1.480000e+02	-2.000000e+00
3	506n	349	2	447.0000	-6.686515e-01	3	1.490000e+02	-1.000000e+00
3	502n	349	2	449.0000	-2.472055e+02	3	1.520000e+02	2.000000e+00 (-250+2.0 = -148)
3	502n	349	2	450.8000	-2.432791e+02	3	1.540000e+02	4.000000e+00
3	502n	349	2	452.8000	-2.348919e+02	3	1.660000e+02	1.600000e+01
3	502n	349	2	454.6000	-2.237324e+02	3	1.760000e+02	2.600000e+01
3	502n	349	2	456.6000	-2.102665e+02	3	1.890000e+02	3.900000e+01
3	502n	349	2	458.4000	-1.950237e+02	3	2.040000e+02	5.400000e+01
3	502n	349	2	460.4000	-1.784835e+02	3	2.210000e+02	7.100000e+01
3	502n	349	2	462.2000	-1.669302e+02	3	2.340000e+02	8.400000e+01
3	502n	349	2	464.2000	-1.490338e+02	4	0.000000e+00	-1.500000e+02
3	502n	349	2	466.0000	-1.306553e+02	4	1.800000e+01	-1.320000e+02
3	502n	349	2	468.0000	-1.119543e+02	4	3.800000e+01	-1.120000e+02
3	502n	349	2	469.8000	-9.303931e+01	4	5.500000e+01	-9.500000e+01
3	502n	349	2	471.8000	-7.398327e+01	4	7.200000e+01	-7.800000e+01
3	502n	349	2	473.6000	-5.483452e+01	4	9.300000e+01	-5.700000e+01
3	502n	349	2	475.6000	-3.562499e+01	4	1.100000e+02	-4.000000e+01
3	502n	349	2	477.4000	-2.279558e+01	4	1.200000e+02	-3.000000e+01
3	502n	349	2	479.4000	-3.527747e+00	4	1.450000e+02	-5.000000e+00

In the simulation, all information polled from the AVL-equipped buses is recorded and comprises:

the vehicle identification number, cell-id, the service route number; the time of polling; the code of the last beacon passed; the odometer reading and the estimated distance to the stop-line. In the example shown in Table 1, SIGSIM has generated an AVL equipped bus serial number 349. The bus is in cell 3 and route 2. From 443.4 s to 447.0 s, the bus is in lane 506_n, at 449.0 s, the vehicle crossed the intersection and moved to lane 502_n. From 449.0 s to 479.4 s the bus is in lane 502_n. The bus is polled at an average of two seconds. From the bus telegram, the code of the last beacon passed, odometer reading, the distance travelled since last beacon, and the estimate distance to the stopline at a given time has been extracted. Note the difference between this distance and the odometer reading i.e. the error, increases as the bus proceeds, until the odometer has been reset by passing a beacon at poll time 464.2 s. SIGSIM uses measurements of distance from the next stop-line (as negative number) for vehicle positions, and this is included in the Table 1. An increase in this distance between poll time 447 s and 449 s implies that the vehicle has crossed the intersection and is heading for the next junction. The junctions modelled in the simulation are 250 metres apart and beacons are placed 200 metres upstream of the next stop-line. The odometer readings are converted to distance from the stop-line and this is shown in the final column of Table 1. Estimated distances from the stop-line that are positive would suggest that the bus has passed the stop-line and is approaching the next junction. The bus is stationary during poll times 443.4 s to 445.2 s, and this corresponds to the stopline of the junction.

5.   IMPROVEMENT OF URBAN TRAFFIC SIGNAL CONTROL USING CELL-BASED AVL REAL-TIME INFORMATION

Ways to use the real-time information contained within vehicle telegrams to update estimates of traffic parameters of selected signal control systems have been considered. When the AVL equipped-bus being polled, the bus telegram includes the information on the position and time when the bus arrived at the back of the queue. These real-time information was used to update the SCOOT-like controller's parameters such as queue profiles at a short time interval. The majority of runs suggest that the contribution of AVL data to this type of signal controller could be insignificant. For AVL information with discrete time signal controller, AVL information is used to estimate the arrival and departure rates of vehicles at a junction, then combining with estimates derived from detector information. It is suggested that this combination of estimates used by the discrete time controller would result in an improvement in signal control.

An accurate position of the back of queue and the delay time of the AVL equipped-bus are supplied when the buses are polled. This could possibly enable early identification of congestion, and allow compensatory measures. The delay time of AVL equipped buses on the routes can be use to indicate which routes in the network has minimum delay hence support the vehicle's route choice.

6. CONCLUSIONS

There are considerable benefits to be gained in being able to extract and make use of real-time information from AVL equipped buses. A number of problems remain. In most AVL systems, data are collected by the control centre through the sequential polling of buses. The frequency with which the buses can be polled, therefore limits the frequency with which the position of any bus may be recorded centrally. In many systems, the low rate of radio data transmission is a bottleneck; London COUNTDOWN system polls any bus on average around once every 30s (Atkins, 1994). Cell-based AVL has opened up new possibilities for dramatically increasing the capacity of the data communication between AVL equipped buses to the traffic signal controller and provide real-time traffic parameters for supporting public transport priority, vehicle monitoring and improving urban traffic control.

REFERENCES

Anagnostopoulos P., Papapanagiotakis, G. and Gonos, F. (1992). PAN-DRIVE: "A vehicle navigation and route guidance system". *Proc. Vehicle Navigation and Information Systems, IEEE International Conference*, Oslo, Sept. 2-4, pp. 14-19.

Atkins, S.T., (1993) "Real-time Passenger Information At Bus Stops - A New Attribute of Service Quality", *Proc. 25th Annual public Transport Symposium, University of Newcastle upon Tyne*, April 1993

Bell, labs. (1979). Special issue on "Advanced mobile phone service". *Syst. Tech. J.* January 1979.

Bell M.G.H, Brookes, D W. (1992). "Discrete time adaptive traffic signal control: the calculation of expected delay and stops". *Transportation Research B.*

Cheong, M. S. (1992). "Integrated Land and Coastal Navigation Using a Differential GPS Receiving System". *Proc. Vehicle Navigation and Information Systems, IEEE International Conference*, Oslo, 2-4 September 1992, pp. 380-386.

Crosta D A (1998) *PART C: Parallel SIGSIM User Guide*. Centre for Transport Studies (CTS), University College London, London, England.

Department of Transport (1984). 'Microprocessor based traffic signal controller for isolated, linked and urban traffic control installations'. *Specification MCE 0141*, London

Gipps P.G. (1981). "A behaviourial car-flowing model for computer simulation". *Transportation Research 15B*, pp. 105-111.

Gipps P.G. (1986). "A model for the structure of lane-changing decision". *Transportation Research 20B (5)*, pp. 403-411.

Hounsell, N.B. (1994). Bus priority at traffic signals: U.K. developments. *Proceeding of the International conference on Advanced Technologies in Transportation and Traffic Management*. Singapore, 18-20 May 1994.

Law M, Nelson J D and Bell, M G H (1997). "Cell-based Approaches to Automatic Vehicle Locationing". *The 2nd Conference of Eastern Asia Society for Transport Studies*, Seoul, Korea, 29-31 October 1997.

Law M, Crosta D A (1998). *PART A: SIGSIM Theory*. Transport Operations Research Group (TORG), Newcastle University, Newcastle upon Tyne, England and Centre for Transport Studies (CTS), University College London, London, England.

Law M, (1998). *PART B: Serial SIGSIM user guide*. Transport Operations Research Group (TORG), Newcastle University, Newcastle upon Tyne, England.

Scorer, T. (1993). "An overview of AVL (Automatic Vehicle location) technologies". *Proc. Colloquium on Vehicle Location and Fleet Management Systems*, Inst. of Electrical Engineers, London.

Smith M W, Nelson JD, Bell M G H (1994a) "Bus as Probe: The use of AVL information for improved signal control". ERTICO, (Ed), *The Intelligent Transport System*, Vol. 6., London, Artech House, 1994, pp. 3064-3071.

Smith M W, Nelson JD, Bell M G H (1994b) Developing the Concept of Buses as Probes: the Integration of Automatic Vehicle Locationing and Urban Traffic Control Systems. *7th IFAC Symposium on Transportation systems: Theory and Application of Advanced Technology*, Tianjin, China, August 1994.

Stewart, J. (1993) Vehicle locationing and positioning monitoring system using satellite navigation and cellular phone. *Colloq. Vehicle Location and Fleet Management System*, Inst. of Electrical Engineers, London, 8 June 1993.

Walker, J. (1990) Mobile Information Systems. Artech House, Inc., Boston - London.

ONLINE SIMULATION OF URBAN TRAFFIC USING CELLULAR AUTOMATA

L. Santen[1], J. Esser[2], L. Neubert[2], J. Wahle[2], A. Schadschneider[1] and M. Schreckenberg[2]
[1] *Institute for Theoretical Physics, University of Cologne, Cologne, Germany*
[2] *Theoretical Physics, Gerhard-Mercator University, Duisburg, Germany*

ABSTRACT

The modelling and prediction of traffic flow is one of the future challenges for science. We present a simulation tool for an urban road network based on real-time traffic data and a cellular automaton model for traffic flow. This tool has been applied to the inner city of Duisburg. The quality of the reproduced traffic states is investigated with regard to vehicle densities and typical features of urban traffic.

1. INTRODUCTION

Congested road networks are reflected by daily traffic jams which have a negative impact from the individual, environmental and economic point of view. Especially in densely populated regions the road network usually cannot be expanded to relax the situation, so that existing infrastructures have to be used as efficient as possible. Meanwhile a lot of work has been done in developing algorithms and strategies for dynamic - i.e. traffic state dependent - traffic management systems; an exemplary overview is given in (e.g. Ran and Boyce, 1996). The basic assumption for successful applications of these systems is information about the present traffic state in the road network.

Typically, traffic data are collected by locally fixed detectors like counting loops or cameras which are usually not installed on every road in the network, but are positioned mainly in the surroundings of crossings. In order to derive traffic information for those regions which are not covered by measurements, it is necessary to combine local traffic counts with the network structure (i.e., type of roads, priority regulations at the nodes, etc.). This is the basic idea of online-simulations: Local traffic counts serve as input for flow simulations in the considered network to provide network-wide information. An advantage of such a microscopic approach is that all aspects which have to be considered for the interpolation like network structures and traffic lights, are directly incorporated in the simulation dynamics.

The outline of this paper is as follows: First, the cellular automaton approach for the simulation is introduced. The underlying road network and the used data basis are described in the third section. Within this section some technical remarks about the construction of the network are given. The reproduction of traffic states is discussed in section four. The result concerning the whole network are presented in the fifth section. We close with a short discussion and an outlook on to further projects.

2. THE CELLULAR AUTOMATON MODEL

The examination and modelling of traffic flow is of great theoretical and practical interest (for a recent overview of the field, see D.E. Wolf *et al.*, 1996; M. Schreckenberg *et al.* 1998). One basic idea of traffic flow models is to describe the relevant aspects of the flow dynamics as simple as possible. In this spirit very recently cellular automaton models were introduced which give a microscopic description of the vehicular motion using a set of rules. One of the most simple cellular automaton models is the one introduced by Nagel and Schreckenberg (1992) (NaSch-model) which still reproduces important entities of real traffic flow, like the density-flow relation and the temporal and spatial evolution of congestion's. Since cellular automata are by design ideal for large-scale computer simulations they can be used in complex practical applications very efficiently.

The NaSch-model for a single lane road is defined as follows:
The road is divided into cells with a length of 7.5m, which is the average effective length of a passenger car in a jam. A cell is either empty or occupied by a vehicle labelled by i. This vehicle is positioned at cell x_i with its discrete velocity v_i which can vary between 0 and the maximum velocity v_{max}. Therefore a cell can be in $v_{max}+2$ different states. The variable g_i denotes the gap, i.e. the number of empty cells between the vehicle i and its predecessor.

In every time step the system is updated according to the following rules (*parallel update*):

1. Acceleration: $v_i \rightarrow \min(v_i+1, v_{max})$,
2. Avoid accidents: $v_i \rightarrow \min(v_i, g_i)$,
3. Randomisation: with a certain probability p_{dec} do $v_i \rightarrow \max(v_i-1, 0)$,
4. Movement $x_i \rightarrow x_i+v_i$.

Without the third rule the motion of the cars is completely deterministic, i.e. the stationary state only depends on the initial conditions. Therefore the introduction of the braking noise p_{dec} is essential for a realistic description of traffic flow. It mimics the complex interactions between the vehicles and is also responsible for spontaneous formation of jams. A comparison with experimental results (Treiterer, 1970) shows that this simple set of rules is sufficient to reproduce the macroscopic features of traffic flow, see Fig. 1.

This basic model can easily be expanded to describe more complex road networks, e.g. multi-lane traffic and intersections (Rickert *et al.*, 1996; Wagner *et a.*, 1997; Esser and Schreckenberg, 1998).

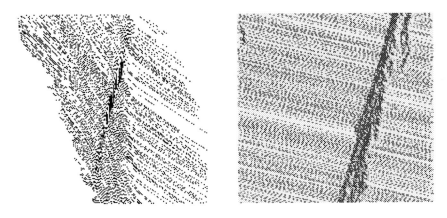

Figure 1. Time-space plots of density waves. The left picture has been generated by using video sequences from an American Highway. Similar structures can be obtained from simulations based on the NaSch-model. Every line describes a trajectory of a single vehicle.

More detailed measurements of highway traffic have shown that the average flow is not a unique function of density. In certain density regimes, either homogeneous states with very high throughputs or phase-separated states where a large jam coexists with a jam free region may exist. The fundamental diagram exhibits a discontinuity, the so-called capacity drop. Intimately related is the phenomenon of hysteresis. Reducing the density from high values, the phase separated state is realised. On the other hand increasing the density from low values the homogeneous state is obtained. In this way one can trace a hysteresis loop. This behaviour (hysteresis loop) cannot be explained within the original model. Thus one has to introduce modified update rules for a description of this effects. This has been done successfully by Barlovic *et al.* (1998). Nevertheless it has been shown that the NaSch-model is sufficient to model traffic flow in urban networks, because they are dominated by the dynamics at the intersection, e.g. traffic lights. Due to this fact the details of the underlying dynamics play a less important rule compared to traffic flow on highways.

3. ROAD NETWORK AND TRAFFIC DATA

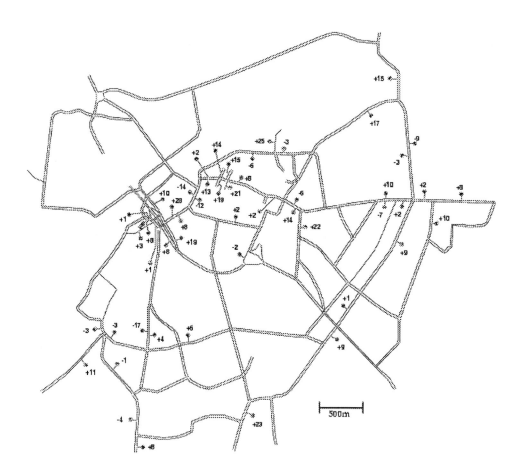

Figure 2. Sketch of the simulated road network with check points (filled circles) and sources and drains (letters). At the check points the data of the whole cross-section are available. The check points are used to tune the local flow. The numbers denote the input rates (cars per minute) averaged over 24 hours.

Typically traffic data are collected by locally fixed detection units. In order to get network wide information, which are necessary e.g., for route guidance systems, one has to interpolate this local data over the whole network. For a precise description of the actual traffic state in the network a realistic description of dynamics must be used. This is where the online simulations come into play. Here the actual measurements serve as input data for numerical simulations of the whole network.

We show results from online simulations of the road network of Duisburg. In order to simulate network traffic one has to consider some basic elements for the construction of the network, edges and nodes (Esser and Schreckenberg, 1998). An edge corresponds to a driving direction on a road. For each edge the realistic number of lanes, turn pockets, traffic lights and priority rules are considered in the simulation. The network consists of 107 nodes (61 signalised, 22 not signalised and 24 boundary nodes) and 280 edges. 138 of them are single-lane roads, the remaining are multi-lane edges. There are 51 check points in the network, where the vehicle flows and densities are adjusted with real traffic data. The check points are marked filled circles in Fig. 2.

As mentioned above we use the NaSch as underlying microscopic model. Due to its small and simple set of rules it is possible to simulate the traffic flow in the complete road network faster than real-time. This is a basic requirement for the prediction of the temporal evolution and for establishing a route guidance system. In our case, the tools have been implemented in C, the programmes are running on a common Pentium PC 133 MHz with Linux operating system, including the data transfer handler, the traffic light manager and the network simulation itself. We are able to simulate a typical day within 20 minutes.

In order to tune the flow at the check points all arriving cars at the check point are deleted and then cars are put in the network with a given source rate corresponding to the measured flow. Obviously, it is also possible to tune only flow differences, but it is not clear whether one has to enlarge or reduce the density in order to obtain the predefined value. Moreover, simulations have shown that both methods lead to the same results for the quantities we took into account.

Since no origin-destination information with sufficient resolution in time and space is available, vehicles drive randomly through the network. This means that the turning direction is assigned to each vehicle randomly (with a realistic probability for each direction) in the so-called marking area, which is placed on every edge in certain distance to the nodes. Currently, turn counts can be derived directly for 56 driving directions; in addition, turn counts at the nodes were collected manually to have at least typical values for directions which are not covered by measurements.

4. REPRODUCTION OF TRAFFIC STATES

A sensitive test for the quality of the online simulations is the ability to reproduce given traffic states. Because no network wide information can be obtained from the online measurements, we have to compare the results of the online simulations with artificial states (reference states) generated by an independent simulation run. Concisely, we perform two simulation runs with two independent sets of random numbers, but the same set of simulation parameters, e.g. the input rates at the boundary nodes. After an initial equilibration of the system to the stationary state, we estimated the reproduction rate of the local density via the following quantity:

$$R_\rho = \sum_{e=1}^{N_e} z_e \left[1 - \left| \rho_e - \overline{\rho}_e \right| \right]$$

The weight of each edge is given by z_e the number of cells on the edge. The density of an edge in the second run ρ_e is compared with the local density of the same edge $\overline{\rho}_e$ in the reference run. Z denotes the total number of cells of all N_e edges. Each measure point serves as possible check point in the reference run.

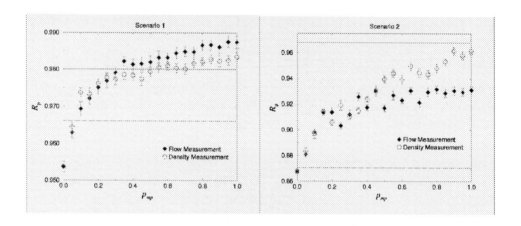

Figure 3. Relation $R_\rho(p_{mp})$ resulting from reproducing traffic states of two scenarios. Simulations in scenario 1 (2) are performed with an input rate $r_s = 0.1$ (0.5) at the boundary nodes. Additionally the values of R_ρ^{ran} (dotted line) and R_ρ^h (dashed line) are shown.

In Fig. 3 we show the results of the reproduction rate dependent on the probability p_{mp} that a measure point is actually used as check point. The results are shown for two different values input rates at the boundary nodes (scenario 1, 2). For obvious reasons the reproduction rate increases for higher values p_{mp}. But for small values of p_{mp} the reproduction rate does not increase monotonously. This is due to the fact that the reproduction rate strongly depends on the position of the check points in the network. In order to show this effect we used a completely new check point configuration for each value of p_{mp}. It should be mentioned that even for the empty network (without any check points) high reproduction rates can be obtained, if the input rates of the simulation and the reference run agree. Therefore it should be possible to extrapolate future states of the network with a reasonable accuracy using online-simulations.
Comparing both scenarios we can see that the reproduction rate is higher for smaller input rates.

This means that it is more difficult to reproduce high density states. Finally, we consider two special cases which are depicted in Fig. 3 as dashed and dotted line. R_p^{ran} denotes the reproduction rate that we obtain if the simulation and reproduction only differ by the set of random numbers used during the time interval where the <u>measurements</u> are performed., i.e. we <u>start</u> the measurements with two *identical* copies of the system. For large value of p_{mp} a higher reproduction than R_p^{ran} can be achieved, because the check points also store the fluctuations in the reference run leading to an extremely high value of R_p. The reproduction rate R_p^h denotes the results obtained from an initial homogeneous state in the reference run.

In Fig. 2 the net rates of vehicles that have to be put into the network averaged over one day are shown. Obviously, at most of the check points vehicles have to be put into the network meaning that there are additional sources in the network which are not covered by check points. This result is not surprising, because of the fact that the roads in the northern part of the city are not well equipped with induction loops. Presently, reliable results of the online simulations can only be obtained in the city centre of Duisburg, where the density of check points is sufficient.

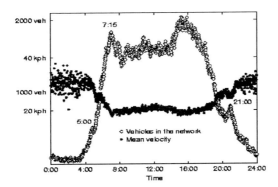

Figure 4. A typical weekday in the network. There are two pronounced peaks indicating the rush hours in the morning and in the afternoon. During day-time the mean velocity is below 20 km/h, whereas its fluctuations strongly increase during the night.

5. RESULTS OF THE ONLINE SIMULATIONS

The number of vehicles in the network and their average velocities during a typical weekday are shown in Fig. 4. It can be seen that the rush hours are reproduced well and that the maximum number of cars in the network is about 2,100, i.e. ten percent of the cells are occupied. But even smaller perturbations like traffic resulting from shift-workers can be resolved, see peak at 5 a.m. and 9 p.m. The mean velocity during day-time is about 20 km/h and is nearly doubled at night. An interesting characteristic of the velocity is its variance which is small during day-time but increases with decreasing amount of cars. This is a typical quality of the free flow regime which exists at smaller densities. The actual traffic state is published in the Internet (OLSIM, 1997) and updated every minute after receiving a new record of data.

6. SUMMARY

Our investigations gave insight to the complex problem of reproducing traffic states with locally measured data. We described the underlying model, the road network and the data basis. We showed that the online simulations can be performed faster than real-time, a basic requirement for traffic predictions. The simulation tool is able to reproduce main features of urban traffic flow.

The experience with the online data enables us to validate and calibrate the model. Furthermore, the influences of static and dynamic changes of the underlying network can be investigated. Beside the task of planning the benefits of route guidance systems and different strategies of routing can be examined. In order to improve the quality of the simulation the number of check points in study area will be increased. The area itself will be enlarged and highways will be included.

7. ACKNOWLEDGEMENTS

The authors would like to thank K. Froese for digitising the Duisburg network and J. Lange, acting for the traffic control centre of the municipal of Duisburg, for providing essential road network data and especially the online traffic data.

We are also thankful to C. Gawron and M. Rickert from the traffic group at the Centre for Parallel Computing of the University of Cologne, S. Krauß and P. Wagner from the *German Aerospace Research Establishment* DLR in Cologne and K. Nagel from the TRANSIMS project at the Los Alamos National Laboratory for many hints and fruitful discussions with regard to the network simulation. This work was developed in the framework of the Northrhine Westfalia Cooperative *Traffic Simulation and Environmental Impact* NRW-FVU (FVU see references).

REFERENCES

Barlovic, R., L. Santen, A. Schadschneider and M. Schreckenberg (1998). Metastable states in cellular automata for traffic flow. *EPJ B* (accepted).

Esser, J. and M. Schreckenberg (1997). Simulation of urban traffic based of cellular automata. *Int. J. Mod. Phys. C* **8**, 1025.

FVU Home Page. Traffic Simulation and Environmental Impact. http//:www.zpr.uni-koeln.de/Forschungsverbund-Verkehr-NRW/.

OLSIM (1997). The simulation of the traffic in the urban road network of Duisburg: http//:www.comphys.uni-duisburg.de/OLSIM/.

Nagel, K. and M. Schreckenberg (1992). A cellular automaton model for freeway traffic. *J. Phys. I France* **2**, 2221.

Ran, B. and D. Boyce (1996). *Modeling Dynamic Transportation Networks*. Springer, Berlin.

Rickert, M., K. Nagel, M. Schreckenberg and A. Latour (1996). Two lane traffic simulation using cellular automata. *Physica* A 231, 534.

Schreckenberg, M. and D.E. Wolf (eds.) (1998). *Traffic and Granular Flow 1997*. Springer, Singapore.

Treiterer, J. *et al.* (1970). Investigation of traffic dynamics by aerial photogrammetic techniques, Interim Report Ees 278-3,Ohio State University, Columbus Ohio (unpublished).

Wagner, P., K. Nagel and D.E. Wolf (1997). Realistic multi-lane traffic rules for cellular automata. *Physica* A **234**, 234.

Wolf, D.E. , M. Schreckenberg and A. Bachem (eds.) (1996). *Traffic and Granular Flow*. World Scientific, Singapore.

MODELLING A DRIVER'S MOTIVATION AND OPPORTUNITY TO CHANGE LANES ON DUAL CARRIAGEWAY ROADS

Fung-Ling Leung and John Hunt, Cardiff School of Engineering, Cardiff University

ABSTRACT

The paper considers the application of neural networks to model driver decisions to change lane on a dual carriageway road. The lane changing process is treated as consisting of two decisions, namely motivation and opportunity. Separate backpropagation neural networks are applied to represent each of the two decisions. The trained motivation and opportunity neural network models are linked to produce a layered network which represents the complete lane changing process. Separate models are developed to represent the nearside to offside lane changing decision, and the offside to nearside lane changing decision. This paper describes the development of the model of the nearside to offside lane changing decision.

For model development, data were collected from several subject vehicle drivers. The results are presented and the implications considered. Selected data were applied to train the neural networks and then an independent subset of data were used to assess performance. When the complete nearside lane changing neural network model was presented with the unseen test examples, 93.3% of the examples were correctly predicted as a lane change or no lane change. These results are shown to be a considerable improvement on those obtained previously.

1. INTRODUCTION

Lane changing is one of the manoeuvres available to drivers on a multi-lane one-way carriageway. Typically a driver will change lane either to join or leave a particular road or to maintain a desired speed. Many factors influence a driver's decision to implement a wish to change lanes. For example a study by McNees (1982) has shown that, on multi-lane dual carriageway roads, the lane changing manoeuvre distance is influenced by traffic volume and the number of lanes. Sparmann (1979) concluded that on multi-lane dual carriageway roads, it is the presence of opportunities which allow drivers to change lanes that influences traffic operation and safety.

Research has shown that highway capacity and safety are affected by driving behaviour such as lane changing and selection of desired speed (Ferrari, 1989; Chang and Kao, 1991). Traffic models are often applied in the development of layouts, assessing capacity and delays, and in evaluating alternative operating strategies for a road section. Microscopic simulation models, which are frequently used, offer the flexibility to represent behaviour for a wide range of highway and traffic conditions and produce measures of performance as output. A possible disadvantage of simulation models is that they represent an individual's interpretation (i.e., the model developer's interpretation) of the behaviour of the real traffic system. During development, assumptions and generalisations concerning the traffic system are made and these typically vary from model to model.

With the advance in technology over recent years, model developers have considered the use of artificial intelligence techniques to model driver behaviour. Intelligent systems, using artificial neural networks (ANNs), have been developed that are capable of autonomously controlling vehicles when presented with information concerning it's driving environment (Niehaus and Stengel, 1991; Pomerleau et al, 1991). In a study by Lyons (1994), ANNs were applied to model lane changing decisions from the nearside lane (lane 1) to the offside lane (lane 2) on dual carriageways. Input data network were presented as a series of binary input patterns that represented the changes in traffic patterns over time, and the speed of a subject vehicle. The ANN was required to determine (based on presented data) whether the subject vehicle was to change lanes or to remain in the current lane. The results, based on observations of lane changing on a section of 2 lane dual carriageway obtained by Lyons (1995), show that 70% of unseen test examples were correctly classified by an ANN model. With the benefit of hindsight it now appears that the approach adopted by Lyons (1994) was probably based on an overestimate of the ability of an ANN to interpret data. This paper considers an alternative neural network approach with the aim of improving performance.

2. ARTIFICIAL NEURAL NETWORKS

2.1 Background

Artificial neural networks (ANNs) are parallel distributed information processing structures that are made up of processing units known as processing elements (PEs). Each PE has a single output connection that carries an output signal. The processing that occurs within each PE depends on the current values of the input signals arriving at the PE and the values stored in the PE's local memory. Within an ANN, the PEs are arranged in layers known as:
- the input layer,
- the hidden layer(s), and
- the output layer.

Like the human brain, ANNs learn by experience. Information from the outside world is presented to the ANN through the PEs in the input layer. Signals leaving the input PEs are then passed into the hidden layer PEs. The internal activity level in each PE in the hidden layer is represented by the summation of the inputs, and the corresponding connection weights. The combined input in each PE is then modified by a transfer function. This transfer function transforms the combined input into a potential output value. The PE will only pass information onto other PEs if the combined activity level within the PE reaches the specified threshold level. The process of modifying connection weights is known as learning. This learning process is commonly referred to as training. Training ANNs involves presenting the ANN model with training examples (or vectors). Having trained the network, unseen test examples are presented to assess how well the ANN model performs. During training care must be taken to ensure that the training examples are not simply memorised by the ANN.

The backpropagation neural network is the most commonly applied ANN paradigm of the many which have been developed. This is primarily because it is simple to implement and produces a solution to most problems. A two step sequence is typically used during training. In the first step the networks response to a particular input pattern is compared with the desired response and the error calculated. The error is propagated back through the network with the weights on the connections adjusted accordingly. The rate at which the errors modify the weights is controlled by a specified learning rate. A momentum term is used to assist in preventing the updating of weights from being caught up in a local minima. The architecture of a backpropagation neural network is determined by the number of input variables presented to the network and the number of responses required from the network. The number of hidden layers and the number of PEs in each hidden layer may be varied.

2.2 Application to model lane changing

Dougherty (1995) has reported an explosion of interest in neural networks applications in the transport field in the period since 1990. Intelligent systems, using neural networks, have been developed that are capable of autonomously controlling vehicles when presented with information concerning its driving environment (Niehaus and Stengel, 1991; Pomerleau et al, 1991). Such systems have the ability to perform driving tasks including lane changing. Lyons (1994) applied ANNs to model lane changing decisions from lane 1 to lane 2 on dual carriageways. Both the backpropagation and the LVQ (Learning Vector Quantisation) were applied. The LVQ paradigm was evaluated on the basis that it was potentially more specific to classification tasks such as that represented by the lane change decision.

Data presented to the ANN by Lyons (1994) were collected by observation of vehicle interaction on the A48, a dual carriageway in Cardiff. The results obtained (Lyons, 1995) show that 70% of unseen test examples were correctly classified by the LVQ ANN model. This performance, while encouraging, suggested that further work was required. In the period since 1994, experience in applying ANNs has developed substantially and it now appears that the approach adopted by Lyons was based on an overestimate of the ability of an ANN to interpret raw data. The approach used also has other disadvantages including the large size of the model which followed from the data representation which was used. Large ANNs are inefficient and take a long time to train.

Here an alternative approach is used in which the task of lane changing is split into two tasks;

- deciding that a lane change is required
- deciding that a lane change can be made

The lane change decision is thus considered as being made up of a motivation decision and an opportunity decision. The motivation decision and the opportunity decision are represented by separate ANNs. These ANNs can later be linked to produce a layered network which represents the lane changing process. In developing these ANNs, data are considered and pre-processed before assembling the input to the neural network.

3. MODELLING LANE CHANGING USING NEURAL NETWORKS

3.1 Basis of method

The proposed approach involves considering the lane change manoeuvre as being made up of a motivation decision and an opportunity decision. Separate models are required for lane 1 to lane 2 changes, and lane 2 to lane 1 changes. This paper considers the model of lane 1 to lane 2 changes. Each decision is represented by a separate ANN. Figure 1 illustrates the structure of the complete lane changing neural network model (LCNNM) which combines the motivation and opportunity neural network decision models.

The driver initially decides if he/she is motivated to change lanes based on measures such as his/her current speed and the positions of surrounding vehicles. This is represented by the motivation neural network model (MNNM). If there is no motivation to change lanes then the driver will remain in his/her current lane. If the driver decides that he/she is motivated to change lanes then he/she needs to determine if there is a safe opportunity to change lanes based on, for example, his/her current speed and the presence of vehicles in lane 2. This is represented by the opportunity neural network model (ONNM).

Initially the variables required to develop each model are assessed. By considering the lane change decision as two separate decisions, it is possible to present each network with fewer input variables. This should result in a more efficient lane changing model. A preliminary investigation using data derived from microscopic simulation provided results which were sufficiently promising to justify an evaluation based on data collected from site observations of lane changing.

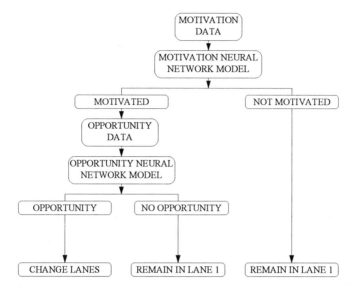

Figure 1. The complete lane changing model

3.2. Data collection

Data were collected from a number of drivers (known as the subject vehicle drivers). Each driver was required to drive for a two hour period along a dual carriageway. The driving characteristics of each driver was recorded visually by three video cameras mounted within the vehicle (known as the subject vehicle). The motivation state of the driver and the subject vehicle speeds throughout the whole trip were orally recorded.

After the subject vehicle drivers had completed their trips, data were extracted from the video recordings. So that the driver motivation and opportunity decisions may be modelled, data associated with the following variables were extracted while the subject vehicle drivers were in the lane 1 (adjacent to the hard shoulder) or lane 2 (adjacent to the central reservation):

- the speed of the subject vehicle driver, subspd (km/h),
- the motivation state of the subject vehicle driver,
- the opportunity situation of the subject vehicle driver,
- the distances from the subject vehicle to the preceding and following vehicles in lane 1 at time t, disfr1(t) and disbck1(t) (m),
- the distances from the subject vehicle to the preceding and following vehicles in lane 2 at time t, disfr2(t) and disbck2(t) (m),
- the relative speeds of the preceding and following vehicles in lane 1, spdfr1 and spdbck1 (km/h), and
- the relative speeds of the preceding and following vehicles in lane 2, spdfr2 and spdbck2 (km/h).

Observations from the recordings and subsequent conversations with the subject vehicle drivers showed that the subject vehicle drivers selected lane 1 as their preferred lane. Table 1 summarises the number of examples collected.

Table 1. A summary of the number of examples collected for lane 1 to lane 2 changes

Example type	No. of examples
Motivated with opportunity	325
Motivated with no opportunity	203
Non motivated	370
Total	898

To enable the ANN to train and test properly, all training examples (or vectors) should contain the same number of input variables. In practice, it is often difficult to obtain the data for each variable; for this data set a distance of 120m and a speed of 0.1km/h were allocated to the 'out of view' vehicle, when no vehicles were in view of the subject vehicle.

3.3 Model development

3.3.1 Pre-processing of data. Relationships between variables were examined initially by plotting and visual inspection. This process enabled the preliminary selection of the most suitable input variables for each of the sub models. Input data for the selected variables were then pre-processed using the software package PREDICT (NeuralWare, 1995). PREDICT has the ability to manipulate, select and prune data for ANN applications. As part of this process, the use of transformations to modify the distribution of the input variables, so that they better match the distribution of the output variable(s) is examined. Overall, PREDICT attempts to find and improve the relationship between the input variables, selected from the raw data, and the output variables. NeuralWorks Professional II/Plus, which offers more detailed diagnostics, was then used to train and test each sub model (Neuralware, 1993).

3.3.2. The motivation sub model. All the motivation examples collected from the subject vehicle drivers while in lane 1 (Table 1) were presented to PREDICT to determine which variables influence a driver's motivation decision. An ANN containing 4 input PEs and 4 hidden PEs was shown to model motivation successfully. The architecture selected by PREDICT is shown in Figure 2. Data associated with the selected variables and transformations were extracted from the lane 1 motivation data set (Table 2). This data set was then separated into a training set (75% of the examples) and a test set (25% of the examples). Several backpropagation ANNs were trained and tested with the data described above. The results showed that the MNNM correctly predicted the outcome for 94.8% of the unseen test examples.

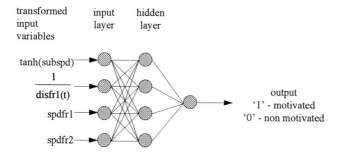

Figure 2. The motivation neural network model

3.3.3. <u>The opportunity sub model</u>. It was assumed that only motivated drivers will consider an opportunity situation. Therefore, all data collected for motivated subject vehicle drivers in lane 1 were presented to PREDICT to determine which variables influence a motivated driver's opportunity decision. The procedure was similar to that used for the motivation sub model. PREDICT found that a backpropagation ANN containing 6 input PEs and 10 hidden PEs would successfully predict a driver's opportunity decision. Figure 3 illustrates the structure of the ONNM.

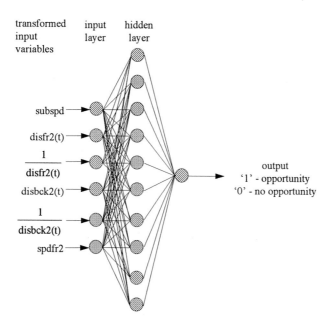

Figure 3. The opportunity neural network model

Following training, the ONNM was presented with the unseen test data. The ONNM correctly predicted the outcome for 92.7% of the unseen test examples.

4. THE OVERALL PERFORMANCE OF THE LANE CHANGING MODEL

The performance of the LCNNM was examined by considering the output responses made by each component of the LCNNM based on the unseen test data. The motivation data is initially passed into the MNNM. The example is passed onto the ONNM if the MNNM finds that the example is a motivated example. The performance of the MNNM therefore has an effect on the overall performance of the LCNNM. Figure 4 shows the results obtained from the lane 1 LCNNM based on the unseen lane 1 test examples.

The unseen lane 1 test set originally consisted of 225 driving examples (of which 132 examples were motivated examples, and 93 examples were non motivated examples). These examples were initially passed through the lane 1 MNNM. The lane 1 MNNM correctly predicted 93.9% of the motivated examples (124 examples) and 95.7% of the non motivated examples (89 examples).

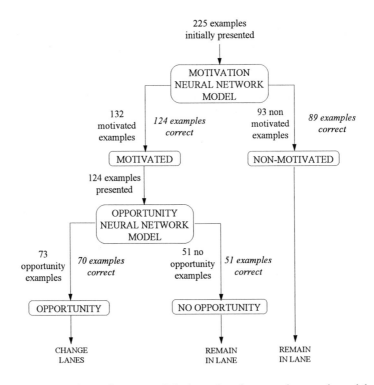

Figure 4. The performance of the lane changing neural network model

As it has been assumed that the subject vehicle driver would only look for an opportunity to change lane when he/she was motivated to do so, only the correctly predicted motivated examples were passed through to the lane 1 ONNM. In total, 124 motivated examples were passed into the lane 1 ONNM (of which 73 were opportunity examples and 51 were no opportunity examples). The lane 1 ONNM was able to correctly predict 95.9% of the opportunity examples (70 examples) and 100% of the no opportunity examples (51 examples).

If all the test examples were correctly classified by both the lane 1 MNNM and ONNM, there would have been 81 lane changing decisions and 144 no lane change decisions (51 motivated with no opportunity examples plus 93 non motivated examples). From the results obtained, the lane 1 LCNNM correctly predicted 68 lane changing examples and 136 no lane change examples. The lane 1 LCNNM therefore correctly predicted 93.3% of the lane changing decisions while the subject vehicle drivers were in lane 1.

The performance of the lane 2 LCNNM which was also developed correctly predicted 98.7% of the unseen lane 2 test examples.

CONCLUDING REMARKS

From the results it can be concluded that, based on data collected from the subject vehicle drivers, backpropagation ANNs can be successfully applied to model driver motivation and opportunity decisions and hence lane changing decisions.

The lane 1 LCNNM was capable of predicting 93.3% of the unseen lane 1 examples and the lane 2 LCNNM was able to predict 98.7% of the unseen lane 2 test examples. A summary of developed components of the LCNNM is shown in Table 2.

Table 2. Summary of the MNNMs and ONNM

Neural network model	Variables and transformations	ANN architecture (input,hidden, output)
MNNM	tanh(subspd), 1/disfr1(t), spdfr1 and spdfr2	4,4,1
ONNM	subspd, disfr2(t), 1/disfr1(t), disbck2(t), 1/disbck2(t) and spdfr2	6,10,1

It can also be concluded that by treating the lane changing process as two separate decisions, and by being more selective in providing the ANN with input data, a much improved lane changing model is developed compared with that developed previously using an ANN.

An ANN provides, in theory at least, an unbiased representation of driver behaviour in practice. However, while the study described in this paper has established that ANNs can be applied to model lane changing, the techniques employed continue to demonstrate some disadvantages in comparison with, for example, microscopic simulation. A major disadvantage is the requirement for the availability of comprehensive data sets with which to train and test the ANN. The trained ANN represents a fit, albeit a sophisticated fit, to observed data and can reasonably be expected to interpolate within the range of the observed data. Extrapolation is, however, likely to be hazardous as the behaviour of the ANN is likely to be unpredictable for input data outside the range over which it was trained. In these circumstances, for example involving changes in road layout or driver behaviour, the algorithm in a simulation model developed from human understanding and reasoning is currently more likely to provide intuitively sensible results.

6. REFERENCES

Ahmed, K.I., Ben-Akiva, M.E., Koutsopoulos, H.N., and Mishalani, R.G. (1996). Models of freeway lane changing and gap acceptance behaviour. *13th International Symposium on Transportation and Traffic Theory, 1996, Lyon, France*, 501-515.

Chang, G. and Kao, Y.M. (1991). An empirical investigation of macroscopic lane change characteristics on uncongested multi-lane freeways. *Transportation Research*, **3C**, 375-389.

Dougherty, M. (1995). A review of neural networks applied to transport. *Transportation Research*, **3C**, 247-260.

Ferrari, P. (1989). The effect of driver behaviour on motorway reliability. *Transportation Research*, **23B**, 139-150.

Godpole, D.N. and Lygeros, J. (1994). Longitudinal control of the lead car of a platoon. *IEEE Transactions on Vehicular Technology*, **43**, 1125-1135.

Lyons, G.D. (1995). Calibration and validation of a neural network driver decision model. *Traffic Engineering and Control*, **36**, 10-15.

Lyons, G.D. (1994). *The application of artificial neural networks to model driver decisions*. PhD Thesis, Cardiff School of Engineering, University of Wales, Cardiff.

McNees, R.W. (1982). In situ study determining lane-manoeuvring distance for three- and four-lane freeways for various traffic-volume conditions. Highway Capacity and Traffic Characteristics, *Transportation Research Record 869*, Transportation Research Board, National Research Council, Washington, D.C., 37-43.

Niehaus, A. and Stengal, R.F. (1991). An expert system for automated highway driving. *IEEE Control Systems*, **11**, 53-61.

NEURALWARE (1993). *Reference Guide (software reference manual for NeuralWorks Professional II/Plus)*. NeuralWare Inc., Technical Publications Group, Pittsburgh.

NEURALWARE (1995). *NeuralWorks PREDICT - Complete solution for neural data modelling*. NeuralWare Inc., Technical Publications Group, Pittsburgh.

Pomerleau, D.A., Gowdy, J. and Thorpe, C.E.I. (1991). Combining artificial neural networks and symbolic processing for autonomous robot vehicle guidance. *Engineering Applications of Artificial Intelligence*, **4**, 279-285.

Sparmann, U. (1979). The importance of lane-changing on motorways. *Traffic Engineering and Control*, **20**, 320-323.

Varaiya, P. (1993). Smart cars on smart roads: problems of control. *IEEE Transactions on Automatic Control*, **38**, 195-207.

On Continuum Modelling of Large Dense Networks in Urban Road Traffic

J. Gwinner, Institute of Mathematics, Department of Aerospace Engineering, University of the Federal Army Munich, D - 85577 Neubiberg

Abstract

This contribution discusses a continuum model of large discrete networks in planar domains. For this model, the Kirchhoff law, boundary conditions and capacity constraints lead in a system optimisation approach to a infinite dimensional constrained optimisation problem and to "mixed" variational inequalities. Mixed finite element methods can be formulated for these variational inequalities such that computable discretizations of the continuum problem are obtained.

1. Introduction

This contribution discusses a continuum model of large discrete networks in urban road traffic. Similar to Beckmann and Puu (1985), Dafermos (1980), Iri (1980), Strang (1983), Taguchi and Iri (1982), we embedded the network in a bounded planar domain and describe the unknown flow at each point by a vector field. As in classical network theory (Ford and Fulkerson, 1962), the unknown flow has to satisfy capacity constraints and the Kirchhoff law (conservation of mass) that is modelled by a constraint on the divergence. Moreover, a fixed demand between origin-destination pairs leads to appropriate conditions on the boundary of the domain.

Such a continuum approximation of traffic flow in large, dense networks has the following advantages in comparison to standard finite network models:

- reduction of necessary data assemblage
- reduction of computer time and computer memory
- better and easier analysis and visualisation of the numerical results.

These advantages of the continuum approach and the effectiveness of the numerical method based on the finite-element approximation technique have been demonstrated by the work of Taguchi and Iri (1982). It is the purpose of the present paper to complement and extend the former work in some aspects. Here in contrast to Beckmann and Puu (1985), Dafermos (1980), Taguchi and Iri (1982), we adapt some concepts of functional analysis (up to now applied to the Stokes problem in fluid mechanics, see e.g. Girault and Raviart (1981)) to make precise the understanding of a feasible flow, respectively of an optimal flow, where employing a system optimisation approach. Thus our continuum model leads to a infinite dimensional constrained optimisation problem. Further by introduction of the Lagrangian associated to the divergence constraint we arrive at novel "mixed" variational inequalities. To obtain numerically tractable approximations of these continuum problems, we show how mixed finite element methods (that are standard for variational problems without constraints, see e.g. Roberts and Thomas (1991)) can be formulated for these variational inequalities. By discretization in space we obtain again mixed variational inequalities, but now in finite dimensions, which in view of the simple constraints and for quadratic or linear objective functions reduce to linear complementarity problems for which direct and iterative numerical methods are available (see Cottle et al., 1992 and Murty, 1988).

2. THE CONTINUUM MODEL

In this section we develop our continuum model of large (discrete) networks by embedding the network in a plane domain. As in classical network theory (Ford and Fulkerson, 1962), the unknown flow has to satisfy a conservation law (Kirchhoff law) and capacity constraints. In this section we concentrate on the system optimisation approach and on a single origin-destination pair. However in the next section, we comment on various variants of our approach, namely on the user optimisation approach and on the extension to multi-class flow (see e.g. Smith and van Vuren (1992) in finite dimensions) and to a finite number of origin destination pairs.

To begin with, we embed a given large dense planar network, e.g. a urban street network, in a large enough (simply connected) bounded domain Ω in $\mathfrak{R}^2$ with Lipschitz boundary Γ. In our continuum (transportation) model, similar to Beckmann and Puu (1985), Dafermos (1980), Taguchi and Iri (1982), Strang (1983), the unknown flow at each point $\underline{x}$ of Ω will be described by a vector field $\underline{u}(\underline{x})$, whose components u_1 and u_2, e.g. in the traffic network, represent the traffic density (number of vehicles per unit length and unit time) through a neighbourhood of $\underline{x}$ in the coordinate directions $\underline{e}_1$ and $\underline{e}_2$ respectively. We have to satisfy the conservation law (Kirchhoff law), this leads to the divergence constraint

$$\text{div } \underline{u}(\underline{x}) = 0 \qquad (\underline{x} \in \Omega). \qquad (2.1)$$

Remark. More generally one can prescribe a scalar field $\rho(\underline{x})$ (number of vehicles per unit area and unit time) and demand

$$\text{div } \underline{u}(\underline{x}) = \rho(\underline{x}) \qquad (\underline{x} \in \Omega).$$

Then obviously, $\rho(\underline{x})$ will be positive if net flow is generated in a neighbourhood of $\underline{x}$, respectively negative if net flow is absorbed there. However, this more general case can be reduced to the homogeneous case $\rho = 0$ by the solution of an appropriate auxiliary problem. Indeed: Solve the Poisson problem $\nabla \phi = \rho$ (with the same boundary conditions as the original problem; see later) for the unknown ϕ and replace the flow $\underline{u}$ by $\widetilde{u} = \underline{u} - \nabla \phi$.

In contrast to Beckmann and Puu (1985), Dafermos (1980), Iri (1980), Taguchi and Iri (1982), we make more precise the function analytic setting. The most simple (in contrast to Strang (1983)) and - in view of the divergence constraint - the natural function space is

$$E(\Omega) = \left\{ \underline{u} \in L^2(\Omega, \mathfrak{R}^2) \mid \text{div } \underline{u} \in L^2(\Omega) \right\}$$

on the Lipschitz domain Ω, a well-known space in fluid mechanics (Girault and Raviart, 1981). Here $\psi \equiv \operatorname{div} \underline{u}$ is the $L^2(\Omega)$ - function that satisfies

$$\int_{\Omega} \chi \psi \, d\underline{x} + \int_{\Omega} \nabla \chi^T \underline{u} \, d\underline{x} = 0$$

for all infinitely differentiable functions χ with its support

$$\operatorname{supp} \chi := \overline{\{\underline{x} \mid \chi(\underline{x}) \neq 0\}}$$

strictly contained in Ω, thus vanishing on Γ. Then $\mathrm{E}(\Omega)$ is a Hilbert space endowed with the scalar product

$$(\underline{u}, \underline{w}) = \int_{\Omega} \operatorname{div} \underline{u} \, \operatorname{div} \underline{w} \, d\underline{x} + \int_{\Omega} \underline{u}^T \underline{w} \, d\underline{x}.$$

This Hilbert space approach necessitates that all point wise equations and inequalities on Ω are to be understood almost everywhere.

Remark. Let us comment on the choice of the function space $\mathrm{E}(\Omega)$ which stands between the more standard spaces $L^2(\Omega, \Re^2)$ and

$$H^1(\Omega, \Re^2) = \left\{ \underline{u} \in L^2(\Omega, \Re^2) \mid \frac{\partial u_i}{\partial x_j} \in L^2(\Omega); \, i, j = 1, 2 \right\}.$$

$L^2(\Omega, \Re^2)$ is a too crude space, since for L^2-vector fields we can only apply the divergence operator in the distributional sense and moreover, we do not have enough regularity at the boundary, what is needed as seen below On the other hand, a H^1-setting does not lead to a well-posed optimisation problem, because the objective function (see below) in the optimal flow problem does not involve any partial derivative, so usually the flow problems lack information about all partial derivatives of a flow field $\underline{u}$.

Further we have capacity constraints. We could consider the implicit set constraints of Taguchi and Iri (1982), but we prefer to introduce a subdomain ω of Ω, moreover vector fields $\underline{s}, \underline{t} \in L^2(\omega, \Re^2)$, where $\underline{s} \leq \underline{t}$ holds componentwise, and impose the simple box constraints

$$\underline{s} \leq \underline{u} \leq \underline{t} \quad \text{in} \quad \omega \tag{2.2}$$

This includes the nonnegativity constraint $\underline{u} \geq 0$ in Dafermos (1980) modelling one way streets. On the other hand, one can imagine more general constraints, e.g. constraints on the components separately in different subdomains. However since these more general constraints do not lead to more insight, we do not elaborate on these further here.

Finally we have boundary conditions to model a single origin-destination pair. Thus we introduce connected measurable disjoint non-empty subsets P, Q, R of Γ such that $\Gamma = \overline{P} \cup \overline{Q} \cup \overline{R}$ and the distance between the origin P (e.g. the boundary of a residential area in the morning business traffic) and the destination Q (the boundary of a labour region),

$$\delta(P,Q) = \inf \left\{ \left\| x_p - x_q \right\| : x_p \in P, x_q \in Q \right\} > 0 .$$

Here we use the Green-Stokes formula (see e.g. Theorem 2.2 of the book of Girault and Raviart, (1981)) that asserts with $\underline{n}$ denoting the outer normal of Ω that we have for all $\varphi \in H^1(\Omega)$, $\underline{u} \in E(\Omega)$

$$\int_{\Omega} \varphi \cdot \operatorname{div} \underline{u} \, d\underline{x} + \int_{\Omega} \nabla \varphi^T \underline{u} \, d\underline{x} = \int_{\Gamma} \varphi \cdot \underline{u}^T \underline{n} \, ds \qquad (2.3)$$

and that in particular the trace $\underline{u}^T \underline{n} \in H^{-1/2}(\Gamma)$, while the trace $\varphi_{|\Gamma} \in H^{1/2}(\Gamma)$. So $\int_{\Gamma} \underline{u}^T \underline{n} \, ds$ is well-defined for $\underline{u} \in E(\Omega)$, but not $\int_{P} \underline{u}^T \underline{n} \, ds$, since the characteristic function $1_P \notin H^{1/2}(\Gamma)$ or (equivalently) cannot be extended to a $H^1(\Omega)$-function. To overcome this difficulty we introduce a cut-off function $\chi_P \in H^1(\Omega)$ such that $0 \le \chi_P \le 1$, $\chi_P|P = 1$ and the support supp χ_P is contained in some ε-neighbourhood U_ε of P, where $0 < \varepsilon < \delta(P,Q)$.

On the other hand, there is no flow across the boundary part R. Therefore we are led to the subspace

$$\mathrm{E}_{0,R}(\Omega) = \overline{\left\{ \underline{u} \in C^\infty\left(\overline{\Omega}, \mathfrak{R}^2\right) : \underline{n}^T \underline{u}|_R = 0 \right\}},$$

where the closure is taken with respect to the norm $\|\underline{u}\| = (\underline{u}, \underline{u})^{1/2}$ of $\mathrm{E}(\Omega)$. Let us note that the Green-Stokes formula (2.3) implies that this space is the orthogonal complement to the closed subspace

$$H^1_{0, P \cup Q}(\Omega) = \overline{\left\{ \phi \in C^\infty\left(\overline{\Omega}\right) : \phi|_{P \cup Q} = 0 \right\}}$$

of $H^1(\Omega)$ with respect to the duality form

$$\langle \phi, \underline{n}^T \underline{u} \rangle = \int_{\Gamma} \phi \cdot \underline{n}^T \underline{u} \, ds$$

in $H^{1/2}(\Gamma) \times H^{-1/2}(\Gamma)$.

Now observing the sign of $\underline{n}$ we can impose for $\underline{u} \in \mathrm{E}_{0,R}(\Omega)$ the non-local boundary condition

$$-\int_P \underline{u}^T \underline{n}\, ds = -\int_\Gamma \chi_P \underline{u}^T \underline{n}\, ds = \vartheta \qquad (2.4)$$

where $\vartheta > 0$ is the given inflow rate of the origin P. On the other hand, the choice $\varphi = 1$ in the Green-Stokes formula (2.3) and (2.1) entail for such $\underline{u} \in E_{0,R}(\Omega)$ the balance equation

$$\int_{P \cup Q} \underline{n}^T \underline{u}\ ds = 0.$$

To sum up, all feasible flows $\underline{u}$ are those $\underline{u} \in E_{0,R}(\Omega)$ that satisfy (2.1), (2.2), and (2.4), which gives a convex and closed subset K. In Gwinner (1995 a) the feasibility problem is studied in the case $R = \Gamma$ and equivalent conditions on the data for the nonemptiness of K are given. Here we simply assume, $K \neq \varnothing$. The set K is generally infinite dimensional, as the example in Gwinner (1995 a) section 3 shows.

Therefore to single out a particular flow, we employ an optimisation procedure. There are two different approaches: the system-optimisation model and the user-optimisation model. In this paper we concentrate on the first model, while we only comment on the latter model in the next section. In the system optimisation model one supposes that there is a planning authority (or society) that has a clearly defined target for the network considered, e.g. an "ideal" traffic pattern seeking for a best approximate "real", that is feasible flow, or the target of reduction of costs (e.g. transportation costs) seeking for a feasible flow that minimises a given cost function. Thus one obtains an optimisation problem of the form

$$(P) \begin{cases} \text{minimize } F(\underline{u}) := \int_\Omega f\left(\underline{x}, \underline{u}(\underline{x})\right) d\underline{x} \\ \text{subject to } \underline{u} \in K. \end{cases}$$

Under some appropriate structural assumptions on the integrand f one can prove the existence of an unique minimal flow (Gwinner, 1995 a, Theorem 4.1). Further let us introduce the Lagrangian

$$L\left(\underline{u}, \eta\right) = F(\underline{u}) + \int_\Omega \operatorname{div} \underline{u} \cdot \eta\, d\underline{x} \qquad u \in U, \eta \in L^2(\Omega),$$

where U is simply given by the box constraints, and by the boundary conditions, that is

$$U := \left\{ \underline{u} \in E_{0,R}(\Omega) \,\middle|\, \underline{u} \text{ satisfies (2.2) and (2.4)} \right\}.$$

Then appropriate structural assumptions permit to obtain the saddle point problem

(SP) Find $(\hat{\underline{u}}, \hat{\eta}) \in U \times L^2(\Omega)$ such that

$$\begin{cases} F'\left(\hat{\underline{u}}, \underline{u} - \hat{\underline{u}}\right) + \int_{\Omega} \operatorname{div}\left(\underline{u} - \hat{\underline{u}}\right) \cdot \hat{\eta}\, d\underline{x} \geq 0 & \forall\, \underline{u} \in U; \\ \int_{\Omega} \operatorname{div} \hat{\underline{u}} \cdot \eta\, d\underline{x} = 0 & \forall\, \eta \in L^2(\Omega) \end{cases}$$

The problem *(SP)* can be formulated as a single variational inequality; in particular in the case of a function F that is quadratic in the unknown $\underline{u}$ we arrive at bilinear forms in the variational inequality.

3. SOME VARIANTS OF THE CONTINUUM MODEL

A related continuum maximum-flow problem is considered in the book of Anderson and Nash (1987, chapter 8.4) and also more specialised in the paper of Taguchi and Iri (1982, problem (P1) p. 184). Let as before, $\Gamma = \overline{P} \cup \overline{Q} \cup \overline{R}$. Further, let a nonnegative $L^2(\Omega)$-capacity function c be given and let $|.|$ denote the Euclidean norm in $\mathfrak{R}^2$. Then this maximum-flow problem reads in our framework as follows:

$$\begin{aligned} \text{maximize} \quad & \int_{\varrho} \underline{n}^T \underline{u}\, ds \\ \text{subject to} \quad & \operatorname{div} \underline{u} = 0 \quad \text{in } \Omega \\ & |\underline{u}| \leq 0 \quad \text{in } \Omega \\ & \underline{n}^T \underline{u} = 0 \quad \text{on } R. \end{aligned}$$

Modifying the arguments of (Gwinner, 1995 a, Theorem 4.1)., one can easily show that this optimisation problem is solvable in $\mathrm{E}_{0,R}(\Omega)$, too.

The extension to a finite number of origin-destination (OD) pairs is rather straightforward. Let $P = P_1 \cup ... \cup P_J, Q = Q_1 \cup ... \cup Q_K$; further let $\Theta = \left(\vartheta_{jk}\right)\left(j = 1, ..., J; k = 1, ..., K\right)$ denote the OD matrix of given total flows between P_j and Q_k with $\delta\left(P_j, Q_k\right) > 0$. Then with the boundary part $R_{jk} = \Gamma \setminus \left(\overline{P_j} \cup \overline{Q_k}\right)$ we introduce the subspace $\mathrm{E}_{0,R_{jk}}(\Omega)$ containing the feasible flows $\underline{u}_{jk}$ that satisfy the divergence constraint and the respective inflow boundary condition. A realistic capacity constraint will not be separable in the flows $\underline{u}_{jk}$, but because of interaction effects, will involve all $\underline{u} = \left(\underline{u}_{jk}\right)$; e.g. $\sum_{j,k} \rho_{jk}\, u_{jk} \leq t$ with space-dependant weights ρ_{jk}. Since we do not assume that the origins P_j respectively the destinations Q_k are pair-wise disjoint, by this approach we can also include multi-class traffic, e.g. the two-class traffic of cars and lorries. (see e.g. Smith and van Vuren (1992) in finite dimensions)

To conclude this section let us point out that also the user optimisation model based on the behavioural axioms of Wardrop leads also to such variational inequalities on convex sets defined by simple constraints; see Dafermos (1980) for a heuristic discussion and Gwinner (1995 b) for a derivation of a finite dimensional variational inequality that models a constrained market equilibrium of supply and demand in a bipartite graph.

4. DISCRETIZATION OF THE CONTINUUM MODEL

To obtain numerically tractable approximations of these continuum problems, we show in this section how mixed finite element methods (that are standard for variational problems without constraints, see e.g. Roberts and Thomas (1991)) can be formulated for these variational inequalities. By discretization in space we obtain again mixed variational inequalities, but now in finite dimensions.

Let us use the most elementary setting for finite element methods: Ω and ω are supposed to be (bounded, open) polyhedral subsets of $\mathfrak{R}^2$; $T_h(\supset \hat{T}_h)$ denotes a triangulation of $\overline{\Omega}(\supset \overline{\omega})$ by triangles T of diameter h_T less than h such that

$$\overline{\Omega} = \bigcup_{T \in T_h} T, \quad \overline{\omega} = \bigcup_{T \in \hat{T}_h} T.$$

Moreover, the family of triangulations $\{T_h\}_{h>0}$ is supposed to be regular, that is, with ρ_T the radius of the circle inscribed in T there holds

$$\inf_{h>0} \min_{T \in T_h} \frac{\rho_T}{h} > 0.$$

The direct approach to the discretization of the problem *(P)* would be the construction of a finite dimensional subset K_h of K by polynomial interpolation on each triangle T (this is a successful procedure in the case of scalar unconstrained variational problems like the Dirichlet problem for the Poisson equation) and to seek a minimal element $\hat{u}_h$ of F (or of an appropriate approximation F_h by a quadrature method) on K_h. Leaving aside the unilateral constraints $\underline{s} \le \underline{u}_h \le \underline{t}$ it is difficult to construct a basis of a finite dimensional subspace V_h of the subspace

$$V = \left\{ \underline{u} \in E_{0,R}(\Omega) \middle| \operatorname{div}\underline{u} = 0 \text{ in } \Omega \right\}.$$

Indeed, any $\underline{w}_h \in V_h$ should satisfy the two constraints:

1. Within each triangle $T \in T_h$, the divergence of $\underline{w}_h$ must vanish, thus in particular with $\underline{v}_T$ denoting the outer normal of T,

$$\int_{\partial T} \underline{w}_h^T \underline{v}_T \, ds = 0.$$

2. The normal traces of $\underline{w}_h$ must be "continuous across the interface" between any two neighbouring triangles T_1 and T_2 with common edge, i.e. $\underline{w}_h^T \underline{v}_{T_1}$ and $-\underline{w}_h^T \underline{v}_{T_2}$ must coincide on $T_1 \cap T_2$ (see Roberts and Thomas (1991), Theorem 1.3).

To circumvent this difficulty we use the Lagrangian formulation *(SP)*. Then we can adopt the construction of finite dimensional subspaces that is known from the numerical analysis of mixed finite element methods; for an exposition see section 6 of Roberts and Thomas (1991).

To obtain a finite dimensional subspace E_h of $E(\Omega)$ we can choose the Raviart-Thomas elements $D_k(T)$ $(T \in \mathrm{T}_h)$ as the space of restrictions to T of the functions in $D_k := \mathrm{P}_{k-1}^2 \oplus \underline{x}\,\mathrm{P}_{k-1}$ $(k \in \mathrm{IN})$ where P_{k-1} denotes the space of polynomials (here in 2 variables) of degree $\leq k - 1$. This leads to the subspace

$$E_{h;0,R} = \left\{ \underline{u}_h \in E_h \mid \underline{n}^T \underline{u}_h(\underline{x}) = 0,\, \forall \underline{x} \in \mathrm{N}_h \cap R \right\}$$

and to the non-conforming approximation

$$U_h = \left\{ \underline{u}_h \in E_{h;0,R} \mid \underline{u}_h \text{ satisfies (2.4) and } \underline{s}(\underline{x}) \leq \underline{u}_h(\underline{x}) \leq \underline{t}(\underline{x}),\, \forall \underline{x} \in \mathrm{N}_h \cap \omega \right\},$$

where N_h denotes the finite point set (depending upon the degree k) associated to the triangles $T \in \mathrm{T}_h$. The corresponding finite dimensional subspace of the multiplier space $L^2(\Omega)$ is then

$$M_h = \left\{ \eta_h \in L^2(\Omega) \mid \eta_h \big| T \in \mathrm{P}_{k-1},\, \forall\, T \in \mathrm{T}_h \right\}.$$

Thus neglecting numerical integration, we obtain the following finite dimensional approximate problem (SP_h): Find $\left(\hat{\underline{u}}_h, \hat{\underline{\eta}}_h \right) \in U_h \times M_h$ such that

$$\begin{cases} F'\!\left(\hat{\underline{u}}_h, \underline{u}_h - \hat{\underline{u}}_h \right) + \displaystyle\int_\Omega \mathrm{div}\,(\underline{u}_h - \hat{\underline{u}}_h) \cdot \hat{\eta}_h \, d\underline{x} \ \geq 0 \quad \forall\, \underline{u}_h \in U_h, \\[2mm] \displaystyle\int_\Omega \mathrm{div}\,\hat{\underline{u}}_h \cdot \eta_h \, d\underline{x} \hspace{3.5cm} = 0 \quad \forall\, \eta_h \in M_h. \end{cases}$$

Note that with the approximate problems (SP_h) in the case of a quadratic objective function F one obtains linear complementarity problems for which direct and iterative numerical methods are available (see Cottle et al. (1992), Murty (1988)).

5. SOME CONCLUDING REMARKS

We emphasise that our continuous model merely intends to cover the deterministic static network problem under constraints, e.g. urban traffic flow at some fixed time instant or at the stationary state, or the transport problem of supply and demand. While theoretical issues like existence of optimal solutions can be settled in the described functional analytical framework of the continuum model, the mathematical question concerning convergence of the proposed finite-element discretizations seems to be open.

ACKNOWLEDGEMENT

The author gratefully acknowledges the constructive criticism of the referee.

REFERENCES

Anderson, E. J. and P. Nash (1987). *Linear Programming in Infinite-Dimensional Spaces*. J. Wiley, Chichester.

Beckmann, M. J. and T. Puu (1985). *Spatial Economics: Density, Potential, and Flow*. North-Holland, Amsterdam.

Cottle, R. W., J.-S. Pang. and R. E. Stone. (1992). *The Linear Complementarity Problem*. Academic Press, New York.

Dafermos, S. (1980). Continuum modelling of transportation networks. *Transportation Res.*,; **14B**, 295-301.

Dautray, R. and J.-L. Lions (1993). *Spectral Theory and Applications* In: Mathematical Analysis and Numerical Methods for Science and Technology, Vol. 3. Springer, Berlin-New York.

Ford, L. R. Jr. and D. R. Fulkerson (1962). *Flows in Networks*. Princeton Univ. Press, Princeton, New Yersey.

Girault, V. and P.-A. Raviart (1981*). Finite Element Approximation of the Navier--Stokes Equations*. In: Lecture Notes in Mathematics, 749. Springer, Berlin-New York.

Gwinner, J. (1995 a). A Hilbert space approach to some flow problems. In: *Recent Developments in Optimization*, Seventh French German Conference on Optimization. Dijon, 1994. *Lect. Notes Econ. Math. Syst.* (R. Durier and C. Michelot, eds.), 429, pp. 170 - 182. Springer.

Gwinner, J. (1995 b). Stability of monotone variational inequalities with various applications. In:

Variational Inequalities And Network Equilibrium Problems. International School of Mathematics "G. Stampacchia", Erice, 1994 (F. Giannessi and A. Maugeri., eds.). pp. 123 -142 Plenum Press, New York.

Iri, M. (1980). Theory of flows in continua as approximation to flows in networks. In: *Survey of Mathematical Programming* (A. Prekopa, ed.),Vol. 2, pp. 263-278. North-Holland, Amsterdam.

Murty, K. G. (1988). *Linear Complementarity, Linear and Nonlinear Programming*. In: Sigma Series in Applied Mathematics, 3. Heldermann, Berlin.

Roberts, J. E. and J.-M. Thomas (1991). Mixed and Hybrid Methods. In: *Handbook of Numerical Analysis: II 1*, (P. G. Ciarlet and J. L. Lions, eds.) pp. 523 - 639. North-Holland, Amsterdam

Smith, M.J. and T. van Vuren, (1989). Equilibrium traffic assignment with two vehicle types. In: *Mathematics in Transport, Planning and Control* (J. D. Griffiths, ed.) pp. 159 - 168. Oxford Press.

Strang, G. (1983). Maximal flow through a domain. *Mathematical Programming*, **26,** 123 - 143.

Taguchi, A. and M. Iri (1982). Continuum approximation to dense networks and its application to the analysis of urban road networks. *Mathematical Programming Study*, **20**, 178 - 217.

ON THE INTEGRATION OF CAR-FOLLOWING EQUATIONS

Dirk Heidemann, Heilbronn University of Applied Sciences (Fachhochschule), Heilbronn/Kuenzelsau, Germany

1. ABSTRACT AND INTRODUCTION

A common principle of obtaining speed-flow-density relationships and thus capacities is by integrating car-following equations. However, the method of integration which is usually found in traffic-theory literature, is incorrect from a mathematical point of view. The error is due to ignoring the statistical *distribution* of the spacings between consecutive vehicles. In the paper it is shown how this distribution can be incorporated into the integration procedure. The results obtained are analysed and interpreted.

2. THE GENERAL CAR-FOLLOWING EQUATION AND ITS INTEGRATION

The general car-following equation is

$$v_{n+1}'(t+T) = c \cdot v_{n+1}^m(t+T) \cdot \frac{v_n(t) - v_{n+1}(t)}{\left(x_n(t) - x_{n+1}(t)\right)^k}, \qquad (2.1)$$

where: v_n and v_{n+1} denotes speed of vehicle no. n and n+1, respectively,

x_n and x_{n+1} denotes position of vehicle no. n and n+1, respectively,

t is continuous time,

T is driver's reaction time,

c ("sensitivity parameter"), m and k are constants, and

the prime denotes the time derivative.

For the purpose of integrating this equation is re-formulated as

$$\frac{v_{n+1}'(t+T)}{v_{n+1}^m(t+T)} = c \cdot \frac{v_n(t) - v_{n+1}(t)}{\left(x_n(t) - x_{n+1}(t)\right)^k}. \qquad (2.2)$$

Integration of the right-hand side of (2.2) yields

$$\beta_1 + c \cdot f_k(x_n - x_{n+1}), \tag{2.3}$$

where β_1 is a constant and $f_k(x_n - x_{n+1}) = \begin{cases} \log(x_n - x_{n+1}), & \text{if } k = 1 \\ (x_n - x_{n+1})^{1-k} / (1-k), & \text{if } k \neq 1 \end{cases}$

Integration of the left-hand side of (2.2) yields

$$\beta_2 + f_m(v_{n+1}), \tag{2.4}$$

where β_2 is a constant and $f_m(v_{n+1}) = \begin{cases} \log(v_{n+1}), & \text{if } m = 1 \\ (v_{n+1})^{1-m} / (1-m), & \text{if } m \neq 1 \end{cases}$

Since the integrals of both sides can only differ by a constant, we finally obtain

$$f_m(v_{n+1}) = \beta + c \cdot f_k(x_n - x_{n+1}), \tag{2.5}$$

where β is constant.

Applying the inverse function f_m^{-1} and taking expectations (E) on both sides of (2.5) leads to

$$E(v) = E\left(f_m^{-1}(\beta + c \cdot f_k(x))\right), \tag{2.6}$$

where $x = x_n - x_{n+1}$.

This is the correct integral of the car-following equation (note that on the left-hand side of (2.6) v_{n+1} has been replaced with v, for in statistical equilibrium the vehicle number is redundant).

The flaw in traffic-theory literature[1] is due to interchanging expectation and function f_m^{-1} and f_k on the right-hand side of (2.6) which yields

$$E(v) = f_m^{-1}(\beta + c \cdot f_k(E(x))). \tag{2.7}$$

Since $E(x) = 1/d$, where d is traffic density, from (2.7)

---

[1] The monograph of Ashton (1966) is just one example. Any other standard textbook on traffic flow theory could have been quoted as well.

$$E(v) = f_m^{-1}\left(\beta + c \cdot f_k\left(1/d\right)\right) \tag{2.8}$$

is obtained. (2.8) is the speed-density relationship which is usually reported to result from integration of the car-following equation.

From (2.6) or (2.8) speed-flow and flow-density relationships can be obtained by the well-known formula

$q = d \cdot E(v)$,

where q is the flow rate. The maximum flow rate is called "capacity" of the road under consideration.

Let us consider the case $m=0$ in some more detail. In this case f_m is the identity mapping and the flaw merely lies in the use of the equation $E\left(f_k(x)\right) = f_k\left(E(x)\right)$. The correct relationship is given by Jensen's inequality as

$E\left(f_k(x)\right) \geq f_k\left(E(x)\right)$, if f_k is a convex function, and
$E\left(f_k(x)\right) \leq f_k\left(E(x)\right)$, if f_k is a concave function.

Thus (2.7) and (2.8) are only correct for $k=0$ (provided $m=0$), whereas for all other values of k they are incorrect. (2.6), however, is correct in all cases.

3. COMPARISONS OF CORRECT AND ERRONEOUS INTEGRATION

In order to compare the correct with the incorrect formula numerically, the distribution of the spacings of consecutive vehicles must be given. It is common and reasonable to assume a Gamma distribution with small shape parameters α (α is confined to the positive integers, for the sake of simplicity) . Thus the probability density function (pdf) of the spacings is assumed to be

$$g_{\alpha,d}(x) = \frac{(d \cdot \alpha)^\alpha}{\Gamma(\alpha)} \cdot x^{\alpha-1} \cdot \exp(-d \cdot \alpha \cdot x), \tag{3.1}$$

if density d prevails. For $\alpha=1$ the exponential distribution is obtained as a special case. Notice that for all values of α the mean spacing is the inverse density, $1/d$, and that the variance of the spacing is $1/(\alpha \cdot d^2)$. Thus, if α tends to infinity, the Gamma distribution converges to the Dirac distribution concentrated on $1/d$. In this limiting case the correct and the incorrect method of integration, i.e. (2.6) and (2.8), coincide.

An analytical integration of the car-following equation is possible in some of the most interesting cases, namely:

- m=0, all k and all positive integers α such that $\alpha+1-k>0$, and
- m=1, k=1 and all positive integers α (provided $\alpha+c$ is positive which is no restriction in practice).

For m=0 and k=1

- the correct method of integration (cf. 2.6) yields

$$E(v) = \beta - c \cdot \left(\log(\alpha \cdot d) - \frac{\Gamma'(\alpha)}{\Gamma(\alpha)} \right), \tag{3.2}$$

- the incorrect method of integration (cf. 2.8) yields

$$E(v) = \beta - c \cdot \log(d). \tag{3.3}$$

Γ denotes the Gamma function. It is well-known that for integer-valued $\alpha>0$ we have $\Gamma(\alpha)=(\alpha-1)!$. Notice that $\Gamma'(1)/\Gamma(1)$ is $-C$ and $\Gamma'(2)/\Gamma(2)$ is $1-C$, where C is Euler's constant (≈ 0.5772).

For m=0 and k$\neq$1

- the correct method of integration (cf. 2.6) yields

$$E(v) = \beta - \frac{\alpha^k \cdot \Gamma(\alpha+1-k)}{(k-1) \cdot \Gamma(\alpha+1)} \cdot c \cdot d^{k-1} \text{, if } \alpha+1-k>0, \tag{3.4}$$

- the incorrect method of integration (cf. 2.8) yields

$$E(v) = \beta - \frac{1}{k-1} \cdot c \cdot d^{k-1}. \tag{3.5}$$

For m=1 and k=1

- the correct method of integration (cf. 2.6) yields

$$E(v) = d^{-c} \cdot \exp(\beta) \cdot \frac{\Gamma(\alpha + c)}{\alpha^c \cdot \Gamma(\alpha)}, \qquad (3.6)$$

- the incorrect method of integration (cf. 2.8) yields

$$E(v) = \exp(\beta - c \cdot \log(d)) = d^{-c} \cdot \exp(\beta). \qquad (3.7)$$

The latter case (m=1 and k=1) is of no practical interest, since it implies monotonic flow-density curves which are implausible.

Comparing (3.2) and (3.3) or (3.6) and (3.7), respectively, one might argue that there is no relevant difference between these formulas, because the difference on the right-hand sides may be absorbed into the constant of integration, β. And β does not even occur in the car-following equation (2.1). However, β has a physical meaning of its own as can be seen from (2.5), e.g.. Moreover, both methods of integration as such can only be compared if everything else including the constant of integration is not changed. And from (2.6) and (2.8) it is clear that the difference between both these formulas will in general depend on the traffic density, and thus cannot be completely absorbed into the constant of integration. This is confirmed by comparing (3.4) and (3.5).

The erroneous integration based on (2.8) leads to two of the most popular speed-density relationships, namely:

- for m=0 and k=1: $E(v) = \beta - c \cdot \log(d),$ (3.8)
- for m=0 and k=2: $E(v) = \beta - c \cdot d.$ (3.9)

In (3.8), which is the model due to Greenberg (1959), c is the "optimum" speed, i.e. the mean speed at capacity, and β is the product of the optimum speed and the logarithm of the maximum density, i.e. the density at which E(v) is zero.

In (3.9), which is the model due to Greenshields (1935), β is the "free" speed, i.e. the mean speed at zero density, and c is the ratio of the free speed and the maximum density.

In figures 1 and 2 (3.2) and (3.3) as well as (3.4) and (3.5) are compared graphically. k is set to 1 or 2, and the shape parameter of the Gamma distribution of the spacings, α, is varied. As already mentioned, if α tends to infinity, then (3.2) and (3.3), (3.4) and (3.5) as well as (3.6) and (3.7), respectively, must coincide.

The shape parameter of the Gamma distribution, α, was assumed to be independent of the traffic density, d, in the discussion above. This means that the coefficient of variation of the Gamma distribution, $\dfrac{1}{\sqrt{\alpha}}$, does not depend on d. However, empirical evidence shows that α increases as d increases. Of course, the dependence of α on d can only be taken into account if the correct method of integration is applied, because the incorrect method ignores the distribution of the spacings. In figure 3 both methods of integration are compared for m=0 and k=1 in the car-following equation, if the relationship between α and d is given by

$$\alpha = \alpha(\mathrm{d}) = \frac{\mathrm{d}_{max}}{\mathrm{d}_{max} - \mathrm{d}}. \qquad\qquad 3.10)$$

From (3.2), (3.3) and (3.10) it is clear that the results of the correct and incorrect methods of integration do not just differ by the values of constant parameters. Instead, the overall structure of the flow-density relationships is different.

4.   INTERPRETATIONS AND CONCLUDING REMARKS

The dependence of the coefficients of the flow-density relationships on the sensitivity parameter of the car-following equation, c, is in general affected by the erroneous way of integrating. In the cases analyzed in this paper, for the same value of c the flow-density-relationships exhibit smaller capacities when the integration is done correctly. Of course it can hardly be imagined that with respect to the correct method of integration the values of the parameters of the car-following equation are consistent with the parameters of the corresponding speed-flow-density relationship given that they are consistent with respect to the erroneous method of integration which has always been used in practice up to now.

The erroneous integration method is correct if the spacings are assumed to be constant, because in this case they have a Dirac distribution which is concentrated on their mean value. As already mentioned this distribution may be considered as a limit of Gamma distributions if the shape parameter α tends to infinity. Since the variance of the Gamma distribution is inversely proportional to α, $\alpha=1$ (exponential distribution) is a case of very heterogeneous spacings. Our investigations thus have shown that *ceteris paribus* low values of α decrease road capacity. This means that the more heterogeneous the spacings are the lower the capacity comes out. Thus the plausible and common opinion is confirmed that traffic flow should be as homogeneous as possible in order to avoid capacity problems and congestion. This was also found by Heidemann (1996) who used a queueing theory approach.

REFERENCES

Ashton, W.D. (1966). *The theory of road traffic flow.* Wiley, New York.

Greenberg, H. (1959). An analysis of traffic flow. *Opns. Res.,* **7**, 79-85.

Greenshields, B.D. (1935). A study of traffic capacity. *Proc. High. Res. Bd.,* **14**, 448-474.

Heidemann, D. (1996). A queueing theory approach to speed-flow-density relationships. In: Transportation and Traffic Theory. *Proceedings of the 13th International Symposium on Transportation and Traffic Theory,* Lyon 1996, pp. 103-118. *Elsevier Science.*

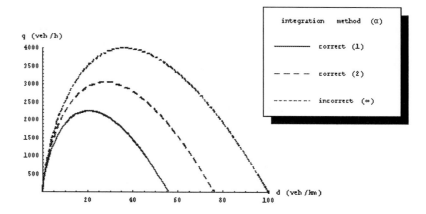

Figure 1.   Flow-density relationship resulting from <u>correct</u> and <u>incorrect</u> integration of car-following equation (2.1). Parameters used: m=0, k=1, Gamma distributions of spacings with shape parameters α=1, α=2, c=108.731, β=500.726 (such that q_{max}=4000 veh/h and d_{max}=100 veh/km for curve from incorrect integration). q=flow rate, d=density.

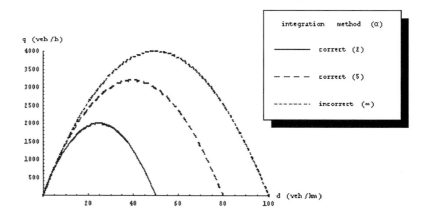

Figure 2.   Flow-density relationship resulting from <u>correct</u> and <u>incorrect</u> integration of car-following equation (2.1). Parameters used: m=0, k=2, Gamma distributions of spacings with shape parameters α=2, α=5, c=1.6, β=160 (such that q_{max}=4000 veh/h and d_{max}=100 veh/km for curve from incorrect integration). q=flow rate, d=density.

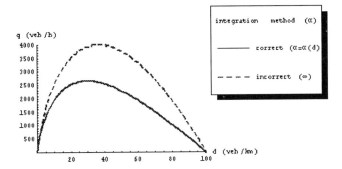

Figure 3.   Flow-density relationship resulting from <u>correct</u> and <u>incorrect</u> integration of car-following equation (2.1). Parameters used: m=0, k=1, Gamma distributions of spacings with shape parameters $\alpha = \alpha(d) = \dfrac{d_{max}}{d_{max} - d}$, c=108.731, β=500.726 (such that q_{max}=4000 veh/h and d_{max}=100 veh/km for curve from incorrect integration).
q=flow rate, d=density.

THE COMPLEX DYNAMICAL BEHAVIOUR OF CONGESTED ROAD TRAFFIC

David J Low and Paul S Addison, Napier University, Edinburgh, UK

ABSTRACT

The mathematical models used to describe the dynamical behaviour of a group of closely-spaced road vehicles travelling in a single lane without overtaking are known as car-following models. This paper presents a novel car-following model, which differs from the traditional models by having an equilibrium solution that corresponds to consecutive vehicles having not only zero relative velocity, but also travelling at a certain desired distance apart. This new model is investigated using both numerical and analytical techniques. For many parameter values the equilibrium solution is stable to a periodic perturbation but, for certain parameter values, chaotic motion results. This shows that in congested traffic, even drivers attempting to follow a safe driving strategy, may find themselves driving in an unpredictable fashion.

1. INTRODUCTION

Roads have provided a great deal of transportation freedom to individuals and businesses but they are themselves restrictive, constraining the motorist to a limited number of lanes and with a limited number of alternative routes for a given journey. This means that when a large number of vehicles travel between the same two points they must share the same finite amount of road space, and congestion occurs. In this paper we will investigate the dynamical behaviour of such traffic. When a vehicle has sole possession of a section of road, then its motion is generally quite simple. The driver will usually attempt to maintain a certain desired constant velocity. However, in the absence of a 'perfect' driver the velocity of the vehicle will oscillate about the desired value. This situation changes quite dramatically when other road vehicles are present. The driver will be required to modify his strategy to avoid collision with other vehicles. This is particularly true on a congested road. On a carriageway where overtaking is not allowed, or is not possible, the driver's main concern will be to avoid colliding with the vehicle immediately ahead. The driver must carefully observe this vehicle, and react appropriately to any changes in its motion. The mathematical models used to describe the motion of an isolated group of closely spaced vehicles (a platoon) travelling in a single lane without overtaking are known as *car-following models* (Chandler et al., 1958; Gazis et al., 1959, 1961; Kometani and Sasaki, 1958; Pipes, 1953). They describe how each vehicle in the platoon, except the lead vehicle, responds to a change in the relative motion of the vehicle ahead by accelerating or braking in a prescribed manner. This approach is in contrast to the main alternative which treats the traffic flow as a continuous fluid.

2. A MODIFIED CAR-FOLLOWING MODEL

The no-overtaking condition restricts the vehicles to motion in one spatial dimension, thus they can be illustrated as in Figure 1. The position on the x-axis of the front of vehicle n at time t, is denoted by $x_n(t)$ and L_n denotes the length of vehicle n.

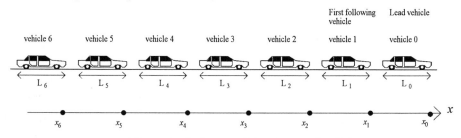

Figure 1. A platoon of closely spaced vehicles.

The traditional car-following model assumes that each vehicle attempts to match the velocity of the vehicle ahead. This makes no allowance for the effect of the inter-car spacing independently of the relative velocity. Vehicles are allowed to travel arbitrarily close together provided their velocities are identical. We believe that it is more realistic to assume that drivers will attempt to achieve a desirable inter-vehicle separation, as well as attempting to minimise the relative velocity. This is modelled by adding a nonlinear spacing-dependent term to the equation traditionally used to model car-following (Addison and Low, 1996,1997). It is convenient to eliminate the vehicle lengths from our considerations by defining new co-ordinates $u_n(t) \equiv x_n(t) + \sum_{j=0}^{n-1} L_j$ which represent each vehicle as a point on the u-axis. The resulting modified car-following equation is expressed as (2.1) where a and b are positive real numbers and vehicle N is the final car in the platoon.

$$\ddot{u}_n(t) = a \frac{\dot{u}_{n-1}(t-\tau) - \dot{u}_n(t-\tau)}{u_{n-1}(t-\tau) - u_n(t-\tau)} + b\left(u_{n-1}(t-\tau) - u_n(t-\tau) - D_n\right)^3 \qquad n = 1,\ldots,N \qquad (2.1)$$

$$\underbrace{\qquad\qquad\qquad}_{\text{Traditional Term}} \qquad \underbrace{\qquad\qquad\qquad}_{\text{Additional Spacing Dependent Term}}$$

The desired separation that the driver of vehicle n attempts to achieve from the vehicle ahead is denoted by D_n. The reaction time of the driver and vehicle is represented by including a delay-time τ. A cubic additional term is chosen as it is the simplest nonlinear term that attempts to readjust the inter-vehicle spacing back to the desired value. It produces an acceleration if the separation is greater than D_n and a deceleration if the separation is less than D_n. The desired separation D_n is most naturally taken to be a constant multiple of the mean velocity, v, of the lead vehicle, $D_n \equiv \lambda_n v$. For simplicity we shall assume that we have a platoon of identical vehicles and drivers, each having the same values of a, b, λ_n. In this paper we deal with the case where the time delay τ is taken to be zero. Numerical solution of the non-zero delay case is dealt with by the authors in Low and Addison (1998).

3.   NUMERICAL SOLUTION OF THE MODIFIED CAR-FOLLOWING EQUATIONS

In order to study the dynamical behaviour of a platoon of vehicles governed by our modified car-following model we investigate the response of the system, to a small sinusoidal perturbation to the platoon's equilibrium state. We assume that prior to time $t = 0$ all the vehicles are travelling at constant velocity v, and corresponding desired separation $D \equiv \lambda v$. Then at time $t = 0$ we introduce a sinusoidal perturbation to the velocity of the lead vehicle, $\dot{u}_0 = v + \omega\, p \sin \omega t$. This perturbation represents the fluctuation in velocity of a driver, with a clear road ahead, attempting to maintain a constant velocity. The strictly positive real number p is the spatial amplitude of oscillation of the lead vehicle. We convert to a system of equations with dimensionless variables by introducing the length scale D, the desired inter-vehicle separation at equilibrium, and the time scale $2\pi/\omega$, the period of oscillation of the lead vehicle's velocity. The new time and distance variables can then be defined as $T \equiv \omega t/2\pi$ and $U_i(t) \equiv (u_i(t) - vt)/D$ for $i = 0,1,\ldots,N$. Thus the dynamical behaviour of the vehicles in the platoon is governed by (3.1) where $A \equiv (2\pi/\omega D)a$ and $B \equiv (2\pi D/\omega)^2 b$.

$$U_i''(T) = A\frac{U_{i-1}'(T) - U_i'(T)}{U_{i-1}(T) - U_i(T)} + B\big(U_{i-1}(T) - U_i(T) - 1\big)^3 \qquad i = 1, \ldots, N \qquad (3.1)$$

The initial conditions for $T \le 0$ are $U_i(T) = -i$ and $U_i'(T) = 0$ where $i = 0, 1, \ldots, N$. The motion of the lead vehicle is described by $U_0'(T) = 2\pi P \sin 2\pi T$ where $P \equiv p/D$. This coupled system of nonlinear differential equations is then solved numerically in Fortran 77 using a traditional 4th order Runge-Kutta routine. This numerical investigation reveals two characteristic types of behaviour.

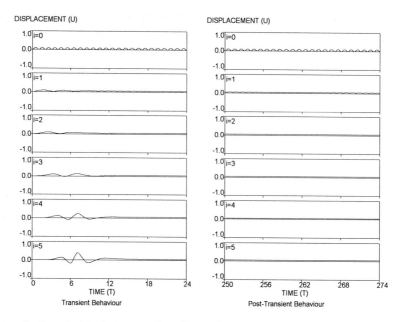

Figure 2. Transient and post-transient time series corresponding to the parameter values
$P = 0.05$, $A = 0.70$, $B = 700.0$.

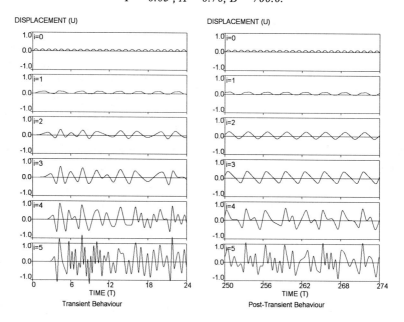

Figure 3. Transient and post-transient time series corresponding to the parameter values
$P = 0.05$, $A = 0.20$, $B = 700.0$.

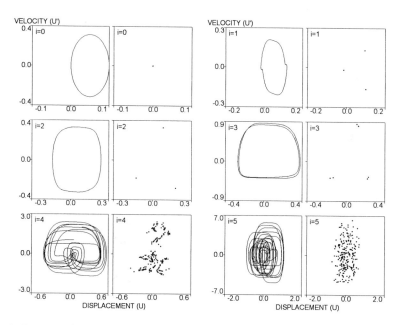

Figure 4. Post-transient phase portraits and corresponding Poincare sections for the parameter values *P = 0.05, A = 0.20, B = 700.0.*

The first characteristic type of solution is illustrated in Figure 2. The steady-state response is a periodic oscillation about the equilibrium position, with frequency identical to that of the lead vehicle. The amplitude of this oscillation decreases as we move along the platoon until it is no longer visible. The second characteristic type of solution is illustrated in Figures 3 and 4. The amplitude of the steady-state now increases rather than decreases. Periodic oscillations may be found with periods equal to that of the lead vehicle or integer multiples of it. Figures 3 and 4 illustrate oscillations with periods of three and six times that of the perturbing oscillation. Further along the platoon we observe, oscillations that are chaotic, that is, not visibly periodic consisting of a broad band of frequency components. Standard techniques (Addison 1997) are used to prove that these solutions are indeed chaotic, and that the degree of chaos increases as we move along the platoon.

4. ANALYTICAL INVESTIGATION OF THE CAR-FOLLOWING MODEL

In order to improve our understanding of the modified car-following model we shall carry out a nonlinear analysis of the model equations. The numerical results show that a common route to chaos is via the periodic oscillation of one vehicle producing a periodic oscillation in the motion of the following vehicle with a period three times as long. Therefore, we shall attempt to determine under what conditions our car-following model admits period three solutions. To make our analysis somewhat easier we shall introduce a new time scale $T = \omega t$ (as opposed to $T = \omega t/2\pi$ used in the numerical simulation). The only change to (3.1) is that A and B are replaced by $\overline{A}$ and $\overline{B}$, where $\overline{A} = A/2\pi$ and $\overline{B} = B/4\pi^2$. Without loss of generality we shall consider the interaction between the lead vehicle and the first following vehicle. Thus we define the new variable $X \equiv U_1 - U_0 + 1$, and the car-following equation becomes

$$(1 - X)X'' + \overline{A}X' + \overline{B}(1 - X)X^3 + (1 - X)P \cos T = 0 \qquad (4.1)$$

We now seek a solution of (4.1) of the form $X = \alpha \cos T + \beta \sin T + \gamma \cos\frac{1}{3}T + \delta \sin\frac{1}{3}T$. We proceed with the analysis, as in Jordan and Smith (1987), by substituting this expression for X in (4.1), and neglecting higher harmonics in our approximations. In order to find steady-state solutions to the system we consider $\alpha, \beta, \gamma, \delta$ to be constant, and define $r^2 \equiv \alpha^2 + \beta^2$ and $s^2 \equiv \gamma^2 + \delta^2$. If we then compare the coefficients of $\cos T$, $\sin T$, $\cos\frac{1}{3}T$, $\sin\frac{1}{3}T$ and the constant term in the resulting expression we find that (4.1) implies

$$P + \alpha\left(\tfrac{3}{4}\overline{B}r^2 + \tfrac{3}{2}\overline{B}s^2 - 1\right) + \overline{A}\beta = -\tfrac{1}{4}\overline{B}\gamma\left(\gamma^2 - 3\delta^2\right) \qquad (4.2)$$

$$-\overline{A}\alpha + \beta\left(\tfrac{3}{4}\overline{B}r^2 + \tfrac{3}{2}\overline{B}s^2 - 1\right) = -\tfrac{1}{4}\overline{B}\delta\left(3\gamma^2 - \delta^2\right) \qquad (4.3)$$

$$\gamma\left(\tfrac{3}{2}\overline{B}r^2 + \tfrac{3}{4}\overline{B}s^2 - \tfrac{1}{9}\right) + \tfrac{1}{3}\overline{A}\delta = -\tfrac{3}{4}\overline{B}\alpha\left(\gamma^2 - \delta^2\right) - \tfrac{3}{2}\overline{B}\beta\gamma\delta \qquad (4.4)$$

$$-\tfrac{1}{3}\overline{A}\gamma + \delta\left(\tfrac{3}{2}\overline{B}r^2 + \tfrac{3}{4}\overline{B}s^2 - \tfrac{1}{9}\right) = -\tfrac{3}{2}\overline{B}\alpha\gamma\delta - \tfrac{3}{4}\overline{B}\beta\left(\gamma^2 - \delta^2\right) \qquad (4.5)$$

$$-\tfrac{1}{2}\alpha P + \tfrac{1}{2}r^2 + \tfrac{1}{18}s^2 = -\overline{B}\left(\tfrac{3}{8}r^4 + \tfrac{3}{2}r^2s^2 + \tfrac{3}{8}s^4 + \tfrac{1}{2}\alpha\gamma\left(\gamma^2 - 3\delta^2\right) + \tfrac{1}{2}\beta\delta\left(3\gamma^2 - \delta^2\right)\right) \qquad (4.6)$$

Note that (4.6) is a linear combination of (4.2-4.5). If $\overline{B} = 0$ then these imply $\gamma = \delta = 0$, hence no period three solution exists. If $\overline{B} \neq 0$ then $(4.2)^2 + (4.3)^2$ and $(4.4)^2 + (4.5)^2$ imply

$$0 = P^2 + 2\alpha P\left(\tfrac{3}{4}\overline{B}r^2 + \tfrac{3}{2}\overline{B}s^2 - 1\right) + 2\overline{A}\beta P + r^2\left(\tfrac{3}{4}\overline{B}r^2 + \tfrac{3}{2}\overline{B}s^2 - 1\right)^2 + \overline{A}^2r^2 - \tfrac{1}{16}\overline{B}^2s^6 \qquad (4.7)$$

$$0 = s^2\left(\tfrac{3}{2}\overline{B}r^2 + \tfrac{3}{4}\overline{B}s^2 - \tfrac{1}{9}\right) - \tfrac{9}{16}\overline{B}^2r\ s^4 + \tfrac{1}{9}\overline{A}^2s^2 \qquad (4.8)$$

By considering $3\alpha(4.2) + 3\beta(4.3) - \gamma(4.4) - \delta(4.5)$ and $3\beta(4.2) - 3\alpha(4.3) + \delta(4.4) - \gamma(4.5)$ we find that $\alpha P = \overline{B}\left(-\tfrac{3}{4}r^4 - r\ s^2 + \tfrac{1}{4}s^4\right) + r^2 - \tfrac{1}{27}s^2$ and $\beta P = -\overline{A}\left(r^2 + \tfrac{1}{9}s^2\right)$. If, in addition, we define $R \equiv \tfrac{3}{4}\overline{B}r^2$ and $S \equiv \tfrac{3}{4}\overline{B}s^2$ and $W \equiv \tfrac{3}{4}\overline{B}P^2$ then (4.7) and (4.8) become

$$W - R^3 - \tfrac{8}{3}R^2S - \tfrac{2}{3}RS^2 + \tfrac{11}{9}S^3 + 2R^2 + \tfrac{70}{27}RS - \tfrac{22}{27}S^2 - \left(1 + \overline{A}^2\right)R + \tfrac{2}{27}\left(1 - 3\overline{A}^2\right)S = 0 \qquad (4.9)$$

$$S\left(4R^2 + 3RS + S^2 - \tfrac{4}{9}R - \tfrac{2}{9}S + \tfrac{1}{81} + \tfrac{1}{9}\overline{A}^2\right) = 0 \qquad (4.10)$$

If $S \neq 0$ then we introduce the new variables $Y \equiv \tfrac{3}{2}R + S - \tfrac{1}{9}$ and $Z \equiv R - \tfrac{2}{63}$, resulting in

$$\tfrac{11}{9}Y^3 - \tfrac{37}{6}Y^2Z + \tfrac{91}{12}YZ^2 - \tfrac{21}{8}Z^3 - \tfrac{38}{63}Y^2 + \tfrac{112}{27}YZ - \tfrac{169}{54}Z^2$$
$$+ \tfrac{2}{9}\left(\tfrac{53}{189} - \overline{A}^2\right)Y - \tfrac{2}{3}\left(\tfrac{232}{189} + \overline{A}^2\right)Z - \tfrac{2}{567}\left(\tfrac{3707}{567} + 13\overline{A}^2\right) + W = 0 \qquad (4.11)$$

$$Y^2 + \tfrac{1}{4}Z^2 = \tfrac{1}{9}\left(\tfrac{1}{63} - \overline{A}^2\right) \qquad (4.12)$$

If the two curves described by (4.11) and (4.12) do not intersect then we must have $S = 0$ and no period three solution would be possible. We can see immediately from (4.12) that $\overline{A}^2 < 1/63$ is a necessary condition for a period three solution to exist. This corresponds to

$A < 2\pi/\sqrt{63} \approx 0.7916$ and is in agreement with Figure 6. If $\overline{A}^2 < 1/63$ then the curve described by (4.12) is an ellipse. As A decreases this ellipse gets larger, reaching its largest size when $A = 0$. Figure 5 shows that as W increases the curve described by (4.11) sweeps across the ellipse. The smaller the value of A the larger the range of W values for which this curve and the ellipse intersect. All of these agree with the results illustrated in Figure 6. In fact these equations predict the possibility of period three solutions for more parameter values than actually occur in the numerical simulation. A stability analysis may reveal which of these actually become period three. The only dependence of these solutions on $\overline{B}$ and P is via the term $W = \frac{3}{4}\overline{B}P^2$. This agrees with the $B \propto P^{-2}$ relationship indicated by Figure 7. Although the behaviour of vehicle 1 does not depend on $\overline{B}$ independently of P, the amplitude of the solution does, and this will in turn affect the behaviour of the next vehicle in the platoon.

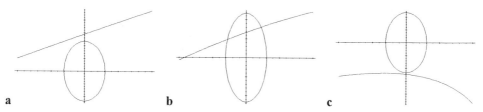

Figure 5. Graphs of Z (on the vertical axis) against Y (on the horizontal axis) with $\overline{A} = 0.085$ (corresponding to $A \approx 0.534$) and **a:** $W = 0.05$, **b:** $W = 0.035$, **c:** $W = 0.005$. With $P = 0.05$ these correspond to $B \approx 1052.8, 736.9, 105.3$ respectively.

We can also use the above analysis to investigate the occurrence of periodic solutions of the same frequency as the input oscillation. In this case we have $\gamma = \delta = 0$ and hence $S = 0$. Then (4.9) and (4.10) reduce to (4.13) which has at least one real positive root and thus such a periodic solution is always possible (although not necessarily stable).

$$R^3 - 2R^2 + \left(1 + \overline{A}^2\right)R - W = 0 \tag{4.13}$$

In Section 3 we discussed the two characteristic types of behaviour that are admitted by our car-following model. In fact, our analytical investigation reveals a third possible behaviour. In this new case the post-transient response of the following vehicle U_1 is a periodic oscillation of exactly the same frequency and amplitude as the lead vehicle U_0 oscillation, although they may be out of phase. This occurs when $(\alpha - P)^2 + \beta^2 = P^2$, and using the expression for αP found earlier this implies $R(2R - 1) = 0$. Taking $R = 1/2$ in (4.13) implies $W = \frac{1}{8} + \frac{1}{2}\overline{A}^2$, and using the original parameters a, b, p, ω, D we can rewrite this as $D^2 = 4a^2/\left(6bp^2 - \omega^2\right)$. This gives the desired separation that allows vehicles to have the same amplitude of oscillation. For a platoon of identical vehicles this is the same desired separation between each pair of vehicles.

Parameter B

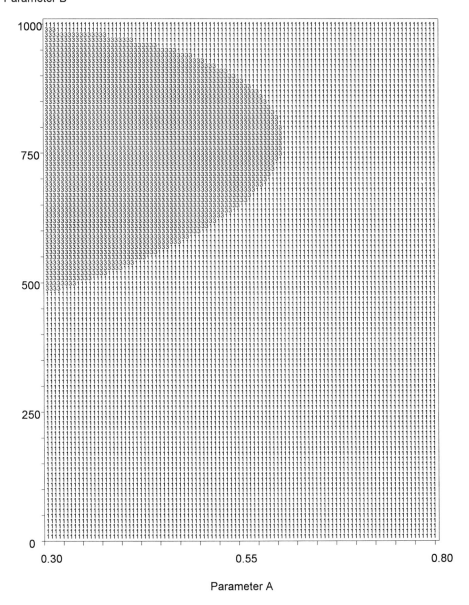

Parameter A

Figure 6. Results of the numerical simulation of the behaviour of vehicle 1, with parameter *P = 0.05* fixed. The numbers indicate the period of the solution as a multiple of that of the lead vehicle.

Figure 7. Results of the numerical simulation of the behaviour of vehicle 1, with parameter $A = 0.3$ fixed. The numbers indicate the period of the solution as a multiple of that of the lead vehicle. A dash indicates that the vehicle has collided with the one ahead.

5. CONCLUSIONS

In this paper we have extended our previous research on our modified car-following model to include an analytical as well as numerical investigation. The combination of the two approaches helps to reveal some of the complex behaviour of this highly nonlinear system. For certain parameter values our car-following model exhibits chaotic behaviour. This leads to the possibility that drivers, attempting to follow a safe driving strategy, may still find themselves within a relatively complex driving environment containing unpredictable behaviours. In this paper we have considered the case where the reaction time is zero. Numerical solution of the non-zero delay case is dealt with in Low and Addison (1998) and reveals that the qualitative shapes of the B–A and B–P parameter spaces are unchanged. The main differences are that, as the time delay increases, the non period one region of the parameter space gets larger; chaotic solutions occur earlier in the platoons; and the degree of chaos is greater for a given vehicle.

REFERENCES

Addison, P. S. (1997). *Fractals and Chaos: An Illustrated Course*. Institute of Physics Publishing, Bristol.

Addison, P. S. and D. J. Low (1996). Order and Chaos in the Dynamics of Vehicle Platoons. *Traffic Engineering and Control*, **37(7/8)**, 456-459.

Addison, P. S. and D. J. Low (1997). The Existence of Chaotic Behaviour in a Separation-Distance Centred Non-Linear Car Following Model. In: *Road Vehicle Automation II*. (C. Nwagboso, ed.), pp. 171-180, Wiley, Chichester.

Chandler, R. E., R. Herman and E. W. Montroll (1958). Traffic Dynamics: Studies in Car-Following. *Opns. Res.*, **6**, 165-184.

Gazis, D. C., R. Herman and R. B. Potts (1959). Car-Following Theory of Steady-State Traffic Flow. *Opns. Res.*, **7**, 499-505.

Gazis, D. C., R. Herman and R. W. Rothery (1961). Nonlinear Follow-the-Leader Models of Traffic Flow. *Opns. Res.*, **9**, 545-567.

Jordan D. W. and P. Smith (1987). *Nonlinear Ordinary Differential Equations*. Clarendon Press, Oxford.

Kometani, E. and T. Sasaki (1958). On the Stability of Traffic Flow (Report-I). *J. Op. Res. Japan*, **2(1)**, 11-26.

Low, D. J. and P. S. Addison (1998). A Nonlinear Temporal Headway Model of Traffic Dynamics. *To appear in Nonlinear Dynamics*.

Pipes, L. A. (1953). An Operational Analysis of Traffic Dynamics. *J. Ap. Phys.*, **24(3)**, 274-281.

THE EFFECT OF SPEED CONTROLS ON TRAFFIC

A D Mason, Department of Applied Mathematics and Theoretical Physics, University of Cambridge, CB3 9EW

A W Woods, School of Mathematics, University of Bristol, BS8 1TW (corresponding author)

ABSTRACT

We use a combination of continuum and car-following models to explore the potential impact of speed-controls on (i) decreasing travel times at times of congested flow; and (ii) increasing the safety of motorway flow approaching the site of an accident.

1. INTRODUCTION

There has been considerable interest in developing means to control or regulate the flow of traffic owing to increasing congestion on freeways. One approach, which has been adopted in the UK on the M25, is to introduce mandatory speed controls when the flow exceeds a particular value. However, the implications of such controls have not been described in detail, and full understanding of such controls is crucial for developing optimal strategies for maximising the flow.

Here we attempt to develop some understanding of speed controls on a freeway by combining the analytical continuum approach of Lighthill and Whitham (1955) with a numerically implemented car following model. The continuum model is strictly valid for traffic in which the drivers are able to make rapid corrections of their speed to a desired speed-headway relation; some important effects associated with drivers' finite response time do emerge from the car-following approach. However, the value of the continuum model lies in the relatively simple description of the phenomena.

We focus on a single lane road, and assume that over some fixed region of the road the speed limit is gradually reduced to a smaller value, while downstream, the speed limit is returned to the original value. The car-following model which we adopt herein follows from that proposed by Bando et al. (1995) in which the cars tend to adjust to a desired speed as a function of the headway, at a rate proportional to the difference in actual speed with this speed. This model has been successfully shown to predict various forms of behaviour recorded on Japanese highways, and appears to capture a number of the features of real traffic (Bando et al. 1995).

First, we use the continuum approach to describe the different effects that a localised decrease in speed can have on the flow, identifying how a large reduction in speed can cause a shock wave to propagate upstream if the flow is overcontrolled, while smaller adjustments to the speed limit simply produce a local perturbation to the flow. We then examine the results of some numerical simulations using a car-following model. This identifies in more detail the deceleration and subsequent acceleration of the vehicles through the speed control zone. In this short contribution, we describe the key results of our modelling; we are presently preparing an extensive paper on the full details of calculating shock speeds and critical fluxes, based on the original models of Lighthill and Whitham.

2. GLOBAL IMPLICATIONS OF SPEED CONTROLS: CONTINUUM MODELS

Continuum models typically specify that there is a regime of low concentration (fast free-flow) and one of high concentration (slow congested flow) in which the speed of the vehicles gradually falls as concentration increases (Figure 1). As a result, the flux of vehicles attains a maximum, $Qm(c1)$ say, at some intermediate concentration, where $c1$ represents the speed limit in the free-flow regime.

If the free-flow speed limit is reduced to $c2$, then the fluxes associated with the free-flow regime also decrease. As a result, the maximal flux which may move along the road falls to a lower value, $Qm(c2)$. Conservation of vehicles immediately identifies there are two types of behaviour depending on the actual flux upstream, Q say (Figure 1). If $Q>Qm(c2)$ then the flow is over-controlled and an upstream propagating shock forms. This shock travels upstream because arriving cars accumulate before passing through the controlled flow region. However, if $Q<Qm(c2)$ then the flow is under-controlled and the flow simply adjusts as it passes through the region of low speed flow.

Woods (1995) has discussed the implications of this result in the context of regulating traffic. In particular, he identified that it may be possible for the vehicles to undergo a transition from congested slow to fast free flow if the speed is lowered to exactly that value such that the flow is controlled on passing through the region of low speed limit. However, such models do not account for the inertia of drivers and the finite response time to changes in the speed. Here we examine the analogous behaviour of the flow using a car following model. We also extend the ideas to investigate strategies for minimising shock-propagation upstream of accidents or roadworks; the sudden deceleration required at such shocks tends to be very hazardous.

3. SIMULATIONS USING A CAR-FOLLOWING MODEL

We follow the model of Bando et al (1995) and specify that the speeds of the vehicles obey laws of the form

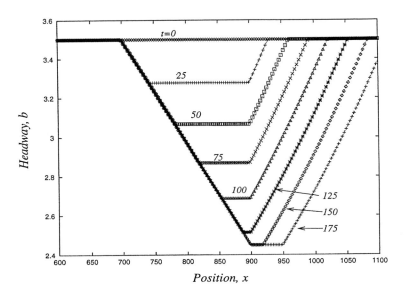

Figure 1. The concentration flow diagram illustrating slow and fast flow regimes. As the speed limit is lowered, the fast flow branch also becomes lowered and the maximum flux decreases.

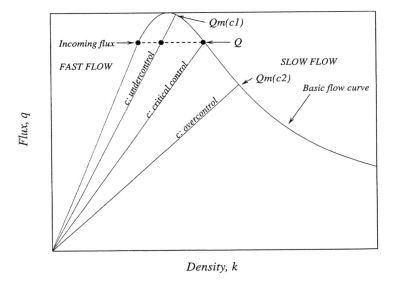

Figure 2. Computer solution of the variation in headway as a function of position along the road for an undercontrolled speed restricted zone. Numerical annotations represent non-dimensionalised time.

$$dv/dt = a(U(b) - v)$$

where v denotes position and $U(b)$ is the speed-headway relation. a is a constant representing a reciprocal reaction time. We specify

$$U(b)=\tanh(b-2) + \tanh2$$

so that the speed limit has value $c=U(1+\tanh2)$ except in regions in which the speed limit, $u(x)$, takes a smaller value than c, in which case we specify that

$$U(b) = \min(\tanh(b-2)+\tanh2, u(x)).$$

The model is solved numerically with a long platoon of vehicles, and the speed control is imposed at time $t=0$ in the centre of the platoon. Prior to that time, the flow is assumed to move downstream at a fixed speed.

3.1   UNDERCONTROL

Figure 2 illustrates how a platoon of vehicles moving along the freeway adjust, according to the above model, when they move into a region in which the speed limit falls uniformly with distance to a lower value. In all diagrams, position is in non-dimensional units relative to an arbitrary point $x=0$ far upstream of the speed control region. The speed limit is imposed at time $t=0$. In this example the flow is undercontrolled so that all the oncoming flux can pass through the speed controlled region. In the region of deceleration, the cars slow down and become more closely packed, eventually following the local speed limit in a form of stationary deceleration wave. The cars originally downstream of this region, where the speed limit takes the new smaller value, all adjust to the new speed, but since they are all initially equi-spaced, they remain equi-spaced, and therefore the flux decreases. However, the cars which emerge from the region of deceleration are much more closely spaced. A region of transition therefore develops as the first cars which pass through the deceleration zone arrive downstream. These cars spend a progressively greater time in the deceleration region, and hence their spacing becomes progressively smaller. Downstream, all vehicles have the same speed, equal to the speed limit, and so there is no further evolution of the inter-car spacing. Hence, this 'headway adjustment wave' travels downstream with the traffic. The oncoming vehicles are then able to progress smoothly along the region of deceleration and the region of lower speed limit, simply by reducing their spacing.

3.2   OVERCONTROL

In the case of overcontrol, the downstream speed limit becomes too small to admit all the oncoming flux of traffic on the downstream speed-controlled section of the road. As a result, a region of congestion accumulates upstream of the speed control with a jump in headway propagating upstream (Figure 3).

Meanwhile, downstream, the headway adjustment wave propagates downstream as in the undercontrol case, with the headway adjusting to the maximum flux subject to the lower speed limit. This behaviour is also seen clearly in a space-time (x-t) diagram (Figure 4) which illustrates how the shock develops upstream owing to the congestion while the headway adjustment wave advances downstream.

4. ACCELERATING THE FLOW THROUGH A SPEED CONTROL

The models of overcontrol and undercontrol above suggest that the use of a localised reduction in the speed limit may serve to transform the flow from slow congested flow to a fast, free flow (cf. Woods, 1995). We have explored this possibility using the car-following model. A congested train of vehicles moving at low speed approaches a region where the speed limit is reduced and then increased in a region downstream of the deceleration zone (Figure 5), so that overall, the speed limit profile is 'V-shaped'.

The minimum speed is chosen so that the flow is overcontrolled. Therefore a shock wave forms and propagates back upstream as the oncoming flux of vehicles exceeds that which can pass through the speed control section. However, as the vehicles pass downstream of the control section, and encounter the region of increasing speed limit, they accelerate towards the fast flow regime. These cars rapidly catch up with the downstream slow, congested flow which originates from before the speed control was imposed. Since the original flux exceeds the flux of oncoming fast vehicles from the control region, a downstream propagating shock wave develops, across which the flow adjusts back to the original speed (Figure 5).

Once this downstream propagating shock has passed through the system, the flow downstream of the speed control will become quasi-steady, and remain in the fast flow regime. By increasing the minimum speed in the speed control section, the degree of overcontrol can be reduced and thereby the size of the shock is also reduced. However, since the flux difference between the standing wave and the downstream congested flow becomes small, the downstream propagating shock wave travels slowly. This is seen in Figure 6, which corresponds to near-critical control.

It is interesting to note how the travel time changes as a result of the imposition of the speed control; in the speed control section, the vehicle speed is reduced and so more time is spent in

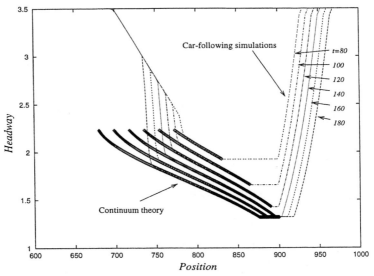

Figure 3. Computer simulation of the overcontrol scenario. Numerical annotations represent non-dimensionalised time.

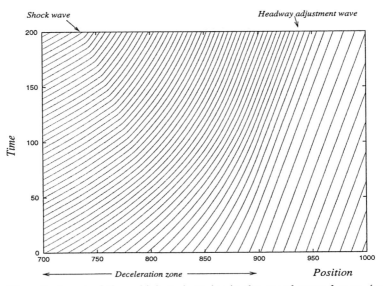

Figure 4. The adjustment of the vehicle trajectories in the speed control zone (overcontrol case). Trajectories are plotted for every third vehicle, for clarity.

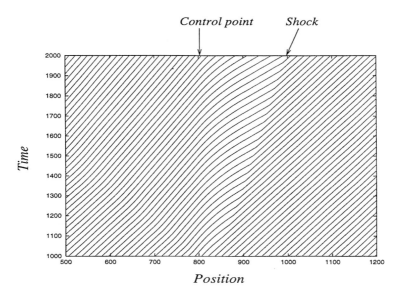

Figure 5. Effect on vehicle trajectories of speed control which is first lowered, then raised. Speed limiting imposed at t=1000.0. Trajectories are plotted for every tenth vehicle, for clarity.

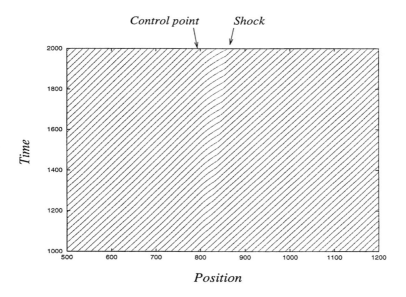

Figure 6. As for Figure 5, but nearer critical control.

that part of the road. However, downstream, the vehicles will be in the fast free-flow regime and so the vehicles spend less time than in the original slow congested flow. Indeed, the

characteristic (*x-t*) diagram illustrates how the flow adjusts across the speed control. As the flow enters the speed control section, the velocity is reduced and initially the car headway decreases. However, on reaching the minimum speed (Figure 5), the vehicles are then able to accelerate downstream and the headway increases, allowing further acceleration towards the downstream speed limit.

Once the downstream propagating shock wave has moved sufficiently far downstream, it is clear that the time to travel a given distance (significantly greater than the distance over which the speed controls are imposed) from upstream to downstream of the speed control is much smaller than in the original congested flow (Figure 7). The benefits appear to be more marked for the case of critical control. However, the downstream propagating shock wave moves slowly in this case, and there is a long time to wait before the shock passes through the system. Moderate overcontrol is therefore the better strategy.

5. IMPLICATIONS FOR SAFETY NEAR THE SITE OF AN ACCIDENT

At an accident site, the maximum flow along the road is often reduced owing to closure of one or more lanes. This typically leads to controlled flow past the accident, and the formation of an upstream propagating shock. This shock requires the rapid deceleration of the vehicles and can lead to further accidents. One possible strategy to avoid such an intense shock is to introduce a zone of speed control upstream of the accident. In this region, the speed limit is gradually reduced so that the speed of vehicles falls and the headway increases. In this way, the magnitude of the deceleration of the vehicles across the shock will be much smaller, leading to increased safety in the oncoming stream of traffic. However, since the shock continues to propagate upstream, the benefit of such a speed control strategy will be limited since the shock will eventually pass through the region of reduced speed and will return to the original large shock upstream. If the accident can be cleared up rapidly, or the speed reduction zone extends over a large stretch of the highway, then this simple speed control strategy would mitigate the problem. However, a superior solution would be to allow the speed control region to propagate upstream at a speed equal to or in excess of the shock speed (Figure 8). In this way the shock would continue to involve a small decrease in speed, and thereby maintain safer driving conditions upstream of the accident.

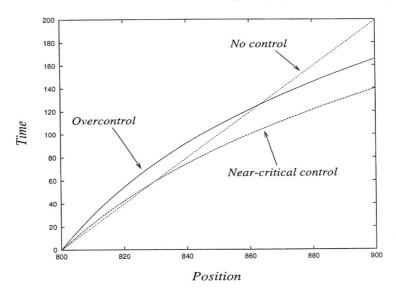

Figure 7. The effect on journey times of a speed control which is first lowered, then raised.

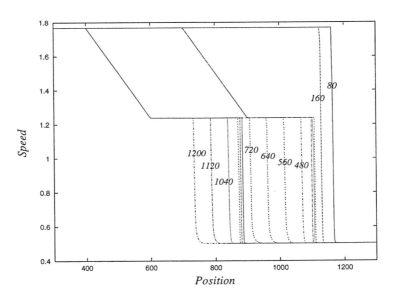

Figure 8. A safe deceleration zone upstream of an accident.

6. DISCUSSION

We have illustrated how a car-following model may be used to examine the potential impact of speed controls in both (i) effecting a transition from congested to free flow and (ii) reducing the magnitude shocks which tend to develop and propagate upstream of an accident site. The modelling identifies that both objectives are possible if sufficient information about the flow and headway is known about the traffic in order to calculate critical speed limits and shock propagation speeds. The ideas presented herein suggest some interesting principles by which motorway flow data, acquired from modern technology, can be applied to control and enhance flow along motorways.

ACKNOWLEDGEMENTS

This work is partially supported by the Smith Institute. We are grateful to Prof. Allsop of U.C.L. for some useful suggestions.

REFERENCES

Bando, M., K. Hasebe et. al. (1995). Dynamical model of traffic congestion and numerical simulation,'' *Phys. Rev. E,* **51**, no. 2, 1035-1042.

Lighthill, M. J. and G. B. Whitham (1955). On kinematic waves II: A theory of traffic on long crowded roads, *Proc. Roy. Soc. Lond. A,* **229**, 317-345.

Woods, A. W. (1995). Effect of speed controls in a continuum model of traffic flow, in *Workshop on traffic and granular flow* (D. Wolf, M. Schreckenberg, and A. Bachem, eds.), HLRZ Forschungszentrum Jülich, Germany, October 1995.

RELIEF SERVICE QUEUES

J.D. Griffiths, School of Mathematics, Cardiff University

ABSTRACT

This research was prompted by work undertaken by the author on the efficiency of shipping operations in the Suez Canal. The physical limitations of the Canal allow only one-way movement of ships for the greater part of its length, and thus ships are organised in convoys. These convoys have fixed starting times, with normally just one convoy per day operating in each direction. When traffic is heavy in the southbound direction, a second (smaller) relief convoy is organised to reduce waiting times which can otherwise exceed 24 hours. The process can be analysed by means of a bulk-service queueing model, where convoys of ships correspond to service batches of customers.

The model has application in the many other fields of transport where relief services are supplied. For example, a coach or train operator will often provide a relief service when customer demand is high. The process may be extended to cover cases where relief is provided for the relief service, resulting in a "cascade" of relief service queues.

1.   INTRODUCTION

The provision of relief service facilities is commonplace in many areas of transport management. For example, a long-distance coach may depart from its origin at a fixed time each day. If the number of passengers exceeds the capacity of the coach, then the coach operator may decide to provide a relief coach departing a short time later. Similar situations arise in air, sea and other public transport systems.

The operator of the transport system is faced with a number of decisions. For example, is it absolutely necessary to provide a relief service? Is it profitable to do so? Will providing relief service improve customer satisfaction? What should be the capacity of the relief service? To answer such questions it is necessary to model the particular situation under consideration in order to estimate such quantities as the number of customers who remain following the departure of the main service.

The work reported in this paper arose initially from a study undertaken by the author relating to shipping operations in the Suez Canal, and the methodology will be introduced by reference to this scenario.

2.   TRAFFIC CONTROL IN THE SUEZ CANAL

From a commercial aspect the Suez Canal is the most important waterway in the world. Reductions in shipping distances, with a consequent saving in transport costs, are typically of the order 40%-50% compared with the alternative route via the Cape of Good Hope. The toll charges imposed by the Suez Canal Authority form a major source of revenue for Egypt, amounting to nearly $2 billion per annum. Currently the number of ships using the Canal rarely rises above 60 per day, i.e. about 30 vessels per day in each direction.

The Suez Canal lies in a North-South direction, connecting the Mediterranean and Red Seas. The Canal is 160 km long and 180-200 m wide, but the navigable channel is only about 110 m wide. This means that there is insufficient width to allow ships to pass one another either in the same direction or in opposing directions, and consequently ships are arranged in convoys. There are two areas of the Canal where opposing convoys may pass one another, viz. Ballah by-pass (km 51-60) and the Bitter Lakes (km 103-113), see Figure 1.

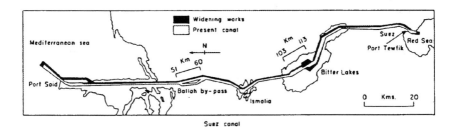

Figure 1. Layout of the Suez Canal

When traffic levels are low, just one convoy per day in each direction is mounted. The southbound convoy (S1) starts from Port Said at 0100 hours, and the northbound convoy (N) from Suez at 0600 hours. These times are fixed and are chosen specifically so that convoy S1 can reach the Bitter Lakes area before the arrival of convoy N. Convoy S1 will then moor at Bitter Lakes until convoy N has passed through, before continuing its journey to the Red Sea. Convoy N is given precedence in the sense of a straight-through passage, since this convoy usually contains tankers carrying crude oil or other hazardous cargoes, and on safety grounds it is advisable not to require this convoy to stop at Bitter Lakes. Figure 2 illustrates this traffic control system diagrammatically.

When traffic levels are high in the southbound direction (greater than about 20 ships per day), it is usual to introduce a second southbound convoy (S2). Indeed, this relief convoy is often used even when traffic levels are low if it will lead to a substantial saving in waiting time for ships which arrive too late for inclusion in the main S1 convoy. Relief Convoy S2 departs from Port Said at 0700 hours, and travels to the western loop of Ballah by-pass where it ties up. Convoy S2 waits for convoy N to pass through the eastern loop of Ballah by-pass before continuing its journey through the Canal without further stoppage, see Figure 2.

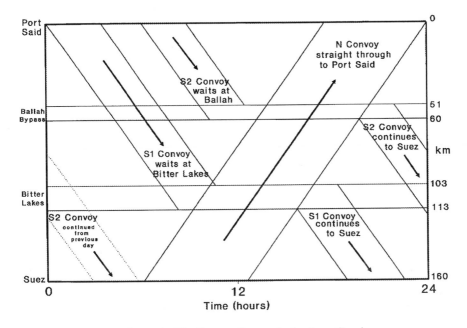

Figure 2. The Convoy System in the Suez Canal

The traffic control system described above is repeated each day, and shipping operations may thus be considered to work on a 24-hour cycle. There is a main service batch (S1) and a possible relief service (S2) operating in the southbound direction, and a main service batch (N) operating in the northbound direction. It has been shown, Griffiths and Hassan (1977), that maximum throughput of shipping in the Canal occurs when the convoys have the following sizes: S1 22 ships, S2 17 ships, N 39 ships. Relief convoy S2 is restricted to 17 ships because Ballah by-pass has only sufficient space to accommodate this number of ships. A more detailed description of the traffic control system in the Canal is given in Griffiths (1995), and a fascinating account of the construction and history of the Canal is provided in Burchell et al (1966).

3.   SHIPPING QUEUES IN THE CANAL

It has been shown that ship arrivals at the Suez Canal may be considered to occur in a random fashion, Griffiths (1995). The decision whether or not to operate the relief convoy depends crucially on the number of ships queueing at the starting times of the southbound convoys (0100 hours for the main convoy and 0700 hours for the relief convoy). Vessels are not allowed to join the tail-end of a convoy after the convoy starting time.

The situation described above is very similar to a bulk-service queue with random arrivals, where the maximum batch size corresponds to the largest allowable convoy size. The major difference between this system and the standard bulk-service queueing model is that ships in a convoy are not served simultaneously, but depart at intervals of time according to their position in the convoy. This situation is analogous to the departure of vehicles from a traffic signal during the green phase of the cycle. However, as the measure of primary interest here is the number of ships queueing at the times when convoys start their transit of the Canal, the manner in which individual ships are processed is not of particular concern. A further difference compared with standard queuing models is that the primary cycle (here 24 hours) may be interrupted by the provision of relief service.

Consider a cycle of operations for convoys S1 and S2 over a 24-hour period as shown in Figure 3. Let A and A' denote the times at which the main convoys start their passage on successive days, and let B denote the time at which the relief service convoy starts its passage. Thus A, A', B represent typical regeneration points using the imbedded Markov Chain technique.

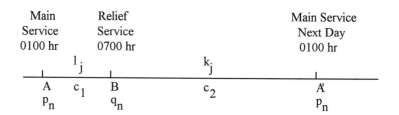

Figure 3. Regeneration Points for Batch-service Queueing Model

Let p_n be the probability that there are n ships queueing to join the main convoy at points A, A', and let the maximum size of this main convoy be denoted by M. Let q_n be the probability that there are n ships queueing to join the relief convoy at point B, and let the maximum size of this relief convoy be denoted by R. Let l_j denote the probability that j ships arrive in the interval AB, and let k_j be the probability that j ships arrive in the interval BA'.

Then we may relate probabilities at A' to those at B by:

$$p_n = k_n \sum_{j=0}^{R-1} q_j + \sum_{j=0}^{n} q_{R+j} k_{n-j} \qquad \text{for } n = 0, 1, 2, ... \qquad (3.1a)$$

and we may relate probabilities at B to those at A by

$$q_n = l_n \sum_{j=0}^{M-1} p_j + \sum_{j=0}^{n} P_{M+j} l_{n-j} \quad \text{for } n = 0, 1, 2, \ldots$$

(3.1b)

We define the following four probability generating functions.

$$P(z) = \sum_{n=0}^{\infty} z^n p_n, \qquad \qquad Q(z) = \sum_{n=0}^{\infty} z^n q_n$$

$$L(z) = \sum_{j=0}^{\infty} z^j l_j, \qquad \qquad K(z) = \sum_{j=0}^{\infty} z^j k_j$$

Multiplying the nth equation of (3.1a) by $z^n (n = 0,1,2,...)$ and summing gives

$$z^R P(z) / K(z) = Q(z) + \sum_{j=0}^{R-1} (z^R - z^j) q_j$$

(3.2)

Using the same technique on (3.1b), we have

$$z^M Q(z) / L(z) = P(z) + \sum_{j=0}^{M-1} (z^M - z^j) p_j$$

(3.3)

Solving (3.2) and (3.3) simultaneously for $P(z)$ and $Q(z)$ we find

$$P(z) = \left[\sum_{j=0}^{M-1} (z^M - z^j) p_j + \{z^M / L(z)\} \sum_{j=0}^{R-1} (z^R - z^j) q_j \right] / \left[z^{M+R} / \{L(z)K(z)\} - 1 \right]$$

(3.4)

$$Q(z) = \left[\sum_{j=0}^{R-1} (z^R - z^j) q_j + \{z^R / K(z)\} \sum_{j=0}^{M-1} (z^M - z^j) p_j \right] / \left[z^{M+R} / \{L(z)K(z)\} - 1 \right]$$

(3.5)

We may write down expressions for $L(z)$, $K(z)$ directly. If ships arrive at random at mean rate λ and if c_1, c_2 denote the (constant) time intervals AB, BA', then

$$l_j = (\lambda c_1)^j \exp(-\lambda c_1) / j! \quad \text{and} \quad k_j = (\lambda c_2)^j \exp(-\lambda c_2) / j! \text{ for } j = 0,1,2,...$$

Hence, $L(z) = \exp\{-\lambda c_1(1-z)\}$ and $K(z) = \exp\{-\lambda c_2(1-z)\}$.

The denominators of (3.4) and (3.5) then become

$$z^{M+R} e^{\lambda(c_1+c_2)(1-z)} - 1 \tag{3.6}$$

It may then be shown in the usual way using Rouché's theorem, see for example Saaty (1961), that (3.6) has exactly $M+R$ distinct zeroes within and on the unit circle provided $\lambda(c_1 + c_2) < M + R$.

Since $P(z)$ and $Q(z)$ are probability generating functions, they must be analytic within and on the unit circle. To preserve this analyticity, it follows that the numerators of (3.4) and (3.5) must vanish at the $M + R$ zeroes of the denominator.

Let the $M + R$ zeroes of (3.6) be denoted by $1, z_1, z_2, ..., z_{M+R-1}$. Then the numerator of (3.4) is zero at these $M + R$ values, leading to $M + R$ equations in the $M + R$ unknowns $p_j (j = 0,1,..., M-1)$ and $q_j (j = 0,1,..., R-1)$. Unfortunately the zero at $z = 1$ leads to the numerator being identically equal to zero, and we thus have to employ a limiting procedure such as L'Hôpital's Rule. This results in the condition

$$M + R - \lambda(c_1 + c_2) = \sum_{j=0}^{M-1}(M-j)p_j + \sum_{j=0}^{R-1}(R-j)q_j \tag{3.7}$$

When this process is complete, in theory $P(z)$ and $Q(z)$ are completely determined. However, the apparent simplicity of (3.4) and (3.5) belies the fact that determination of the $(M + R)$ unknown probabilities in their numerators is a rather difficult task when $M + R$ is large. For example, in the Suez Canal scenario we have $M = 22$, $R = 17$. Thus we need to determine the 38 zeroes (plus $z = 1$) of (3.6) which lie within the unit circle, and then invert the 39×39 matrix to determine the unknown $p_j (j = 0,1,...,21)$ and $q_j (j = 0,1,...,16)$. Readers who have attempted inversion of large matrices will have first-hand knowledge of the manner in which rounding errors build up, and although various computer packages claim to invert matrices up to size 100×100 say, such claims need to be treated with caution. In the present example, it was necessary to evaluate the 39 zeroes of (3.6) to 16 decimal places, and to use double-length arithmetic throughout the inversion process. Efficient methods of calculating the zeroes of expressions such as (3.6) have been given by Griffiths et al (1990). An example of the location of the zeroes of (3.6) for $M = 22$, $R = 17$, and a traffic intensity of 0.84 (i.e. $\lambda = 1.365$ ships/hr, $c_1 = 6$ hr, $c_2 = 18$ hr) is shown in Figure 4.

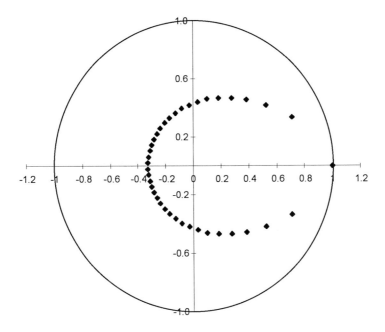

Figure 4. The zeroes of $z^{M+R} - e^{w(z-1)}$ for $M = 22$, $R = 17$, $w = \lambda(c_1 + c_2) = 32.76$.

The queue lengths at times A and B are given by $P'(1)$ and $Q'(1)$ respectively. These expressions are

$$2[M + R - \lambda(c_1 + c_2)]P'(1) = \sum_{j=0}^{M-1}[M(M-1) - j(j-1)]p_j + \sum_{j=0}^{R-1}[R(R-1) - j(j-1)]q_j$$

$$+2(M - \lambda c_1)\sum_{j=0}^{R-1}(R - j)q_j$$

$$-[(M + R)(M + R - 1) - 2\lambda(c_1 + c_2)(M + R) + \lambda^2(c_1 + c_2)^2]$$

$$(3.8)$$

$$2[M + R - \lambda(c_1 + c_2)]Q'(1) = \sum_{j=0}^{M-1}[M(M-1) - j(j-1)]p_j + \sum_{j=0}^{R-1}[R(R-1) - j(j-1)]q_j$$

$$+2(R - \lambda c_2)\sum_{j=0}^{M-1}(M - j)p_j$$

$$-[(M + R)(M + R - 1) - 2\lambda(c_1 + c_2)(M + R) + \lambda^2(c_1 + c_2)^2]$$

$$(3.9)$$

Graphs of these queue lengths against the arrival rate of ships are provided in Figure 5. As may be expected, the graphs have an asymptote at an arrival rate of 39 ships per day, corresponding to the total capacity of the two southbound convoys. The mean queue length for the main convoy reaches the capacity of that convoy (22 ships) when the arrival rate is about 28 ships per day, and the mean queue length for the relief convoy reaches the capacity of that convoy (17 ships) when the arrival rate is about 35 ships per day. The queue lengths initially increase linearly with the arrival rate of ships, being respectively 75% and 25% of the average daily number of ship arrivals, reflecting the arrival windows of $c_2 = 18$ and $c_1 = 6$ for the main and relief convoys.

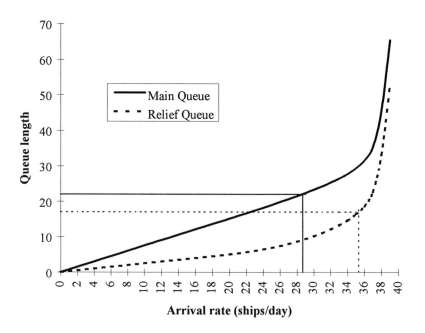

Figure 5. Queues for Main and Relief Convoys

It is worth mentioning that there is an alternative technique available for calculating the probabilities p_n, q_n, and hence the mean sizes of the two convoys. We may consider (3.1a) as a relationship linking the p_n on day m to the q_n on day $(m-1)$, and (3.1b) as relating the q_n on day m to the p_n on day m. Hence, by using any feasible set of probabilities for the p_n on day 1 we can calculate recursively the q_n on day 1, then the p_n on day 2, then the q_n on day 2, and so on. The p_n and q_n settle down to the steady state values quickly when the arrival rate is not too large, but rather slowly as the rate approaches the limit $M + R = 39$. For example, an arrival rate of 30 ships per day (equivalent to a traffic intensity of 0.77) requires only about 10 iterations to produce probabilities correct to 6 decimal places, whereas an arrival rate of 38 ships per day (traffic intensity of 0.974) requires about 3000 iterations for the same degree of accuracy.

To answer the questions originally proposed regarding the advisability of mounting a relief service or not is relatively straightforward in the case of the Suez Canal. The cost of providing the relief convoy is minimal, since this convoy merely requires mooring facilities at Ballah-by-pass, and these facilities are already present. However, there are some minor costs involved associated with the necessity to provide control of the relief convoy (as described earlier). Thus, it would not be sensible to incur these costs for the sake of a single ship, and the recommendation is that the relief convoy should only be provided if the number of ships waiting is about 4 or 5. This occurs at a daily arrival rate of about 20 ships, and the main convoy would be accommodating about 15 ships on average.

4.   DISCUSSION

The model provides an opportunity to quantify queue lengths in many transport situations where relief service may be an option. The measures derived from the model can be used as input to a cost analysis if it is required to justify the provision of relief service on financial grounds. For example, a long distance coach operator may only be willing to provide a relief coach if the cost of doing so were at least offset by the extra revenue received. It is also possible to use the model as a design tool, since it is feasible to run the model with a variety of different batch sizes for the relief service, thereby enabling a decision to be made on the appropriate size of coach, train, ferry, etc. which should be on stand-by for relief duties. This is an interesting aspect of the Suez Canal study, since there seems no very good reason why exactly 17 mooring places should have been created in Ballah by-pass, thus determining the maximum batch size of the relief convoy. Indeed, Figure 5 shows clearly that maximum batch size would be rarely achieved for the relief convoy, since the arrival rate would need to be about 35 ships per day for this to occur. Since the southbound arrival rate is usually 30 or less ships per day, Figure 5 shows that the maximum size of the relief convoy need only be about 10 ships.

It is possible to extend the ideas of the model to provide multiple relief service. These additional relief facilities could occur at the same time as the initial relief service when it is known beforehand that the initial relief service cannot provide the required level of backup, or it could occur sequentially with relief of a relief service, i.e. a "cascade" of relief services is provided. Conceptually this is not difficult to envisage, but the algebraic manipulation required to produce the desired queueing formulae becomes a little cumbersome.

REFERENCES

Burchell, S.C. and C. Issawi (1966). *Building the Suez Canal*, Cassel, London.

Griffiths, J.D. and E.M. Hassan (1977). The maximum shipping capacity of the Suez Canal, *Maritime Policy and Management*, **4**, 235-248.

Griffiths, J.D., W. Holland and J.E. Williams (1990). Efficient solutions to a well-known transcendental equation, *Bull. IMA*, **26**, 156-157.

Griffiths, J.D. (1995). Queueing at the Suez Canal, *J. Op. Res. Soc.*, **46**, 1299-1309.

Saaty, T.L. (1961). *Elements of Queueing Theory*, McGraw-Hill, London.

TRANSIENT DELAY IN OVERSATURATED QUEUES

B.G. Heydecker and N.Q. Verlander, University of London Centre for Transport Studies, University College London

ABSTRACT

The estimation of queue length and delays in queues that are oversaturated for some part of a study period is of substantial importance in a range of traffic engineering applications. Whiting's co-ordinate transformation has provided the basis for several approaches to this. We analyse this approach and present an explicit form for the derivative of queue length with respect to time, which we then use to establish various properties. We also report the results of numerical comparisons with exact formulae for certain special cases and show that these offer little or no advantage over the co-ordinate transformation approximations and can be computationally impractical in study periods of moderate duration.

1. INTRODUCTION

Transport systems often become overloaded for some part of peak periods, and evaluation of their performance under these conditions is therefore important in assessing and improving their design. However, no simple analytical approach is available to estimate queue lengths and delays that is appropriate in both undersaturated and oversaturated conditions. In this paper, we develop the analysis of the heuristic co-ordinate transformation approach devised by Whiting (see, for example, Kimber and Hollis, 1979) for the estimation of queue lengths and hence delays under transient overloaded conditions in the presence of stochastic effects. This approach provides a convenient means of estimation, and is especially useful when traffic intensity is close to unity because in that case neither deterministic nor steady-state stochastic approaches is suitable. Although the co-ordinate transformation method is widely used and indeed has been adopted by the UK and other national government departments of transport (see for example, Vincent, Mitchell and Robertson, 1980; AkHelik, 1980; Semmens, 1985a, b; Burrow, 1987; Transportation Research Board, 1997), it has not been analysed fully. In this paper, we investigate the dynamic properties of the resulting formulae and consider their consequences for practical application.

We establish an exact expression for the derivative with respect to time of queue length formulae that result from co-ordinate transformation. We show that the rate of change of the queue length estimated according to this approach can be expressed conveniently as a function of traffic intensity and current queue length, but also depends on the time since the queueing process was initialised. An immediate consequence of this is that the estimate of queue length varies in a way that depends on the frequency of reinitialisation of the queueing process. The special case of instantaneous re-initialisation gives rise to a remarkably simple formula which we interpret in terms of equilibrated queues.

We investigate the special case of queues with Poisson arrivals and exponential service times, for which several other analyses are available. When the traffic intensity is unity, Whiting's estimate of queue length grows asymptotically in proportion to the square root of capacity times time in each of the cases where the queueing process is initialised once only and is re-initialised instantaneously, but with a greater rate in the latter case. We compare these results with those of Newell's diffusion approximation which is intermediate between them. We also make comparisons with numerical solutions to exact analyses of this case, which shows that the exact analyses can be computationally expensive and can also be subject to numerical difficulties.

2. THE CO-ORDINATE TRANSFORMATION APPROACH

2.1 Introduction

Non-zero queue lengths arise when for some period of time, the mean arrival rate of traffic exceeds the capacity of a system. Two special cases of this are amenable to analysis: the deterministic case in which the arrival rate and capacity vary over time in some known way and there are no stochastic variations, and the steady-state stochastic case in which the arrival and service processes have constant mean rates but are subject to stochastic variations. We present a preliminary analysis of these two cases because they are the principal components of the co-ordinate transformation method. The analysis that we present is for queue length including any customer in service; delay can be deduced from this in a straightforward manner.

Consider first the deterministic model. In order for a non-zero queue length to arise, the arrival rate must exceed capacity for some period of time. Let the capacity of the queue be Q, the length of the queue at time $t = 0$ be L_0, and the arrival rate be q for some time after that so that the traffic intensity is $\rho = q/Q$. We can calculate the queue length $L_d(\rho, t)$ at time $t > 0$ according to the deterministic model in the case $\rho \geq 1$ as

$$L_d(\rho,t) = \left[L_0 + (\rho - 1)Qt\right] \tag{2.1}$$

The same expression holds if $\rho < 1$ up to time $t = L_0 / \left[Q(1 - \rho)\right]$ when the queue dissipates. From this, the traffic intensity $\rho_d(L, t)$ that gives rise to a queue length L after time t is

$$\rho_d(L,t) = \left[1 + (L - L_0)/(Qt)\right] \quad \left(L \geq Max(0, L_0 - Qt)\right) \tag{2.2}$$

According to this model, the instantaneous rate of change of queue length is

$$\frac{dL}{dt} = (\rho - 1)Q \quad (\rho \geq 1) \quad \text{or} \quad (L > 0) \tag{2.3}$$

Consider next the steady-state stochastic case. In order for a steady-state to exist, the traffic intensity must be less than unity. In this case, non-zero queue lengths arise because of the stochastic variations in arrivals and departures. Suppose that the traffic intensity is ρ and that the variance of service times is σ^2; we characterise the irregularity of service by the quantity $C = \left[1 + (Q\sigma)^2\right]/2$ which takes the value 0.5 for regular service, 1.0 for exponentially distributed service times, and for which a typical value at traffic signals is 0.55 (Catling, 1977; Branston, 1978). When arrivals form a Poisson process, the mean equilibrium queue length including any customer in service is (see, for example, Cox and Smith, 1961)

$$L_e(\rho) = \rho + C\rho^2 / (1 - \rho) \quad (\rho < 1) \tag{2.4}$$

The traffic intensity $\rho_e(L)$ that gives rise to a mean equilibrium queue length L is then

$$\rho_e(L) = \begin{cases} \left(L + 1 - \sqrt{(L-1)^2 + 4LC}\right)/\left[2(1-C)\right] & (C \neq 1) \\ L/(L+1) & (C = 1) \end{cases} \tag{2.5}$$

This expression can be evaluated for any $L > 0$ and will always give a value in the range [0, 1).

2.2 Whiting's co-ordinate Transformation

As it stands, the deterministic analysis of queue length represented by equations (2.1 - 2.3) is useful in practice only when either the queue is overloaded, *ie* when $\rho > 1$, or a substantial queue remains; otherwise it yields a null estimate of queue length after a certain amount of time has elapsed. On the other hand, the stochastic analysis represented by equations (2.4 - 2.5) can only be applied when $\rho < 1$; it yields estimates of queue length that diverge as $\rho \to 1^-$ and the time taken for a queue to approach equilibrium also diverges in this limit. These two approaches are mutually incompatible, and neither gives a satisfactory result when $\rho = 1$.

Whiting observed that for any strictly positive value of t, the steady-state stochastic queue length estimate (2.4) could be sheared so that its asymptote is given by the deterministic estimate (2.1); he interpreted the resulting function $L(\rho, t)$ as an estimate of the queue length that would arise at time t in a stochastic queue with initial queue length L_0 and traffic intensity ρ. This co-ordinate transformation can be expressed conveniently in terms of the traffic intensities that would give rise to a certain queue length L according to the various models. Thus Whiting's transformation specifies the traffic intensity ρ in terms of the deterministic (2.2) and stochastic equilibrium (2.5) ones as

$$\rho = \rho_d(L,t) - 1 + \rho_e(L) \tag{2.6}$$

Several authors (including Catling; Kimber and Hollis; AkHelik; and Han, 1996) have developed this analysis by inverting (2.6) to give expressions for queue length as a function of ρ and t for various cases, giving estimates as the solution to certain quadratic equations. Following Han's rearrangement to eliminate the possibility of a singularity at time $t = (1-C)/Q$, the solution to the present form (2.2), (2.5) and (2.6) is

$$L(\rho,t) = \frac{B}{2\left[A + \sqrt{A^2 + (Qt+1-C)B}\right]} \tag{2.7}$$

where $\qquad A = (1-\rho)(Qt)^2 + (1-L_0)Qt - 2(1-C)(L_0 + \rho Qt)$

and $\qquad B = 4(L_0 + \rho Qt)\left[Qt - (1-C)(L_0 + \rho Qt)\right]$

This heuristic approach provides convenient estimates that have been found (Kimber and Daly, 1986) to give reasonably accurate estimates. It has the important property of providing a smooth transition between those provided by two models presented above in the region of practical importance of traffic intensity close to unity. We note, however, that the limiting value of the deterministic estimate of queue length (2.1) as $t \to 0$ is L_0 independent of the traffic intensity ρ, so that the co-ordinate transformation is singular in this limit.

2.3 Derivatives of Co-ordinate Transformed Formulae

We now establish our main result, which is an exact expression for the derivative with respect to time of the estimate of queue length given by the inverse of (2.6). We note that (2.6) expresses the queue length L as an implicit function of t, so that the implicit function theorem can be applied (see, for example, Dieudonn9, 1969, p 270). This approach is found to be advantageous by comparison with direct differentiation of expressions such as (2.7). Thus for a certain value of ρ, we have

$$\frac{dL}{dt} = -\left(\frac{\partial \rho}{\partial t}\right)\left(\frac{\partial \rho}{\partial L}\right)^{-1} \tag{2.8}$$

From (2.6), we see that the sole dependence of ρ on t is through ρ_d given by (2.2), so that

$$\begin{aligned} \frac{\partial \rho}{\partial t} &= -(L-L_0)/(Qt^2) \\ &= -(\rho - \rho_e(L))/t \end{aligned} \tag{2.9}$$

The dependence of ρ on L is through each of ρ_d and ρ_e given by (2.2) and (2.5) respectively, so

$$\frac{\partial \rho}{\partial L} = \frac{\partial}{\partial L}(\rho_d + \rho_e)$$

$$= \left(\frac{1}{Qt} + \frac{\partial \rho_e}{\partial L}\right) \qquad (2.10)$$

Using (2.9) and (2.10) in (2.8), and rearranging gives our main result:

$$\frac{dL}{dt} = \frac{(\rho - \rho_e(L))Q}{\left(1 + \frac{\partial \rho_e}{\partial L} Qt\right)} \qquad (2.11)$$

3. ANALYTICAL PROPERTIES

3.1 Introduction

Equation (2.11) for the derivative with respect to time of the co-ordinate transformation estimate of queue length is expressed in terms of the current traffic intensity ρ and capacity Q, and the value and derivative with respect to queue length of the equilibrium traffic intensity that is associated through (2.5) with the current queue length. We now establish some properties of the co-ordinate transformation approach by analysis of expression (2.11). We consider in turn the dependence of the derivative on the value of t, its limiting value as $t \rightarrow 0$, and its limiting behaviour as $t \rightarrow \infty$ for certain configurations.

3.2 Non-transitivity of queue-length calculations

The expression (2.11) for the derivative of mean queue length with respect to time shows an explicit dependence on the time t since the queueing system was initialised with queue length L_0. This approach to queue-length estimation is generally applied for time-slices throughout which the mean arrival rate is taken to be constant: it provides an estimate of the final queue length on the basis of the initial one and the traffic intensity. The duration of these time-slices is arbitrary, though values that are used in practice are generally in the range 5 - 15 minutes. Equation (2.11) shows that for any combination of values of the current queue length L, the traffic intensity ρ, and the capacity Q, the queue length will change less rapidly as t increases. An immediate consequence of this is that the estimate of queue length changes more rapidly if the queueing process is re-initialised more frequently by subdivision of time-slices. This then quantifies the known property of non-transitivity of queue length estimates given by the co-ordinate transformation approach.

3.3 Limiting Behaviour for Small *t*

At the instant of re-initialisation, (2.11) gives rise to a remarkably simple formula for the rate of change of queue length:

$$\frac{dL}{dt} = \left(\rho - \rho_e(L)\right)Q \tag{3.1}$$

This shows that at the instant of initialisation, the derivative of queue length is indeed well determined, although the co-ordinate transformation itself is singular. Furthermore, in the case that the traffic intensity is equal to that indicated by (2.5) for which the initial queue length L_0 is an equilibrium value, the queue length will be stationary. This supports the interpretation of the initial queue length as the mean of an equilibrated distribution rather than a value that is known exactly. This provides an explanation of the results observed by Olszewski (1990) who showed that substantial differences can arise between results from the co-ordinate transformation approach and a probabilistic treatment in which the initial queue is supposed to have a point distribution corresponding to a known value.

The content of (2.11) is a differential equation, the integral of which is provided for constant traffic intensity ρ by the solution to (2.6). However, if the queueing process is reinitialised instantaneously, the differential equation reduces to (3.1) which offers some scope for analytic integration, depending on the way in which the traffic intensity ρ varies with time.

3.4 Limiting Behaviour for Large t when $\rho = 1$

Consider the special case of a queue with Poisson arrivals, exponential service times, and a single server, which is denoted by the Kendall notation *M/M/1* and for which the irregularity parameter takes the value $C = 1$. In the case that the mean arrival rate is identical to the capacity, so that $\rho = 1$, no straightforward analysis is suitable. Consider first the co-ordinate transformation estimate of queue length for large values of time t after initialisation in the form (2.7). Identifying the greatest power of t in each of the numerator and denominator of (2.7) and simplifying gives that in the case $C = 1$ and $\rho = 1$, for large t

$$L(t) \approx \sqrt{Qt} \tag{3.2}$$

so that the queue length increases asymptotically like the square root of the number of customers that could have been served.

Consider now the effect of supposing that the queue is reinitialised instantaneously. Substituting the equilibrium traffic intensity $\rho_e(L) = L/(L+1)$ from (2.5) that is associated with a certain queue length L for the case $C = 1$ into the formula (3.1) for the rate of change of queue length, integrating and rearranging gives that in the case $C = 1$ and $\rho = 1$,

$$L(t) = -1 + \sqrt{(1+L_0)^2 + 2Qt} \tag{3.3}$$

so that for large t

$$L(t) \approx \sqrt{2Qt} \tag{3.4}$$

Thus in this case, instantaneous reinitialisation of the queueing process in the co-ordinate transformation gives rise to estimates queue lengths that increase at a rate that is asymptotically $\sqrt{2}$ greater than if the process were never reinitialised. Expressions (3.2) and (3.4) provide respectively lower and upper bounds on the asymptotic rates of change of estimates of queue length arising from the co-ordinate transformation approach for the *M/M/1* queue at $\rho = 1$ with differing time slice duration.

We note with interest that Newell (1971, p 115) presents asymptotic results for this case based upon a diffusion approximation. Newell gives

$$L(t) \approx \sqrt{4Qt / \pi} \tag{3.5}$$

which is rather closer to (3.2) than it is to (3.4): the ratio of the asymptotic rates given by (3.5) and (3.2) is $2 / \sqrt{\pi} \approx 1.13$ whilst that of (3.4) and (3.5) is $\sqrt{\pi / 2} \approx 1.25$.

4. NUMERICAL INVESTIGATION

4.1 Introduction

Consider now the numerical evaluation of the co-ordinate transformation estimate of queue length (2.7) for long time over a single long period and in the special case that arises when the queueing system is reinitialised instantaneously that is characterised by (3.1) at each instant. Certain other analyses are apply to the *M/M/1* queue which can be brought into the comparison.

4.2 Analyses of the *M/M/1* Queue

Morse (1958, p 66) presented an expression for the probability of transition between specified queue lengths over a certain time for the *M/M/1* queue with an upper bound on the queue size and arbitrary traffic intensity. This is expressed as a finite sum involving trigonometric functions. In this analysis, the bound on the sum represent the maximum queue size rather than truncation of an infinite series; changes to the number of terms in the series will change the values of each of the terms themselves. These probabilities can be used to estimate the mean queue length at the end of a study period, conditional on some specified distribution of the initial state.

Analyses of the dynamics of an *M/M/1* queue that is initially empty have been presented by Saaty (1961, p 93), Grimmett and Stirzaker (1992, pp 416-20), and Bunday (1996, pp 184-9). Saaty expressed the probability distribution of the number of customers in the queue after a certain time in terms of an infinite sum involving Bessel functions: in this case, truncation of the series of summation is undertaken on grounds of numerical expediency. Grimmett and Stirzaker, and Bunday each apply combinatorial analysis of the Poisson distribution to establish formulae that include exponentials and factorials.

4.3 Numerical Comparisons

Consider again the special case of an *M/M/1* queue that is empty at time $t = 0$ and has traffic intensity equal to unity: at this traffic intensity, neither the deterministic nor the equilibrium stochastic analyses is satisfactory. We investigate the use of various approaches to estimate the state of the queue after 100 mean service times, which corresponds to about 3 minutes of operation of a single lane at traffic signals.

Using Morse's analysis, the upper limit for the summations was set initially to 500, indicating space for up to 500 customers before balking occurs: this value is used within each of the terms of the summation. Evaluating the queue length probability distribution at time $t = 100 / Q$ caused some numerical difficulties. Some of the calculated estimates of probabilities were negative: the probability of there being 37 customers in the queue at this time was calculated as -3.088×10^{-4}. The sum of the positive probabilities was 0.917, indicating a substantial lack of normalisation. The mean queue length, estimated on the basis of the positive probabilities, was 9.094, and normalising the positive probabilities increased this to 9.915. Increasing the upper limit of summation to $N = 1000$ and adjusting the probability calculations accordingly maintained positive probability estimates up to queue length 40, increased the sum of positive probabilities to 0.955 and increased the estimate of mean queue length to 9.804; normalising the positive probabilities increased this estimate of the mean queue length to 10.263. Beyond these numerical difficulties, this method was found to be computationally expensive.

Using Saaty's analysis, probabilities were calculated for the queue lengths 0-98 using 100 terms in the summations. All probability estimates for time $t = 100 / Q$ were positive, their sum was found to be 0.997 and the estimate of mean queue length was 10.784. Probabilities calculated for time $t = 50 / Q$ were found to sum to 0.99993 and hence were normalised to within 10^{-4}. Because of their use of factorials, we were unable to evaluate satisfactorily the expressions of either Grimmett and Stirzaker or Bunday at time $t = 100 / Q$. However, at time $t = 50 / Q$ it was possible to calculate the first 57 terms of the former distribution using an upper limit of summation of 112, which was normalised to within 10^{-3}, and to calculate the first 165 terms of the latter distribution, which was normalised to within 10^{-8}. We note from this that these formulae present numerical difficulties for study periods that are long by comparison with a mean service time although they remain within a range of practical interest.

Evaluating the co-ordinate transformation estimate of mean queue length (2.7) with $L_0 = 0$, $C = 1$ and $\rho = 1$ at time $t = 100 / Q$ gives the estimate 9.512, whilst the integral of the instantaneous rate of change of queue length (3.1) over this period gives the estimate 13.177. Finally, using the asymptotic estimates of rate of change of mean queue length furnished by (3.2), (3.4) and Newell's estimate (3.5) yielded the values 10.0, 14.14 and 11.28 respectively.

5. DISCUSSION AND CONCLUSIONS

The estimation of mean queue length under conditions of transient overload provides an important element in the evaluation of transport systems. The co-ordinate transformation method provides a versatile and computationally convenient means to do this, but it varies according to the arbitrary specification of a time-slice duration. Two possibilities to obviate this arbitrariness are to use either a single time slice for the whole study period or an implicit time-slice duration of 0 by using the instantaneous dynamics of queue length. Neither of these possibilities is entirely satisfactory: the former removes the facility to model temporal variations in arrival rate and capacity, whilst the latter entails the calculation of queue length as the numerical solution to a differential equation.

Various other analyses of the *M/M/1* queue have been investigated. This showed that Morse's approach is computationally expensive, and that this and the methods developed by Grimmett and Stirzaker, and Bunday are computationally ill-conditioned for study periods with duration corresponding to 100 mean service times, giving rise to negative probability estimates and non-normalised probability distributions. Saaty's approach was found to be preferable to these, being computationally convenient and providing plausible estimates of probability distributions, though it applies only when the initial queue length is 0.

Numerical results from the co-ordinate transformation approach showed that its application in a single time-slice leads to underestimates of queue lengths, whilst the instantaneous application leads to overestimates with errors of rather greater magnitude. These estimates do have the advantage of being relatively convenient to calculate. Finally, coarse estimates can be made for the case tested on the basis of the asymptotic rate of queue length growth, and these appeared to introduce further errors of magnitude that is comparable to those inherent in the method.

The analysis of the co-ordinate transformation approach to queue length estimation has yielded a computationally convenient expression for the rate of change of estimated queue length. This expression is informative in respect of the behaviour and interpretation of the co-ordinate transformation method and for the parameter values that are required for its use.

ACKNOWLEDGEMENTS

This work was funded by the UK Economic and Social Research Council under the LINK programme. The authors are grateful to Puff Addison, Suzanne Evans and to Richard Allsop for their stimulating discussions and to an anonymous referee for his helpful comments.

6. REFERENCES

AkГElik, R. (1980) Time-dependent expressions for delay, stop rate and queue length at traffic signals. Australian Road Research Board, Internal Report **AIR 367-1**.

Branston, D. (1978) A comparison of observed and estimated queue lengths at over-saturated traffic signals. *Traffic Engineering and Control*, **19**(7), 322-7.

Bunday, B.D. (1996) An introduction to queueing theory. London: Arnold.

Burrow, I.J. (1987) Oscady: a computer program to model capacities, queues and delays at isolated traffic signal junctions. Transport Research Laboratory, Report **RR 106**.

Catling, I. (1977) A time-dependent approach to junction delays. *Traffic Engineering and Control*, **18**(11), 520-3, 6.

Cox, D.R., and Smith, W.L. (1961) Queues. London: Chapman and Hall.

Dieudonne, J. (1969) Foundations of modern analysis. London: Academic Press.

Grimmett, G.R. and Stirzaker, D.R. (1992) Probability and random processes. Oxford: Clarendon Press.

Han, B. (1996) A new comprehensive sheared delay formula for traffic signal optimisation. *Transportation Research*, **30A**(2), 155-71.

Kimber, R.M. and Daly, P.N. (1986) Time-dependent queueing at road junctions: observation and prediction. *Transportation Research*, **20B**(3), 187-203.

Kimber, R.M. and Hollis, E.M. (1979) Traffic queues and delays at road junctions. Transport and Road Research Laboratory, Report **LR 909**.

Morse, P.F. (1958) Queues, inventories and maintenance. Chichester: Wiley.

Newell, G.F. (1971) Applications of queueing theory. London: Chapman and Hall.

Olszewski, P.S. (1990) Modelling of queue probability distribution at traffic signals. **In**: Transportation and Traffic Theory (ed M. Koshi). London: Elsevier, 569-88.

Saaty, T.L. (1961) Elements of queueing theory. London: McGraw Hill.

Semmens, M.C. (1985a) Picady 2: an enhanced program to model capacities, queues and delays at major/minor priority junctions. Transport Research Laboratory, Report **RR 36**.

Semmens, M.C. (1985b) ARCADY 2: an enhanced program to model capacities, queues and Delays at roundabouts. Transport Research Laboratory, Report **RR 35**.

Transportation Research Board (1997) Highway Capacity Manual. **Special Report 209**. Washington DC: TRB, National Academy of Sciences.

Vincent, R.A., Mitchell, A.I. and Robertson, D.I. (1980) User guide to TRANSYT version 8. Transport and Road Research Laboratory, Report **LR 888**.

REAL-TIME SIMULATION TRIALS IN AIR TRAFFIC CONTROL: QUESTIONS IN PRACTICAL STATISTICS

D. T. Marsh, National Air Traffic Services Ltd

ABSTRACT

Real-time simulation trials involving operational air traffic controllers are an essential stage in the validation of new ideas and computer assistance tools for air traffic control. This paper describes the business and technical background to such trials and then highlights two statistical issues which continue to complicate the design and reporting of trials:

- Reconciling objective, subjective, quantitative and qualitative data: striking the right balance between controlled measurement and expert opinion;

- Correlation and independence in sequences of data: designing cost-effective trials without over-sampling.

1. BACKGROUND

National Air Traffic Services Limited (NATS) is responsible for air traffic control in UK en route airspace, over a large section of the North Atlantic and at major UK airports.

Air traffic is growing at about 3.8% per year in the UK, see Figure 1. So if NATS is to continue to meet its customers' needs, it must identify ways to increase airspace capacity, whilst at least maintaining current levels of safety.

One of the roles of NATS' Department of Air Traffic Management Systems Research (DASR) is to assess future concepts and tools for air traffic control through the use of small-scale, real-time simulations involving validated air traffic controllers. For example, DASR has recently looked at a graphical tool to help plan the routes of aircraft over the Atlantic, for inclusion in a new ATC system due to be in service shortly after the year 2000.

For reasons dating back to the days when aircraft navigated by flying towards and away from radio beacons, with little radar coverage, aircraft in domestic airspace mostly follow piecewise-linear routes, from named waypoint to named waypoint. These routes are also assigned names. Figure 2 shows the route structure over some of Southern Wales: the Brecon sector of airspace runs from Cardiff in the South to near Llangollen in the North. For instance, the South-North route EXMOR, BCN (near Brecon), RADNO (near Radnor), MONTY is known as UA25.

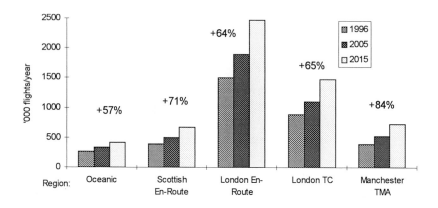

Figure 1. Traffic projections for the UK

Typically, there are two controllers for each sector: one takes tactical responsibility for maintaining separation between aircraft within a sector, whilst a colleague coordinates the planning of how and when the aircraft will enter and leave the sector. Away from airports, the capacity of the airspace is largely determined by the number of aircraft that each such pair of controllers can safely handle. In the past, capacity has been increased by increasing the number of sectors thus reducing the number of aircraft under each controller's responsibility and hence reducing his or her workload. However, there is a limit to how far this can be continued without suffering from diminishing returns, as the workload involved in coordinating the transfer of flights between sectors escalates.

Alternatives to increasing the number of sectors are to change the operating procedures or to provide computer assistance tools to the controller so as to reduce the workload involved in safely handling a given number of aircraft. The early stages of assessment for such changes involve panels of controllers and researchers developing ideas and testing them using fast-time simulations. Ideas which get through these stages need more realistic testing, with controller-in-the-loop, real-time simulations. The NATS Research Facility is a small-scale, real-time simulator comprising two measured sectors plus all of the support facilities needed to ensure realistic operations of those sectors. It is used to evaluate prototype systems through comparisons with a baseline (current day) system using a number of simulated traffic samples.

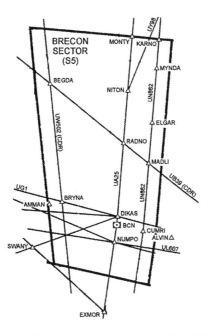

Figure 2. The Brecon airspace sector (simplified)

2.   MEASUREMENTS FROM TRIALS

The types of results gathered and analysed from these simulations include:

(i)   objective workload measurements, such as how much time the controller spends talking to pilots, or how long it takes to plan a safe route for a particular aircraft;

(ii)   subjective workload measurements, such as 'ISA', in which each controller assesses their own workload on a 1-5 scale, every 2 minutes during the simulation run, where 1 represents under-utilised and 5 is overloaded; Figure 3 shows an example of the ISA responses from one trial, plotted against the number of aircraft the controller was dealing with at the time;

(iii)   quality of service to airlines: how many aircraft get the flight level they ask for? For those that don't, how far off are they? How long are the delays?

(iv)   capacity: for example what landing rate can be achieved?

(v)   tool usage: how much of the time are controllers using the new tools? Are they using one in preference to another?

(vi)   system usability: do the controllers find the system easy to learn and to use?

(vii) controller acceptability: how happy are the controllers with the new concepts or tools? What obstacles do they see to using them operationally?

(viii) safety: trial prototypes are not safety-critical as it is a simulation rather than 'real' aircraft, but what would be the safety implications of implementing the tools operationally?

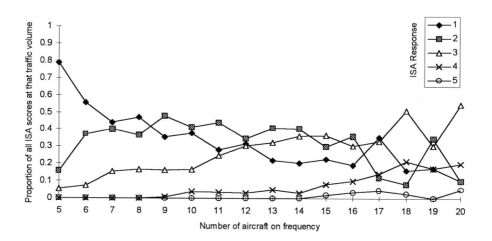

Figure 3. Frequency of ISA response as a function of number of aircraft

3.   RECONCILING THE OBJECTIVE, SUBJECTIVE, QUANTITATIVE AND QUALITATIVE

The views of controllers are an essential part of the evaluation: if the tools or concepts are accepted into operational use, then these are the people who will still be responsible for the safety of the aircraft. Opinions, however, are not sufficiently accurate ways of identifying quantitative differences: the evidence is that controllers can say if quality of service improved or declined, but any estimates by how much are very inaccurate. There is also the problem sometimes of different controllers having widely different opinions.

A decision to change a procedure, or to develop and implement a tool to operational standards involves considerable expense. Such a decision must be based on quantitative information about the benefits that may accrue, even if no-one expects that an improvement measured in the simulation trial will be exactly that achieved in a fully-operational system as no simulation ever exactly matches real life.

It is quantitative analysis, then, that is the principal aim of analysing the results of the trial: that is, descriptive and inferential statistics of the measurements of workload, capacity, safety and quality of service. All controller comments made are logged and classified, but they are used to throw either light or doubt on the quantitative results.

Workload and quality of service are complex ideas which cannot be fully encompassed by one or two measurements. A conclusion of the trial will have to be based on a degree of consistency between a range of quantitative measurements, backed up by qualitative comments and observations. Practically, this makes the conclusions more robust and more persuasive for the customer. A single statistical result from a single workload measurement is not going to make a robust business case for implementing changes to the operational system.

Experimentally however, trusting in a positive result from only three types of measurement out of a possible four, say, increases the likelihood of a type I error. Nevertheless the relatively high significance level of 5% is used without any Bonferroni adjustment, because experience has shown this threshold to be effective at isolating interesting results from the statistical noise in trials on the NATS Research Facility.

In all of this, it is important to remember the provenance of the data: these may be controlled trials, but having the controllers in the loop makes the results noisy, and the subjective data are at best ordinal. The statistical analysis is therefore kept at a level of sophistication which fits the data, and this means largely non-parametric tests.

4.   CORRELATION AND INDEPENDENCE IN TIME SERIES

Data from the simulations are often in the form of sequences, for example observations every two minutes, or of each aircraft as it comes in to land. The aim is to compare sequences to see if one represents an improvement over the other. This section discusses the difficulties of working out how much data it is safe to take from each sequence. There does not appear to be an account of time series (for example Kendall and Ord, 1990; Cox and Lewis, 1966) that addresses the short sequence, non-parametric case, rather than modelling longer sequences of mostly ratio-scale data, or for that matter to be an account of non-parametric statistics (such as Siegel, 1988) that deals adequately with time series.

To run trials, controllers are taken away from operational duties, typically for two weeks at a time. Into that time are packed training, one-off demonstration runs and controlled, measured, matched sets of runs. To strike the right balance, and to be able to make some assessment of what the resolving power of the trial is likely to be, it is necessary to have a reasonable understanding of the independence of the data during the trial's planning phase. To illustrate the difficulties that arise, this section gives two examples of measurements which have caused long discussions: ISA and aircraft spacing.

$$G_{n,n+1}(x_n, x_{n+1}) = F_n(x_n)F_{n+1}(x_{n+1}) \tag{1}$$

where G is the joint distribution of pairs of adjacent scores and F is the distribution of the individual scores. In practice, this is not practicable, for the following reasons:

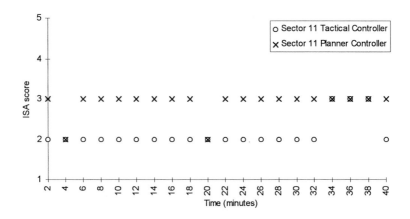

Figure 4. Example sequences of ISA workload responses

1. Workload depends on traffic volume, which varies with time, ie $F_n \neq F_{n+k}$ for some time delay k. Thus within-run frequency of ISA response is not a good indicator of F, which should at least be considered as a function of traffic volume, F_v.

2. The responses are *subjective* workload, so aggregating responses across controllers is unreliable (in practice scores are matched by controller). So F_v is really a function of controller too, $F_{v,c}$. (This assumes a simple design where a controller always fills the same role - ie the tactical or planning position - in the same sector.)

3. Workload depends on the 'organisation', i.e. the combination of operational concept (i.e. the airspace design, the traffic sample, and the duties of the controller) and tools, even if the null hypotheses is that there is no difference. Thus $F_{v,c}$ is also a function of organisation, $F_{v,c,o}$.

Figure 5 shows 12 suchdistributions:'PD1+' is the trial, 'Org0' is the organisation, '10T' is the controller role, 'a' and 'e' are two individual controllers. Traffic volumes have been aggregated, so '9' means the controller was responsible for 7, 8 or 9 aircraft at the time of the ISA response. The lines shown are for ease of reading only, the distributions are, of course, discrete. The differences between controllers are immediately apparent: for *e* the distributions at traffic volumes 9 and 12 are similar; for controller *a*, 9 and 12 are quite different, but 9 is similar to 6.

Each of the distributions is based on relatively few responses: the mean number of data points used is 13, but two are based on a single datum.

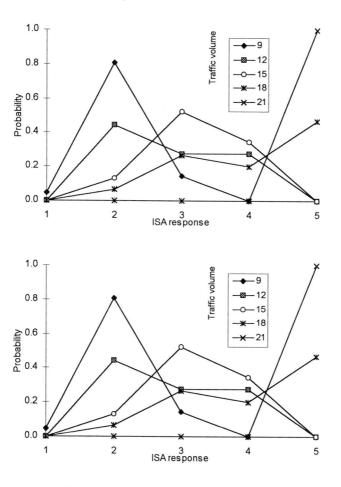

Figure 6. Example ISA response distributions.

(Controllers a and e in role 10T in PD1+ trial Org0)

Under some simplifying assumptions and aggregation, there could be in the order of 200 distributions to estimate. Even given these distributions, the observed and theoretical frequency tables are littered with 0s, so that chi-squared comparison, for testing equation 1, requires manual aggregation of each table. Even then there may be significant numbers of low frequencies, so the test may be of dubious reliability. No estimates can be made before the trial. Even if it were practical to make these estimates after the trial, each sequence of responses would have to be considered separately and a separate decision made about how much of that data could be included. A theoretical solution to the question of independence seems unworkable in practice.

To make any further progress, it is necessary to consider ISA in more detail. 'ISA' stands for instantaneous workload assessment: a response of 2 means the controller feels under-used; 3 means busy, but at a sustainable level; 4 means busy enough for non-essential tasks needing to be put off for the moment, this is not sustainable; 5 means overloaded. The range of ISA is small, and in practice, the upper end of the scale is rarely used (see Figure 3). The lack of variation might, then, be due to ISA's being a coarse-grained measurement, rather than dependence. What other evidence is there?

Figure 7 shows the output from a detailed workload assessment model called 'PUMA'. PUMA looks at the overlapping timings of the overt actions (such as making a phone call) and covert activities (such as thinking of a solution to a potential conflict between aircraft) that together make up the controller's tasks. The underlying model for calculating workload is not relevant here. What Figure 7 shows is the workload for an individual task, which itself lasts only seconds; this workload is changing with a frequency in the order of seconds, ie much more rapidly than the 2 minute interval for ISA sampling.

In other words, controllers' tasks take seconds rather than minutes: if the controller is solving a potential conflict between two aircraft at one ISA sample time, two minutes later the activities will concern a different problem for an essentially different group of aircraft. For this reason, it is assumed that the samples are independent, even if auto-correlated, and all of the data are used.

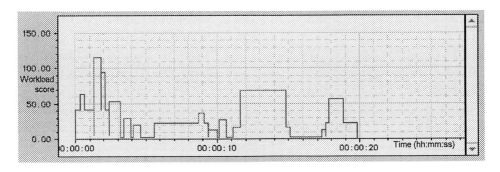

Figure 7. Results from a detailed workload model

The second example is of aircraft spacing on final approach to an airport. The required minimum spacing depends on the relative weights of the pairs of aircraft and the meteorological conditions, so we take deviation from that standard spacing as the measurement for all pairs of aircraft. This is illustrated in Figure 8: as the first aircraft in a pair reaches a 4 nautical mile threshold from the runway, the difference between the distance to the next aircraft and the standard spacing for that ordered pair of aircraft, Δ, is calculated.

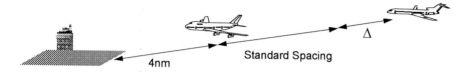

Figure 8. Measuring spacing on approach

In practice, this argument neglects feed-back: the controller can see what is happening to $\Delta(a, b)$ and adjust $\Delta(b, c)$ by delaying or advancing instructions to c. However, examining real-life data before the trial suggested there was still correlation between one measurement and the next, but that the correlation over larger intervals was not significant. The trial and analysis were therefore planned on the assumption that only alternate values from each sequence could be used in the comparison. This in turn determined our expectation of how large a difference the trial might be able to detect and thus the length of trial required.

As it turned out, the actual measurements from the trial were not significantly auto-correlated. So the analysis of the trial results was more powerful than expected. In effect, more trial runs had been devoted to some of the measurements than really necessary.

5. CONCLUSIONS

Real-time trials are the first opportunity in the development of new air traffic control tools or procedures systematically to gather objective measurements in parallel with controller comments. Statistical analysis of these results plays an important part in forming impartial and objective assessments of the value of such developments. A theory of brief time series, especially of non-parametric data, might help make the trials more cost-effective. In its absence, robust results are based on the systematic assessment of all available data sources rather than the statistics of any one measurement.

REFERENCES

Cox, D.R. and P.A.W. Lewis (1966). *The Statistical Analysis of Series of Events*. Methuen, London.

Kendall, M. and J. K. Ord (1990). *Time Series*. Edward Arnold, Sevenoaks.

Siegel, S. and N.J. Castellan (1988). *Nonparametric Statistics for the Behavioural Sciences*. McGraw-Hill, New York.

PLANNING AIRCRAFT MOVEMENTS IN AIRPORTS WITH CONSTRAINT SATISFACTION

Henk H. Hesselink*, Stéphane Paul**,
*National Aerospace Laboratory, NLR, Amsterdam, The Netherlands;
**Alcatel ISR, Massy, France

ABSTRACT

Currently, the European air transport system is experiencing an annual growth of 7%. With an increasing number of flights, airports are reaching their capacity limits and are becoming a bottleneck in the system. Mantea is a European Commission funded project dealing with this issue. This paper focuses on planning decision support tools for airport traffic controllers.

The objective of our planning tools is to achieve a better use of the available airport infrastructure (taxiways and runways). To generate a *safe* plan, many rules must be taken into account that restrict the usage of airport tarmac: international regulations, airport operational procedures, aircraft performance, weather conditions and sometimes even controller "usual practices". To generate a *realistic* plan, extensive monitoring of the traffic situation as well as suitable timing must be achieved. In the life cycle of a flight, 11 out of 15 possible causes of delay occur in an interval of 10-20 minutes, between aircraft start-up request and push-back. This means that precise planning before the end of this period is highly improbable. On the other hand, planning after this period implies the need for fast responses from the system.

In the Mantea project, an architecture is proposed in which a co-operative approach is taken towards planning aircraft movements at the airport. Controllers will be supported by planning tools that help assigning routes and departure times to controlled vehicles, in planning runway allocation (departure sequence) and occupancies, and in monitoring plan progress during flight phases. The planning horizon relates to medium term operations, i.e. 2-20 minutes ahead. The Mantea planning tools implement the following functions: *runway departure planning, routing*, and *plan conformance monitoring*. The tools will reduce the controller's workload, increase the level of safety for airport surface movements, and reduce the number of delays and operating costs for the airliners.

In this paper, we will focus on the constraint satisfaction programming techniques used in Mantea for (1) runway departure planning, (2) itinerary search and taxi planning functions.

The airport tarmac and runway vicinity air routes have been modelled as a graph. Real time constraints have brought us to develop an algorithm linear in complexity for the itinerary search problem. Operational pressure has led us to develop fast search strategies for scheduling (i.e. use of heuristics, hill climbing…).

1   INTRODUCTION AND ARCHITECTURE

With an increasing number of flights, the European air transport system is currently experiencing an annual growth of 7%, the airspace and airports are becoming more and more congested. Figures show that in June 1997 (the busiest month of the year), on average 4,500 flights per day experienced delays, compared with under 3,750 in the same month in 1996 (Jane's Airport Review).

In this paper, we focus on the planning of taxi routes and on departure runway sequencing. The latter includes the allocation of air routes to departing aircraft: SIDs (Standard Instrument Departures). Constraint satisfaction programming techniques are used for:

- the runway departure sequencing and SID allocation function,
- the itinerary search problem,
- the taxi route planning function.

The current chapter shows an architecture for airport surface movement functions in which a co-operative approach is taken towards planning aircraft movements. Chapter 2 describes the runway planning and the taxi route planning problems.

1.1 Congestion at Airports

The airport infrastructure has been identified as becoming a major bottleneck in the Air Traffic Control (ATC) system. The construction of new airports or the expansion of existing ones, however, is extremely expensive and has a strong impact on the European environment. Aviation authorities are seeking methods to increase the airport capacity with the existing infrastructure (in all weather conditions), while at least maintaining the current level of safety. This paper is based on work that was carried out in the Mantea (MANagement of surface Traffic in European Airports) project, which was partly funded by the European Commission.

In this paper, we propose a planning function to support traffic controllers at airports. The planning function provides a decision support tool to reduce the controller's workload, make better use of the available airport infrastructure, reduce the number of delays, and achieve an optimal throughput even under bad weather conditions (viz. low visibility).

1.2 Aircraft Planning

Controllers in airport control towers are responsible for the management of surface traffic. Inbound aircraft are handed over from the arrival controller (often on another location); the sequence and arrival times of inbound aircraft cannot as such be planned in the tower. The tower receives the arrival information a few minutes before the aircraft lands, so that a taxi plan can be made from the arrival runway exit point to the terminal gate.

Departure plans are made in the control tower. Start-up times, taxi plans, and runway plans can be made in advance. Each departing aircraft is assigned a time slot, which is a co-ordinated time interval of about 15 minutes in which the aircraft has to take off. This co-ordination is done with the CFMU (Central Flow Management Unit) in Brussels before the flight starts; the CFMU planning aims at obtaining a constant traffic flow through all sectors in Europe. For the airport controllers, this CFMU restriction ensures that the feeders (i.e. the points where controllers hand over the flight to the next one) are not overloaded.

The planning process is complicated by the fact that even under normal operating conditions, at least three different controllers handle the aircraft over the airfield. Under stress situations, more controllers may be assigned to handle airport traffic. Several controllers act consecutively on each plan. The plans must be established some 2-20 minutes before a flight comes under executive control.

The decision support tool that we propose overcomes most of the problems listed above. We propose a co-operative approach where the planning process is initiated by the runway sequencer, because the runway is usually the scarcer resource. Arrival plans are generated forwards in time towards the assigned gates; departure plans are generated backwards through time, ending with the establishment of a start-up time. To ensure plan achievement during the execution phase, monitoring and re-planning functions need to be part of the proposed architecture.

1.3 Airport Functional Architecture

An airport, or A-SMGCS (Advanced Surface Movement Guidance and Control System), is a complex system involving surveillance, monitoring, planning/routing, and guidance functions (see figure 1 and reference AWOP A-SMGCS).

On an airport tarmac, aircraft and other vehicles are moving independently. These mobiles are sensed by a Surface Movement Radar (SMR) or some other type of sensor. A picture of the traffic situation is built and shown to the airport controller. To support controllers in their job, they are also provided with monitoring tools, which warn them about potential accidents. The controller uses a radio or more sophisticated means (such as the lighting system) to communicate with pilots and guide the aircraft to their destination point on the airport. Coherent guidance instruction is made possible through the use of planning tools, ranging from paper flight plan strips to computer scheduled ground routes and departure sequences.

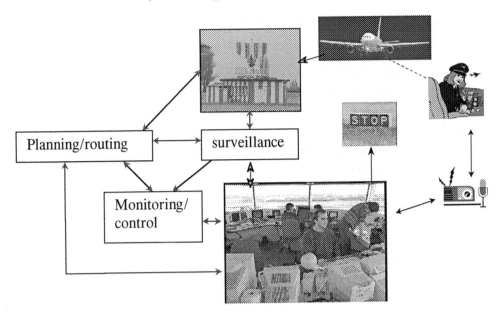

Figure 1: Typical airport functional architecture

1.4 Constraint Satisfaction for Airport Planning

Planning and scheduling have had a long relationship with constraint representation and constraint-based reasoning. A constraint is a formalisation of a restriction to the planning process. Constraints usually specify relationships between different plans and specify how scarce resources can be used. The separation rules that are applicable in air traffic control can best be regarded as restrictions or constraints. Therefore, constraint satisfaction seems an appropriate technique for solving the runway and taxiway planning problems.

The airport tarmac and runway vicinity air routes have been modelled as a graph. On the ground, the links are bi-directional, although preferred routes can be modelled. In the air, the graph is a directed graph.

2 PLANNING

Two planning functions are defined in the architecture: the runway departure planner and taxiway routing management.

2.1 Runway Departure Planning

The objective of the runway planning function is to establish an optimal sequence in which aircraft can depart from the available runways and to plan their initial (climb) flight phase: the SID (Standard Instrument Departure route). Constraints specify that each aircraft takes off within its allocated time slot, that separation criteria between aircraft (both on the ground and in the air) are observed, and that the feeders towards adjacent control sectors are not overloaded.

Standard rules for separation of consecutive aircraft prescribe a two minutes interval. Specific rules exist for consecutive aircraft in different categories. Furthermore for planning SIDs, aircraft on the same air route need to comply with air separation criteria, which are usually larger than on the ground.

In this paper, a decision support function for the runway controller is proposed that may be divided into three sub-functions:
1. Runway allocation. Determine which runway is best to use in the given circumstances (e.g. wind direction and force, aircraft type, distance from gate to runway, etc.).
2. Sequencing. Determine an optimal sequence for departing aircraft, obeying all constraints that apply for one specific airport.
3. Multiple line-up position allocation. Determine which line up position is most suitable for a given aircraft in the given circumstances. For efficiency, aircraft may use an intersection to enter the runway, so that not the whole runway has to be used for the takeoff run.

2.1.1 Constraints

To specify the runway sequencing problem in constraint satisfaction, we need to define variables $V = \{v_1, ..., v_n\}$, their associated domains $D = \{D_1, ..., D_m\}$, and a number of constraints $C = \{C_1, ..., C_k\}$ that restrict certain combinations of values for the variables. The underlying problem translates into the following variables and domains[1]:

$$V = \{R, R_{entry}, t_{takeoff}, SID, F\}$$

$$R \; 0 \; \{25, \; 16R, \; 34L, \; 16L, \; 34R, \; 07)$$

$$R_{entry} \; 0 \; \{Holding \; point, \; A, \; B\}$$

$$t_{takeoff} \; 0 \; \{00.00h \; .. \; 23.59h\}$$

$$SID \; 0 \; \{ELBA \; 5A, \; ELBA \; 5B, \; ELBA \; 5C, \; BASTIA \; 5A \; ...\}$$

$$F \; 0 \; \{AZ123, \; KLM456, \; BA789, \; ...\}$$

where

R is the runway, typically a number and an optional 'L' or 'R'.

R_{entry} is the entry position to a runway, each runway has at least a holding point (the very beginning of the runway) and possibly one or two additional entry points.

$t_{takeoff}$ is the assigned time of takeoff, a one minute interval has been chosen.

SID is the assigned departure route (Standard Instrument Departure).

F are all flights that need to be scheduled during a given period.

---

[1] The example is from Rome Fiumicino airport, one of the validation sites of Mantea.

We have defined five categories of constraints:

- Separation constraints. These concern restrictions on the departure of aircraft at the same runway because of preceding aircraft that may be too close.
- Runway usage constraints. These determine the runway number that will be used, based on the necessary runway length, meteorological conditions, runway surface condition, and runway equipment.
- Line-up constraints. These concern the possibility of lining up other than at the runway holding point and special operations that may be used under good visibility conditions.
- TMA exit point and SID constraints. Separation in the air must be guaranteed and the feeders to the following control sectors must not be overloaded.
- Sequencing and timing constraints. These specify that each aircraft must take off within its time slot and give specific constraints for ordering.

One constraint defines that aircraft in lighter weight categories should be scheduled at least three minutes after a preceding one (separation constraint to avoid wake turbulence effects):

$$\forall F1,\ \forall F2,\ \text{where } F1 \neq F2$$
$$\neg\ (R(F1) = R(F2))$$
$$\omega\ (t_{takeoff}(F1) > t_{takeoff}(F2))$$
$$\omega\ (w(F1) <= w(F2))$$
$$\omega\ (t_{takeoff}(F1) + 3 <= t_{takeoff}(F2))$$

where

$F1$ and $F2$ are flights to be scheduled,
R is a function that provides the allocated runway,
w is a function that provides the aircraft weight category,
$t_{takeoff}$ is a function that provides the takeoff time.

This constraint defines the situation where the aircraft of flight one is heavier than that of flight two and then specifies the four conditions that should not apply (otherwise it *is* allowed to schedule flight one before number two, e.g. when they are on different runways).

SIDs are defined as of a number of waypoints, leading to the feeder to the next sector. SIDs may overlap or cross. For the scheduling of aircraft at waypoints, similar constraints apply as for runway separation, just like for prevention of overloading the feeders to the next sectors. These constraints can be regarded as separation constraints between two following aircraft. Then, we need to specify the relation between the waypoints in the SIDs:

$$\forall P1,\ \forall P2,\ \text{where } P1 \neq P2$$
$$\neg\ (sameSID(P1, P2))$$
$$\omega\ \neg\ (follows(P2, P1))$$
$$\omega\ (t_{over}(P1) + flying\ time(P1, P2) = t_{over}(P2))$$

where

P1 and P2 are points to be overflown,

sameSID is a function that checks if two points belong to one SID,

follows is a function that checks if two points follow each other,

t_{over} is a function that provides the time the flight passes a waypoint.

2.1.2 Algorithm

There are several possible solutions for scheduling a number of departing aircraft at an airport. Constraint satisfaction normally finds just *one* solution to a specified problem. For the runway sequencing function we want to find the optimal sequence out of all possible solutions. In our algorithm, once a complete schedule is found, it will be evaluated against some predefined cost function, after which hill climbing and branch and bound techniques ensure an efficient search process.

2.2 Taxiway Plan Management

Once a runway sequence is proposed, the taxiway planning function needs to get the aircraft from the gate to the runway in time. The main functions of the taxiway plan management function are:

- routing (i.e. the search and choice of a taxi route),
- computation of the approximate time required by an aircraft to go from one point to another point on the airport tarmac,
- scheduling of aircraft taxi movements resulting in taxi plans.

For all of the above functions, an airport tarmac model is required so that routing can be established.

2.2.1 Airport Tarmac Model

Basically, the tarmac is represented as a graph with nodes and edges. However, to model the topology, the edges are grouped in logical areas corresponding to a segmentation dependent on the airfield lighting system. Each geographical way (i.e. taxiway, runway) is cut up in logical segments (called logical areas), see figure 2.

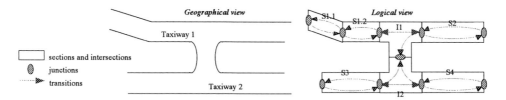

Figure 2: Airport models

The logical areas are terminated by nodes. Each node refers to two logical areas and makes the link between them. Edges are used to go from a node to another one on the same logical area. One edge per direction of traffic is built with operational properties, e.g. a transition is allowed when the traffic regulation authorises to go in this direction; all transitions are maintained since regulations may change dynamically during operation.

2.2.2 Routing

Building a taxi route is basically finding a path between two points of the tarmac (i.e. between two nodes in the graph). The search algorithm we have developed makes intensive use of constraint propagation without needing any generation. Therefore, there will be no backtracking and a search process linear in time can be guaranteed.

A constrained variable called D is associated to each node. D, constrained to be a positive integer, is used to carry the distance of the current node to the route start point. The word "distance" does not necessarily corresponds to the Euclidean distance between intersections but to some cost function that depends on aircraft characteristics and configuration of the taxiways (angles, maximum speed on taxiway…). In figure 3, the constrained variables definition domains are given between square brackets.

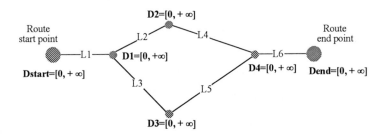

Figure 3: Constrained variables defined for route search

The constrained variable D is set to 0 at the route start point. The following constraints are posted on all other variables: for all adjacent nodes, the constrained variable D on the current node is lower or equal to the sum of the constrained variable D of the adjacent node and the edge length (see figure 4).

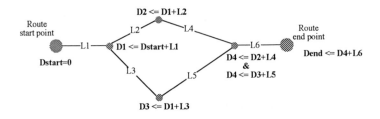

Figure 4: Constraints posted for route search

The above constraints mean that all constrained variables D are lower or equal to the shortest distance to the route start point (see figure 5).

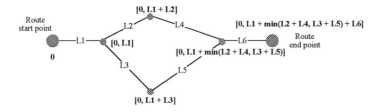

Figure 5: Constrained variables domain definition after constraint posting

When the constrained variable of the route end point is set to its maximum value, constraint propagation will bind all variables on the shortest route. The shortest route will therefore be defined by the set of nodes whose constrained variables are bound (see figure 6).

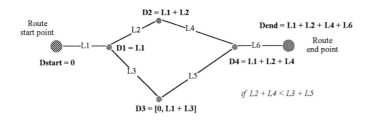

Figure 6: Constrained variables after instantiation of Dend

2.2.3 Scheduling of Aircraft Taxi Movements

After a route assignment for each aircraft, the different aircraft need to be scheduled in time. This scheduling of aircraft taxi movements on the airport tarmac is basically a resource allocation problem. Each edge of the graph is considered as a resource whose capacity is equal to its length. Each aircraft movement on a logical area is considered as a set of two activities: the taxi activity and the reserve activity.

The taxi activity uses the resource corresponding to the edge it is travelling on during the total travel duration. The required capacity is equal to the aircraft length plus some safety margin. The reserve activity uses all the other resources of the logical area covered, at full capacity, during a time sufficient to guaranty the following aircraft from jet blast effects. One side effect of the reserve activity is to prohibit another aircraft entering the logical area in the opposite direction.

To ensure a smooth aircraft movement, all taxi activities on the same route are linked together by time constraints, i.e. taxiActivity(Edge$_{i+1}$) StartsAtEnd taxiActivity(Edge$_i$).

Aircraft cannot overtake each other on a taxiway. This is ensured by a logical operator between constraints specifying that if a taxi activity on an edge starts before another, then it must also end before.

The goal which is set, is made up of two parts: the inbound aircraft travel activities are required to start as soon as possible (because a landing aircraft cannot stop and wait to taxi) and the outbound aircraft travel activities are required to start as late as possible (because we want to bring the aircraft at the runway, just in time for takeoff).

The cost function associated with the scheduling is currently to limit the global taxi duration. It can be easily changed to favour inbound or outbound aircraft depending on the airport regulations and practices.

3   CONCLUSION

In this paper, constraint satisfaction has been applied to a route search and scheduling problem. It is shown that the technique can well be applied to establishing optimal taxi routes and runway sequences for airport planning.

The functions we propose are designed as decision aids to traffic controllers in the control tower. We show an A-SMGCS airport architecture in which a co-operative approach is taken towards the planning of aircraft movements at the airport, and where several controllers act on the same plan. Plans are established 2-20 minutes before an aircraft comes under control. The main components of the planner are a runway sequencing function and a taxiway plan management function.

The objective of the runway departure planning function is to establish an optimal sequence in which aircraft can depart from the available runways and to plan their initial climb flight phase. Constraints are formalised to specify standard separation and additional separation rules, i.e. relations between attributes of the flights. A constraint satisfaction algorithm is extended with hill climbing and branch and bound techniques in order to find the optimal solution efficiently.

Taxi plan management comprises the route search for shortest routes and the assignment of aircraft movements to taxi lanes (resources), which are modelled in a graph. Constraints specify the aircraft's position at the resource and the ordering of aircraft through time.

4   REFERENCES

An. (1996). Proposed Document for Advanced Surface Movement Guidance & Control Systems (A-SMGCS), draft_02, 3rd AWOP A-SMGCS SG-meeting, Frankfurt.

An. (1997). Europe Measures Delays, *Jane's Airport review*, p. 26, December 1997.

Beck, J.C. (1994). A Schema for Constraint Relaxation with Instantiations for Partial Constraint Satisfaction and Schedule Optimization, Masters Thesis, Uni. of Toronto.

Hesselink, H.H. (1997) et. al. Mantea Design, Plan Tools, MANT/ISR-TEC-D5.1-057, Paris.

Markus Stumptner (1997). An Overview of Knowledge-based Configuration, *AI Communications*, Vol.10, pp. 111-125, ISSN 0921-7126.

A GAME THEORETIC MODEL OF ROAD USAGE

Tim James, Department of Economics, University of Sheffield, Sheffield

ABSTRACT

This paper explores the use of the theory of games as a tool for modelling road usage decisions. By explicitly considering the interdependence among road users and their 'payoffs' (utility) from road use, an adapted symmetric form of the 'game of chicken' is developed. The paper explores the various possible equilibria in this game. The mixed strategy symmetric Nash equilibrium is derived and a probability of road usage by a single user is calculated from this. The latter depends explicitly upon the generalised costs of road usage that include both the monetary and time costs of travel. From the single road user probability of usage, measures of the expected number of users are derived. An examination is made of how these vary with variations in the level of the generalised cost of usage. This provides an insight into how the tools of game theory may aid the understanding of the generation and regulation of road congestion.

1. INTRODUCTION

The transport economics and planning literature contains a myriad of approaches that have been used to model the demand for road usage (see, for example, Ort ʃzar and Willumsen, 1990 and Button, 1993). However there seems to be no explicit analysis of the interdependence among individual road users in this literature. This paper seeks to fill this gap by exploring the use of the theory of games as a tool for examining the demand for road usage in the context of interdependent agents.

By explicitly modelling the interdependence among road users and their 'payoffs' from road use, an adapted symmetric form of the 'game of chicken' is developed. The paper explores the various possible equilibria in this game. The mixed strategy symmetric Nash equilibrium is selected and a probability of road usage by a single user is derived. The latter depends explicitly upon the generalised costs of road usage that include both the monetary and time costs of travel. From the single road user probability of usage measures of the expected number of users are derived. An examination of how these vary with variations in the level of the generalised cost of usage. This provides an insight into how the tools of game theory may aid the understanding of the generation and regulation of congestion.

2. INTERDEPENDENCE IN PRIVATE ROAD TRAVEL: AN ADAPTED 'GAME OF CHICKEN'

A scenario in which N mutually aware agents face a one-time simultaneous decision about whether or not to travel independently using private transport using a particular stretch of road is envisioned. Travel has associated with it utility that *may* be positive whilst the alternative of not travelling has an assigned utility of zero. The simultaneous nature of the decision-making means that agents make their individual choices about travel in ignorance of all other like decisions. The two choices open to the ith $(i = 1,...,N)$ agent can be represented as:

$$a_i = \begin{cases} 1 \\ 0 \end{cases} \tag{2.1}$$

where $a_i = 1$ is associated with the choice of private road travel and $a_i = 0$ is the choice not to travel. A decision vector for the agents as can thus be defined as:

$$\boldsymbol{a} = (a_1,...,a_i,...,a_N) \tag{2.2}$$

It is also possible to summarise the N individual agents' decisions with a scalar that gives the number of them that decide to travel:

$$\boldsymbol{a} = \sum_{i=1}^{N} a_i \tag{2.3}$$

In effect (2.2) is just a more extensive expression of the scalar in (2.3). In the discussions that follow use of both of these two definitions is made as necessary without loss of generality.

All agents are assumed to be rational and aware of the number of other agents making similar decisions to them.

It is assumed that the individual utility associated with travel using private transport, u_i, is determined by three variables:

- the private monetary cost of private transport use for the journey in question;
- the price of road usage;
- and, journey time which depends directly upon the number of other road users.

Obviously it is possible to consider the inclusion of other factors which affect the utility associated from the use of private transport by road. The assumption of three defining variables is merely an illustrative abstraction.

The general functional relationship between these variables and an individual's utility from making the journey is:

$$u_i = u_i(t(\mathbf{a}), r, c)$$

where $t(a)$ is the time associated with travel with $a - 1$ other road users, r is the price of road usage and c is the private monetary cost of private transport use. It is logical to expect u_i to be decreasing in each of $t(a)$, r and c. Further, it is also logical to assume that $t(a)$ is increasing in a. It is of course perfectly feasible to consider money metric measures of individual utility rather than the abstract form here if information on the effects of congestion (here in the form of rising a) on journey time and the monetary value of road users' time are available.

The payoff to the ith agent from its strategy choice, $u_i = u_i(t(\mathbf{a}), r, c)$, is determined not only by its own individual choice but also those of all $N - 1$ other agents. If $a_i = 0$ then it is assumed $u_i = u_i(t(\mathbf{a}), r, c) = 0$. If however the agent in question decides to travel then the payoff it receives declines as the number of other agents deciding upon travel increases. That is to say, $u_i(t(\mathbf{a}'), r, c) > u_i(t(\mathbf{a}''), r, c)$ for $a' < a''$. Finally, in relation to the structure of payoffs, it is assumed that there is some level of agent numbers choosing travel which implies negative individual payoffs above that number. Thus $u_i = u_i(t(\mathbf{a}), r, c)$ is separable into $u_i(t(1), r, c), \ldots, u_i(t(a^{max}), r, c) > 0$ and $u_i(t(a^{max} + 1), r, c), \ldots, u_i(t(N), r, c) < 0$ for some a $(N > a^{max} > 1)$ is assumed. The fact that the number of agents who could travel with an associated positive payoff is less than N gives rise to the 'interdependence problem' among these mutually aware agents.

2.1 The Two-Agent Case

The discussion at this juncture is best illustrated by considering the two-agent case. Figure 1 represents the pure strategy choices and payoffs for the corresponding two-agent scenario in matrix form.

Examination of the respective payoffs in this 2x2 matrix reveals that there is no problem of travel interdependence either when private road transport *never* offers positive utility $(0 > u_1(t(1,0), r, c) = u_2(t(0,1), r, c) > u_1(t(1,1), r, c) = u_2(t(1,1), r, c))$ or when it *always* has an associated positive utility $(u_1(t(1,0), r, c) = u_2(t(0,1), r, c) > u_1(t(1,1), r, c) = u_2(t(1,1), r, c) > 0)$. In fact the only interesting case is where the structure of payoffs is such that only one of the two agents can travel with positive utility. That is the case where $u_1(t(1,0), r, c) = u_2(t(0,1), r, c) > 0 > u_1(t(1,1), r, c) = u_2(t(1,1), r, c)$. An alternative statement of this payoff structure is $a^{max} = 1$.

	$a_2 = 0$	$a_2 = 1$
$a_1 = 0$	$u_1(t(0,0),r,c)$ $u_2(t(0,0),r,c)$	$u_1(t(0,1),r,c)$ $u_2(t(0,1),r,c)$
$a_1 = 1$	$u_1(t(1,0),r,c)$ $u_2(t(1,0),r,c)$	$u_1(t(1,1),r,c)$ $u_2(t(1,1),r,c)$

Figure 1. Two Agent Game Matrix

In the circumstance where only one of the two agents may travel with positive utility there is no dominant pure strategy choice for the two agents. To illustrate this argument, consider the best responses of agent 2 to the possible pure strategy choices of agent 1. If agent 1 decides to travel then agent 2's best response is not to travel. However, if agent 1 does not travel then agent 2's best response is to travel. The arguments apply in a symmetric fashion to agent 1's choices. The problem is that both agents make their decisions simultaneously and are thus unaware of the decisions of their rival: no *symmetric* pure strategy Nash equilibrium exists for this particular game. However, there are in fact two *pure* strategy Nash equilibria: $a' = (1,0)$ and $a'' = (0,1)$. These are discussed later but are unappealing because of their asymmetric nature. The search for a symmetric Nash equilibrium solution to this game thus centres on mixed strategies.

2.2 The Two-Agent Mixed Strategy Nash Equilibrium

A mixed strategy for an agent in this game consists of an ordered double (p_{i0}, p_{i1}) where $1 \geq p_{is} \geq 0$ and $p_{i0} + p_{i1} = 1$. p_{is} denotes the probability that the ith agent (here $i = 1,2$) will select the sth pure strategy $(s = 0,1)$. In the context of the particular game we are concerned with p_{i0} is the probability of playing the pure strategy of not travelling and p_{i1} is the complementary probability of playing the pure strategy of travelling. The mixed strategy decision vector for the two-agent case is:

$$p = (p_{10}, p_{20}) \tag{2.4}$$

Allowing the two agents to play mixed strategies means that their goal is now the maximisation of expected payoffs. The expected payoffs to the two agents are:

$$E_1(u_1|p) = (1 - p_{10})\{p_{20}u_1(t(1,0),r,c) + (1 - p_{20})u_1(t(1,1),r,c)\} \tag{2.5}$$

$$E_2(u_2|\mathbf{p}) = (1 - p_{20})\{p_{10}u_2(t(1,0),r,c) + (1 - p_{10})u_2(t(1,1),r,c)\} \qquad (2.6)$$

The conditions for the existence of a symmetric Nash equilibrium in mixed strategies, $\mathbf{p}^* = (p_{10}^*, p_{20}^*)$, are:

$$E_1(u_1|(p_{10}^*, p_{20}^*)) \geq E_1(u_1|(p_{10}, p_{20}^*)) \qquad (2.7)$$

$$E_2(u_2|(p_{10}^*, p_{20}^*)) \geq E_2(u_2|(p_{10}^*, p_{20})) \qquad (2.8)$$

$$p_{10}^* = p_{20}^* \qquad (2.9)$$

In (2.9) we make symmetry of mixed strategy a condition of equilibrium because it has intuitively appealing properties in this context of identical agents. Further justification for the selection and interpretation of the symmetric mixed strategy equilibrium as the appropriate solution to this game can be found in Binmore (1992) and Rubinstein (1991).

Derivation of the symmetric mixed strategy Nash equilibrium for this game follows that of Rasmusen (1989). The relevant first-order conditions that follow from the calculus of maximisation are:

$$p_{20}u_1(t(1,0),r,c) + (1 - p_{20})u_1(t(1,1),r,c) = 0 \qquad (2.10)$$

$$p_{10}u_2(t(1,0),r,c) + (1 - p_{10})u_2(t(1,1),r,c) = 0 \qquad (2.11)$$

Thus the two-agent Nash equilibrium mixed strategy choices to this game are:

$$p_{10}^* = p_{20}^* = 1 - \frac{u_i(t(1,0),r,c)}{u_i(t(1,0),r,c) - u_i(t(1,1),r,c)} \qquad (2.12)$$

In the context of this particular formulation of the game of chicken (2.12) may also be derived by setting $E_i(u_i|(p_{10}, p_{20})) = 0$ (see Rasmusen, 1989). That this constitutes a symmetric Nash equilibrium (satisfying both (2.7) and (2.8)) can be demonstrated by considering agent 1's optimal strategy in the various possible cases. If p_{10}^* and p_{20}^* satisfy (2.12) then from (2.5) it must be true that $E_1(u_1|(\mathbf{p}^*)) = 0$. Suppose that $p_{20} < p_{20}^*$ then $E_1(u_1|(\mathbf{p})) > 0$. Agent 1 then maximises $E_1(u_1|(\mathbf{p}))$ by choosing $p_{10} = 1$. In consequence agent 2's optimal response is to set $p_{20} = 0$. In the case where $p_{20} > p_{20}^*$ $E_1(u_1|(\mathbf{p})) < 0$. Agent 1 then maximises $E_1(u_1|(\mathbf{p}))$ by choosing $p_{10} = 0$ and in consequence agent 2's optimal response is to choose $p_{20} = 1$. Similar arguments follow if the second agent's perspective is considered. However, the resulting Nash equilibria, $p^1 = (1,0)$ and $p^2 = (0,1)$, can be eliminated by the imposition of the symmetry condition in (2.9).

2.3 The N-Agent Mixed Strategy Nash Equilibrium

Generalisation from the two-agent to the N-agent case is straightforward. The mixed strategy decision vector in (2.4) requires extension to:

$$p = (p_{10}, p_{20}, \dots, p_{N0}) \qquad (p_{i1} = 1 - p_{i0}) \qquad (2.13)$$

Omitting the 'i' subscript for notational convenience, we have the following general form for the expected payoff to each of the N agents from the choice of p_1^*:

$$E(u_i | \mathbf{p}^*) = p_1^* \left[\sum_{a=1}^{N} \frac{(N-1)!}{(a-1)!(N-a)!} (p_1^*)^{a-1} (1 - p_1^*)^{N-a} u_i (t(a), r, c) \right] \qquad (2.14)$$

In (2.14) the terms involving the expected payoff to not travelling have been omitted for simplicity since they are equal to zero by assumption. Again the solution that satisfies (2.7)-(2.9) involves solving for a mixed strategy that implies the ith agent's expected payoff is zero. Generally it becomes difficult to write down explicit solutions for p_1^* beyond the three-agent case and so iterative methods must be employed to find (and select from) the roots of the higher-order polynomials generated from setting (2.14) equal to zero.

3. THE PROBABILITY OF ROAD USAGE

3.1 The Individual Probability of Road Usage

p_1^* may be interpreted in a straightforward manner as the probability that an agent will use private road transport. Inspection of an agent's expected payoff function demonstrates that this probability is dependent upon the number of agents involved, N, and the form of the payoff function. Comparative statics are rather difficult given the nature of (2.14) but we can make some general comments as to the effect of changing the nature of the payoff function (holding the number of agents constant) on the individual probability of pursuance. Readers are referred to (2.5) and (2.6) for a simple check on the following propositions.

1. If the positive payoffs to travel increase (decrease) in value ceteris paribus then p_1^* will increase (decrease).
2. If one or more of the negative payoffs associated with travel become(s) more (less) negative ceteris paribus then p_1^* will fall (rise).

3. If the positive payoffs to travel increase (decrease) and the negative payoff(s) to travel fall (rise) then the result is ambiguous and depends upon the relative magnitude of these changes.

If the payoff function to travel is strictly decreasing in a (as would seem reasonable) then p_1^* will be a decreasing function of N. This is because as the number of other agents making the decision to travel increases so does the likelihood of increasingly negative payoffs for each individual agent. The specific form of the relationship between p_1^* and N depends crucially upon the form of the payoff function, $u_i(t(a), r, c)$.

3.2 The Overall Probability of Travel and the Expected Number of Road Users

Having examined the individual probability of travel the overall measures of the probability of travel are affected by varying the numbers of agents is now examined. These overall measures are in fact based on a binomial distribution. The expected number of agents deciding to pursue the prospect is:

$$E(a) = Np_1^*$$
(3.1)

The probability that there will be $\bar{a}$ or more agents choosing to travel is:

$$pr(a \geq \bar{a}) = \sum_{a=\bar{a}}^{N} \frac{N!}{a!(N-a)!} (p_1^*)^a (1 - p_1^*)^{N-a}$$
(3.2)

4. DISCUSSION

4.1 Supply-to-Demand Feedback

One effect that the game theoretic model outlined allows for is that of 'supply creating its own demand'. This is a phenomenon that is discussed (but not explicitly modelled) in Ort´zar and Willumsen (1990), Downs (1992) and Lewis (1994). This is allowed for through the specification of the individual utility function used in the model above. For instance, a road improvement scheme tends to reduce the time costs of travel for all values of the actual number of road users which in turn raises the individual utility associated with travel for any a. The model outlined above predicts a rise in the individual probability of road usage in this circumstance. This implies a rise in the expected number of road users: an increase in demand.

4.2 'Auto-regulation' of Congestion

Another point worth emphasising is that modelling the interdependence among agents making the choice of whether to travel or not explicitly allows for the possibility of the 'auto-regulation' of congestion. This is the argument is that road congestion (whether actual or potential) is to some extent self-regulating. One example of the way in which this may transpire is as follows. As the number of potential road users rises so correspondingly may the number of actual users (thus leading to congestion). However, this potential rise in congestion is tempered by the fall in the individual probability of travel caused by the rise in N (as outlined in the model above). Thus actual levels of congestion will be tempered (to a greater or lesser degree depending upon the form of individual agents' utility functions).

4.3 Road Pricing

The possibilities offered by road pricing in the control of congestion have long been a subject of interest (see, for example, Ministry of Transport, 1964, Walters, 1968, and Vickery, 1969, Morrison, 1986, Hau, 1991, Downs, 1992, Evans, 1992, and Lewis, 1994). Whilst many aspects of road pricing have been examined in this literature there appears to be no analysis of road pricing in a game theoretic framework where the interdependence among road users is explicitly modelled. The model outlined in this paper acts as a basis for the analysis of road pricing effects from an individual agent's perspective.

Road pricing works in the game theoretic model outlined above through its effect upon the individual utility from travel. If the price of road use, r, is raised then the individual utility from travel, $u_i(t(a), r, c)$, falls whatever the actual number of road users, a. This then leads to a fall in an individual agent's probability of travel, p_1^*, for a given total number of potential users, N, and thus the expected number of actual road users in the mixed strategy Nash equilibrium, $E(a)$, will correspondingly fall. If the principal objective of road pricing is to regulate congestion to ensure traffic flow stays at its maximum level, $\bar{a}$, then r should be adjusted for different values of N so that p_1^* is such that $E(a) = \bar{a}$ is maintained. This implies, writing p_1^* as a function of N, that:

$$1 - (1 - p_1^*(N))^{\frac{N}{N+1}} = p_1^*(N-1)$$

must hold for varying values for N.

REFERENCES

Binmore, K. (1992) *Fun and Games*. D.C. Heath and Company: Lexington, MA.

Button, K.J. (1993) *Transport Economics*. Cambridge University Press: Cambridge, England.

Downs, A. (1992) *Stuck in traffic: coping with peak hour traffic congestion*. The Brookings Institution and Lincoln Institute of Land Policy: Washington, D.C..

Evans, A. (1992) "Road congestion pricing: when is it a good policy?" *Journal of Transport Economics and Policy*, 26, 213-43.

Hau, T. (1991) *Economic fundamentals of road pricing – a diagrammatic analysis*. World Bank: Washington, DC.

Lewis, N.C. (1994) *Road Pricing: theory and Practice*. Thomas Telford: London.

Ministry of Transport (1964) *Road pricing – the economic and technical possibilities*. HMSO: London.

Morrsion, S. (1986) "A survey of road pricing." *Transportation Research A*, 20, 87-97.

Ort ∫zar, J. de D. and Willumsen, L.G. (1990) *Modelling Transport*. John Wiley & Sons: Chichester, England.

Rasmusen, E. (1989) *Games and Information*. Blackwell: Oxford, UK.

Rubinstein, A. (1991) "Comments on the interpretation of game theory." *Econometrica*, 59, 909-924.

Vickery, W. (1969) "Congestion theory and transport investment." *American Economic Review (Papers and Proceedings)*, 59, 251-60.

Vorob'ev, N.N. (1977) *Game Theory; Lectures for Economists and Social Scientists*. Springer-Verlag: New York.

Walters, A. (1968) *The economics of road user charges*. World Bank: Washington D.C..

THE APPLICATION OF AN ELASTIC DEMAND EQUILIBRIUM MODEL FOR ASSESSING THE IMPACTS OF URBAN ROAD USER CHARGING

Dave Milne, Institute for Transport Studies, University of Leeds, UK.

ABSTRACT

Road user charging has been proposed as a solution to the uncontrolled growth of traffic and congestion in urban areas. In the absence of evidence from real world applications, modelling techniques provide the best information about the potential impacts and benefits of different charging approaches. This research has employed an elastic demand network equilibrium model, as part of the well established SATURN suite of computer programs, to represent a series of alternative road user charging systems which have been proposed for practical application.

Results have been obtained for both the impacts of charges on the volume and spatial distribution of road travel demand and for aggregate measures of network performance, such as travel distances, times and costs. Some interesting issues have emerged regarding the overall performance of charging systems in comparison with prior expectations and the specific impacts of charges related to travel conditions, which attempt to approximate the economic theories of marginal cost pricing. In addition, doubts are raised regarding the ability of steady-state equilibrium models to provide plausible representations of behavioural responses to charges which may vary significantly in time and space. It is suggested that alternative modelling techniques may provide superior user response predictions.

1.   INTRODUCTION

Most previous modelling work related to road user charging has focused either on very detailed network optimisation issues in a simple, synthetic context or on strategic impacts for transport demand and the economy in a practical environment. This paper reports work carried out at an intermediate level, using a steady-state equilibrium assignment model to provide a detailed representation of route choice and an elastic assignment function to allow travel demand to respond to changes in origin-destination cost.

The research described here focuses on a real-world model application in Cambridge and upon four alternative road user charging systems, which have previously been proposed for implementation. These charging systems are:

(i)   *cordon or point charging*, where fixed fees are levied for passing particular points on the road network;

(ii)   *distance-based charging*, where fees are levied relative to the total distance travelled;

(iii)   *time-based charging*, where payments are related to total travel time; and

(iv)   *delay-based charging*, where payments are made only for time spent in congestion.

The evaluation is in terms of detailed impacts on travel network performance rather than economic benefits. Results presented are for a morning peak situation and include the levels of charge which produce similar global demand impacts from the four systems, a comparison of effects upon the spatial distribution of travel demand and a comparative analysis of network performance based on travel distances, travel times and cost. More detailed coverage of this work is available elsewhere (Milne, 1997).

2.   THEORETICAL BACKGROUND

The justification for road user charging originates from the economic theory of *Marginal Cost Pricing*, under which it is stated that the market for any commodity will only operate efficiently if users pay the full costs of their activities. In road transport, it has long been argued that while users pay their own travel costs in full (the *Marginal Personal Cost*), they contribute nothing towards costs which they impose on others through delays and environmental impacts, encouraging inefficiency by overuse. In a very simple situation, with no spatial or temporal dimensions, it can be shown that achieving full cost coverage (the *Marginal Social Cost*) through a corrective tariff equal to these *External Costs* will reduce travel demand to a level which maximises social benefits, reducing congestion and adverse environmental effects in the process.

It is widely acknowledged that such tariffs could not be imposed precisely in complex reality with existing technologies and that practical applications would need to expand upon the economic and environmental efficiency objective to embrace other goals such as the specific protection of environmentally sensitive areas, the achievement of urban planning targets, maintaining and improving equity and raising revenue (May, 1992). However, the failure of other practical policies to control road travel demand and reduce congestion has prompted considerable interest in road user charging as a travel demand and network management tool. The four systems investigated each involve a trade-off between the extent to which they attempt to replicate marginal cost structures and their simplicity of understanding and implementation. To date practical experience has been biased ⁻ds simplicity, with only cordon or point charges existing in a real urban context (Singapore and three Norwegian cities). Thus, predictive modelling provides the best opportunity to evaluate the technological alternatives.

3. THE MODEL

Modelling work was conducted within the SATURN (Simulation and Assignment of Traffic to Urban Road Networks) computer package, a mathematical model for the analysis and evaluation of traffic patterns on explicit road networks (Van Vliet, 1982). The principal function of the model is to assign traffic (in terms of average flows of *Passenger Car Units*) to routes based on the concept of *Wardrop User Equilibrium*, under which it is assumed that all drivers attempt to minimise their travel cost individually and distribute themselves amongst routes on the network such that:

(i)   no individual can reduce travel cost by changing route;

(ii)   the average cost on all used routes between any two given points is the same; and

(iii)   the average cost on any unused route is greater.

Travel cost is represented by economic *generalised cost*, based on a weighted combination of time and operating cost (approximated through distance), in this case 7.63 pence per minute and 5.27 pence per kilometre for a single average user class. The contribution of travel time to generalised cost is calculated using *speed-flow relationships*, which are associated with each road link and junction turning movement of the modelled network. The model is temporally static, so that all trips experience average conditions, and equilibrium is considered to represent a long-run average situation rather than any particular day.

In the simplest case demand for road travel is fixed. However, where the effects of a policy on generalised costs would be likely to result in significant changes in the pattern of trips, such as under road user charging, demand response may be included through an *Elastic User Equilibrium* assignment procedure. This varies the number of trips using the road network for each origin-destination movement based on changes in total generalised travel cost compared with a reference situation, where the costs and corresponding demand level are known. Thus, if the cost of a particular movement increases fewer people will travel and demand will reduce.

Equally, if total travel cost falls, additional trips will be generated. For this study the relationship between demand and cost was defined by the following constant elasticity function:

$$T_{ij} = T_{ij}^{0}(c_{ij} / c_{ij}^{0})^{(-E)}$$

where:

T_{ij}	=	road travel demand after elastic assignment
T_{ij}^{0}	=	road travel demand in the reference case
c_{ij}	=	road travel costs in the test case
c_{ij}^{0}	=	road travel costs in the reference case
E	=	elasticity parameter

The results reported here have used a single generalised cost elasticity value of -0.5, which was found to be consistent with literature both for elasticity with respect to out of pocket costs (Goodwin, 1992) and for values derived from elasticity with respect to travel time (The Highways Agency et al, 1997).

Charges have been represented in the model as additions to generalised costs in the assignment procedure. Thus, cordon charges were applied as cost penalties on those links which formed part of charging cordons or screenlines, while the other three systems required factors to be applied to the relevant elements of the generalised cost calculation for all road links and turning movements within the charged area. For cordon, distance and time-based charging, the model provides an accurate reflection of charges which would be applied using practical technology. For delay-based charging, the definition of delay within the model (all travel time in excess of free-flow) is rather more rigorous than that possible with current technology (Oldridge, 1990), but may be closer to the concepts of marginal cost pricing. In all cases it was assumed that road users would respond to direct charges in equal weight to the costs of their time and vehicle operation, as no superior information was available on this issue.

4. RESULTS

Figure 1 shows the Cambridge urban road network and illustrates the spatial manner in which charges were applied. For cordon charging, three concentric cordons were defined, as shown, to produce an increasing concentration of charges on the approach to the more congested central areas of the city. The two inner cordons fall immediately either side of a well used orbital route around the centre, while the outer cordon was placed around the perimeter of the developed urban area, immediately inside a ring of orbital bypasses. Six screenlines were also employed between the two outer cordons to separate the main radial routes and discourage orbital rerouteing through residential suburbs. A uniform charge was imposed for all cordons and screenlines in any direction. For distance, time and delay-based charging, charges were imposed at a uniform rate throughout the developed urban area, within the outer cordon.

As levels of charge are expressed in very different units, comparison of the four systems requires the definition of a common base against which performance may be assessed. During this study two particularly useful bases have emerged. First, the level of trip suppression has allowed spatial and route choice issues to be examined independent of total travel demand. Charge levels were identified which reduced overall trip-making by up to 15%. Given that charging was confined to the main urban area, the highest level of suppression is approximately equivalent to trip-making reductions of 24% advocated on environmental grounds, once unaffected journeys are excluded (Royal Commission on Environmental Pollution, 1994). Second, the average journey charge, assuming no behavioural response affecting either travel demand or routes chosen, provides a measure of financial incentives for change from the viewpoint of individual users. Total revenue also provides an

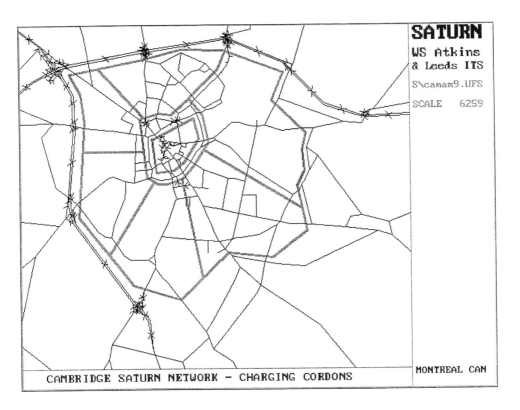

Figure 1: Cambridge urban road network and location of charges

interesting comparator, as it relates to some of the most important practical economic decisions required to implement charging, including ability to cover operating costs, funds available for other projects and calculation of economic benefits. However, in the Cambridge network revenues have been found to correlate very closely with trip suppression.

Table 1 shows the unit charge levels required under the four systems to achieve trip suppression of 5, 10 and 15 per cent.

Table 1: Charge levels required to produce given reductions in total trips

Trip Suppression	5%	10%	15%
CORDON (pence per crossing)	21	45	90
DISTANCE (pence per kilometre)	10	20	37
TIME (pence per minute)	5	11	19
DELAY (pence per minute delay)	60	200	500

Cordon, distance and time-based charging all require an approximate doubling of the unit rates which produce 5% trip suppression to achieve 10% and an increase of around four times to achieve 15%. For delay-based charging, the equivalent factors are three and eight. This shows that all systems allow for significant avoidance of charges through route choice, compared with the shape of the elastic demand function, and that this effect is naturally greatest for delay-based charges because delays tend towards zero as increasing numbers of trips are suppressed. The resulting unit rates of delay-based charge required to suppress total trips by 10% or more are extremely high, suggesting that the use of this approach for reducing road travel demand may be both practically and politically inappropriate. However, the model assumption that drivers know the costs on all feasible routes may allow greater avoidance of delay-based charges within a steady-state equilibrium context than would be expected in reality.

Table 2 shows the impacts of charges which result in 15% overall trip suppression on the distribution of road travel demand.

Table 2: Percentage trip Suppression by area (for 15% reduction overall)

Cordon From/To	1	2	3	Distance From/To	1	2	3
1	9	31	26	1	24	25	19
2	33	20	19	2	25	33	16
3	24	17	2	3	19	15	2
Time From/To	1	2	3	Delay From/To	1	2	3
1	34	31	17	1	28	23	11
2	30	32	13	2	30	28	9
3	16	12	2	3	17	14	2

KEY: **1** : city centre; **2** : remainder of urban charge area; **3** : external areas

In all cases wholly external trips (3-3) are largely unaffected, suggesting that they experience little relative increase in cost due to avoiding the charged area or encountering additional traffic diverted to the bypasses. Trips internal to the city centre (1-1) and within the remainder of the urban charge area (2-2) are reduced to a smaller extent under cordon charging than under the other systems, because many of these journeys can be made without crossing a cordon or screenline. Indeed, some individual origin-destination movements increase due to reduced congestion. By contrast, time and delay-based charges have the greatest impact on trips internal to the city centre, due to their propensity to levy higher charges in congested conditions. Cordon charges also have a strikingly larger impact on longer journeys between the city centre and external areas, because these drivers have no way of avoiding the concentric cordons. However, perhaps the most interesting comparison between systems is by direction of travel. For those movements which are directionally opposed (2-1 & 1-2; 3-1 & 1-3; 3-2 & 2-3), only delay-based charging results in greater trip suppression for the more congested inbound peak direction. This lends support to one of the original arguments in favour of the system: that it may be more equitable than other alternatives by relating the impact more closely to travel conditions encountered (Oldridge, 1990). For cordon, distance and time-based charging, the unavoidable elements of the fees become the dominant component of generalised cost as charge levels rise, causing little variation in impacts related to travel conditions.

Figures 2a-e provide a brief summary of results from a detailed analysis of the impacts of the four charging systems on network performance and travel conditions.

Figures 2a and 2b, show trends in total travel distance within the charge area and on the outer orbital bypasses. They provides the best measure of changes in overall traffic levels and may be one of the most important indicators from an environmental viewpoint. The plots show a significant transfer of traffic from within the charged area to the outer orbital for all systems, although the scale of effect is much lower in the case of delay-based charging. For cordon, distance and time-based charging, the percentage decreases in travelling within the charged area are a little less than double the percentages increases on the orbital: for delay-based charging the levels are similar. As the absolute travel distance total on the Cambridge outer orbital in the current situation is approximately 25% greater than that within the charge area, this implies that delay-based charging causes an increase in total distance travelled over the full network despite trip suppression effects. A detailed analysis of routes chosen within the assignment has revealed that delay-based charging causes very widespread rerouteing within the charge area to avoid congestion and reduce fees paid, involving a significant transfer of traffic from major to minor routes and an increase in the total number of routes chosen by individual origin-destination movements to barely plausible levels. This supports concerns raised earlier that an equilibrium approach may be unable to provide a faithful representation of real behavioural responses to delay-based charges. Nevertheless, model results of this nature provide a useful illustration of the behavioural processes that travel condition dependent charges may encourage. As would be expected, distance-based charging causes the greatest impact on in traffic levels because it charges directly for the measure under assessment.

Figures 2c and 2d show trends in total travel time, which are affected by trip suppression, rerouteing and congestion levels. Comparison of these plots with those for travel distance can yield important information about network effects, separately for the charge area and the outer orbital. In particular, it is possible to infer impacts on average network speeds while maintaining an understanding of the parameters used to generate them. Thus, comparing Figures 2a and 2c, it is clear that time and delay-based charging produce the greatest increases in speeds within the charge area, but it is also possible to note that in the latter case this may be due as much to rerouteing in favour of longer, naturally faster routes as it is to congestion relief in the central area. A more detailed analysis of travel times has show that the average time taken increases under delay-based charging across all geographical journey types (using a disaggregation similar to that adopted for Table 2), as a direct result of the increase in distance travelled apparent in Figures 2a and 2b. Comparison of Figures 2b and 2d shows that all four systems cause average network speed to fall on the outer orbital, with increases in traffic causing negative effects on travel conditions. A detailed investigation has revealed that this is primarily related to increased delays at junctions between the orbital bypasses and the major radial approaches of the city. The fact that these intersections are grade separated on the main bypasses to the north and west explains the negligible impacts upon wholly external trips described in Table 2. Delay-based charging produces a significantly smaller reduction in orbital speeds because of the lower propensity to transfer traffic away from the charge area.

Figure 2e combines outputs for total travel distance, total travel time and trip suppression in the form of average generalised travel cost (excluding charges). This represents an effective measure of network performance which focuses on the spatial and route choice strengths of the model and reduces dependence on the relatively simple treatment of road travel demand, which would limit the power of economic measures such as consumer surplus. It demonstrates clearly that all four of the systems tested produce optimum network benefits of less than 5 per cent and tend towards disbenefits above relatively low charge levels. Average charges of £1 per journey are only around one third of the price that commuters might currently expect to pay for parking in a typical UK city. In the case of cordon charges, there are no benefits at any charge level. The greatest benefit occurs under time-based charging, which reduces average generalised costs by around 4 per cent for an average journey charge of approximately 55 pence. Delay-based charging produces benefits across a wider range of charge levels than the other systems, but requires an average journey charge of approximately 65 pence to generate benefits of the order of 3%.

5.   CONCLUSIONS

The network benefits of the four practical road user charging systems investigated during this study have been significantly smaller than was originally anticipated. In particular, improvements in travel conditions appear much lower has been suggested by practical modelling studies which have used a much coarser treatment of route choice issues (The MVA Consultancy, 1990).

In addition, implementation of all systems within a geographically limited urban boundary has caused transfer of traffic and reductions in network performance in the areas immediately beyond. In the simplest economic case it is assumed that marginal cost pricing would be universal. Modifying this in practice may significantly reduce efficiency benefits.

Delay-based charging, which may provide the closest practical approximation to marginal costs, has been shown to be the least effective alternative for reducing road travel demand and has encouraged significant increases in travel distance, suggesting that it may be inappropriate for achieving environmental goals. It has also generated results within an equilibrium model which are difficult to reconcile with behavioural expectations, suggesting that fully dynamic and microscopic modelling approaches may be required.

6. REFERENCES

Goodwin, P.B. (1992) **A review of new demand elasticities with special reference to short and long run effects of price changes.** *Journal of Transport Economics and Policy* (26)2.

The Highways Agency, The Scottish Office Development Department, The Welsh Office, The Department of the Environment for Northern Ireland and the Department of Transport (1997) **Induced traffic appraisal.** *Design Manual for Roads and Bridges* 12(2) part 2.

May, A.D. (1992) **Road pricing: an international perspective.** *Transportation*, 19(4).

Milne, D.S. (1997) **Modelling the network effects of road user charging.** PhD thesis, Institute for Transport Studies, University of Leeds, unpublished.

The MVA Consultancy (1990) **Joint authorities and environmental study.** Lothian Regional Council, Edinburgh.

Oldridge, B. (1990) **Electronic road pricing: an answer to traffic congestion?** Proc. *Information Technology and Traffic Management* conference, HMSO, London.

The Highways Agency

Royal Commission on Environmental Pollution (1994) **Eighteenth report: transport and the environment.** HMSO, London.

Van Vliet, D. (1982) **SATURN - a modern assignment model.** *Traffic Engineering and Control* 23(12).